Study Guide

for

Carroll's

Sexuality Now: Embracing Diversity

Study Guide

for

Carroll's

Sexuality Now: Embracing Diversity

Ann Kolodji
University of Colorado Health Sciences

Australia • Canada • Mexico • Singapore • Spain • United Kingdom • United States

Printed in the United States of America
2 3 4 5 6 7 07 06 05

Printer: West

0155-06771-0

For more information about our products, contact us at:
Thomson Learning Academic Resource Center
1-800-423-0563

For permission to use material from this text, contact us by:
Phone: 1-800-730-2214
Fax: 1-800-730-2215
Web: http://www.thomsonrights.com

Thomson Wadsworth
10 Davis Drive
Belmont, CA 94002-3098
USA

Asia
Thomson Learning
5 Shenton Way #01-01
UIC Building
Singapore 068808

Australia/New Zealand
Thomson Learning
102 Dodds Street
Southbank, Victoria 3006
Australia

Canada
Nelson
1120 Birchmount Road
Toronto, Ontario M1K 5G4
Canada

Europe/Middle East/South Africa
Thomson Learning
High Holborn House
50/51 Bedford Row
London WC1R 4LR
United Kingdom

Latin America
Thomson Learning
Seneca, 53
Colonia Polanco
11560 Mexico D.F.
Mexico

Spain/Portugal
Paraninfo
Calle/Magallanes, 25
28015 Madrid, Spain

Table of Contents

Preface

This student-friendly study guide for *Sexuality Now: Embracing Diversity* provides human sexuality students with a comprehensive learning resource. The study Guide orients you to each chapter of the textbook by providing an overall view then breaking it down into interactive sections so you can assimilate the information in ways that best suit your learning style. Each chapter includes a chapter summary, learning objectives, and a detailed chapter outline. Each chapter also contains a Test Yourself section, which provides you with the opportunity to test your knowledge with both fill-in-the-blank and short answer questions for topics covered in the chapter. A Post-Test section completes each study guide chapter, providing the opportunity to synthesize and apply your knowledge through true/false, multiple choice, and matching questions for each chapter. The answers and page references to every question for both the Test Yourself and the Post-Test sections appear at the end of each chapter.

For the best use of the study guide in conjunction with the test book, read the chapter summary, learning objectives, and chapter outline prior to reading the chapter in the textbook in order to prepare yourself for key concepts and theories. After you've read the textbook chapter, return to the study guide to begin cementing your knowledge of the research presented in each chapter. First, work on the Test Yourself section by completing the section-by-section fill-in-the-blank and short answer questions grouped by topic. Refer to the textbook as a resource to help you review important information. You can also use the chapter outline to help highlight the structure and major features of each chapter. The answer key for each Test Yourself section will help you assess your progress and provide you with page references to refer back to specific textbook sections. Once you feel like you have assimilated the information sufficiently, test yourself in the Post-Test section. Complete the true/false, multiple choice, and matching questions for the entire chapter as if you were taking an exam for the course. As you respond to these questions, you may want to highlight ones that you find difficult. Once you've finished with these questions, use the answer key at the end of the chapter to evaluate your performance. Use the page references in the answer key to refer back to the textbook to review questions or areas that proved more challenging for you. Most students find that learning about human sexuality is more interactive and enlightening than they thought it would be—so try to take your time and have fun!

About the Author

Ann Kolodji works as a Testing Supervisor and Senior Research Associate in the Program for Early Developmental Studies at the University of Colorado Health Sciences in Denver, Colorado. She received her Ph.D. in Education from the Human Sexuality Program at the University of Pennsylvania and works as a part-time sexuality educator and consultant.

Chapter 1 – Exploring Human Sexuality: Past and Present

CHAPTER SUMMARY

Sexuality is mysterious and exciting. Sexuality for human beings has evolved beyond biological sexual functioning to include customs, laws, and fantasies. Many of the cultural implications of sexuality can be seen in media representations of sexuality that have contributed significantly to sexual understanding. The evolution of the Internet has changed social relationships in relation to sexuality. There is increased access to sexuality information, paraphernalia, pornography, and online relationships.

As the first civilizations were established, the first laws and rituals around sexuality began to develop. The interaction of three ancient cultures, Hebraic, Greek, and Roman, made major contributions to modern Western society's views on sexuality. After the development of Christianity, early Christian views on sexuality evolved and have had much impact on contemporary views of sexuality in the Western world. The legacy of early Christianity was a general association of sexuality with sin. Nonprocreative sex was strictly forbidden, along with contraception, masturbation, and sex for pleasure's sake. In the early Middle Ages, the influence of the Catholic Church slowly began to increase while the views towards various sexual issues fluctuated during this long period of time.

During the Renaissance that began in Italy in the late 1300s, intellectual and artistic thought turned from somberness to sensuality. In the early sixteenth century, Martin Luther founded a movement known as Protestantism, which differed in previous Christian views in that it allowed divorce, saw marital love as blessed, and considered sexuality not just for procreation but as a natural function. The Enlightenment, an intellectual movement of the eighteenth century, considered sexual pleasure as natural and desirable. The Victorian Era, which lasted into the twentieth century, was a time of great prosperity in England during which propriety and public behavior became more important and sexual attitudes became more conservative.

In the United States, the history of sexuality included the Puritans, who were a religious group who fled England and tried to set up a biblically based society in the New World. They set severe sanctions for sexual transgressions, including the death penalty for sodomy, bestiality, adultery, and rape, although they did believe that sexuality was good and proper with marriage. Settlers throughout early American history used the sexuality of minorities, African slaves, Native Americans, and Mexicans, as an excuse to oppress them. After the Revolutionary War, the church's power began to diminish in the United States, and the country entered a period of practical, utilitarian philosophy that stressed the individual's right to pursue happiness.

At the end of the nineteenth century, the medical model of sexuality began to emerge, and Americans became obsessed with sexual health, sexual restraint. The Comstock Act of 1873 was passed, which prohibited the mailing of obscene and indecent writing or advertisements including articles about contraception and abortion. With the twentieth century, sexual reform movements became visible, leading to a birth control movement spearheaded by Margaret Sanger. The Social Hygiene Movement, feminism, and gay and lesbian liberation were other reforms that influenced sexuality in American history.

LEARNING OBJECTIVES

After studying Chapter 1, you should be able to:

1. Provide examples of how sexuality can be contradictory and confusing.
2. Identify three of the six goals the textbook lists as goals for this human sexuality course.
3. Describe the impact of media on sexual development and sexual learning.
4. Discuss the ways the Internet has been changing patterns of communication around sexuality.
5. Compare and contrast different ancient civilizations' attitudes and practices related to sexuality.
6. Identify how the Hebraic, Greek, and Roman civilizations have influenced sexual attitudes in the Western world.
7. Discuss some of the different attitudes and practices that existed in ancient civilizations when it came to same-sex relationships.
8. Explain how religious and spiritual beliefs influenced the sexual attitudes and practices in ancient China and India.
9. Describe the contributions that Christianity has had on contemporary Western sexual attitudes.
10. Provide examples of sexual developments that occurred during the Middle Ages.
11. Compare early Christian views of sexuality with early Islamic views of sexuality.
12. Explain how views of sexuality evolved from the Middle Ages to the Victorian Era.
13. Identify the contribution that Protestantism made to sexual attitudes and practices.
14. Compare and contrast the divergent attitudes around sexuality during the Enlightenment and the Victorian Eras.
15. Describe the sexual practices and attitudes of early American settlers.
16. Identify the sexual conflicts that occurred in American society during the nineteenth century.
17. Provide examples of the sexual reform movements during the twentieth century.

CHAPTER OUTLINE

I. HUMAN SEXUALITY IN A DIVERSE WORLD

A. **Only Human: What is Sexuality?**

1. Only human beings have gone beyond sexual behaviors to create ideas, laws, customs, fantasies, and art around the sexual act, making sexuality a uniquely human trait.
2. Sexuality is an area of human life that can be contradictory and confusing.

B. **Why Are We Here? Goals for Human Sexuality Course**

1. To develop a broad and accurate knowledge base about sexuality.
2. To understand the various influences on the development of your sexual knowledge, attitudes, relationships, and behaviors.
3. To have a clear understanding of society's attempts to regulate your sexuality.
4. To identify trends and changes that have influenced your sexual attitudes and values.
5. To understand the biological basis and the complex political, media-related, and ethical issues of sexuality.
6. To become more comfortable talking about sex.

C. **Sex Sells: The Impact of the Media**

- Many images, from magazines to videocassette packaging, are explicitly or subtly sexual.

1. More television shows now address sexual issues.
    a. *Will and Grace*
    b. *Real World*
    c. *Jerry Springer*
    d. *Taxicab Confessions*
2. The Internet has given many people access to sexual information and the ability to communicate about sexuality despite social norms, embarrassment, and fear.
    a. Websites are available that offer sexual advice, sex toys, photographs, video clips.
    b. More than 8,000 chat rooms devoted to sex exist, allowing people the freedom to talk to others and act out their sexual fantasies.
3. Despite that the media's message about sexuality are often inaccurate, unrealistic, and misleading, many people accept information provided by the media as fact.
    a. No mention of **sexually transmitted infections (STIs)**
    b. How might media images negatively affect adolescents?

II. THE EARLY EVOLUTION OF HUMAN SEXUALITY

A. **Stand Up and Look Around: Walking Erect**

1. Our early ancestors were **quadrupeds** who walked on all fours; the evolution of an upright posture changed the way humans engaged in sexual intercourse.
2. The **phallus**, the male symbol of sex and potency, became associated with displays of aggression.
3. The upright posture of the female emphasized her breasts, hips and rotated the pelvis forward so that the clitoris and the breasts were more easily stimulated.
4. Homo sapiens appeared about 200,000 years ago, where life was difficult and included high infant mortality, disease, malnutrition, and a harsh environment.

B. **Sexuality in the Ancient Mediterranean**

1. As the first civilizations were established, the first codes of law began to develop.
2. Some ancient medical texts discuss cures for sexually transmitted infections.
3. Male circumcision was first performed in Egypt and possibly parts of Africa.
4. Egyptian women inserted sponges and other objects into their vaginas as forms of contraception.

5. Abortion was forbidden because a great value was put on having as many children as possible (especially sons).
6. Prostitution was common, and temple prostitutes often greeted worshippers.
7. History tells us more about men's experiences, since women's voices were often silent.
8. Egyptians condemned adultery, but it appears to still have been relatively common.
9. The interaction of three ancient cultures, Hebraic, Greek, and Roman, made a contribution to modern Western society.
10. The Hebrews
    a. The Hebrew Bible contains explicit rules about sexual behavior concerning adultery, same-sex intercourse, and incest.
    b. The Hebrew attitude toward sexuality has formed the basis of sexual attitudes in the West for centuries.
11. Greece (1000-200 B.C.E.)
    a. The Greeks were more sexually permissive than the Hebrews, with stories and myths full of incest, rape, and **bestiality**.
    b. Greek **pederasty** was considered a natural form of sexuality practiced by Socrates and Plato, where an older man would befriend a post-pubescent boy, helping him with his intellectual, physical, and sexual development. They would engage in sexual activities with each other.
    c. Love in Greece was discussed in **homoerotic** terms where man's nonsexual love for another man was seen as the ideal love as discussed by Plato, which led to the term **platonic** being used to describe nonsexual friendships.
    d. Pederasty was not the only form of Greek sexuality. Also literature describes deep heterosexual love and marriage between women and men.
12. Rome (Fifth Century B.C.E. to Seventh Century C.E.)
    a. Marriage and sexual relations were viewed as a means to improve one's economic and social standing, with love viewed as secondary to fairness, respect, and mutual consideration in marriage.
    b. Romans had permissive attitudes toward same-sex sexual behaviors.

C. **Sexuality in Ancient Asia**

1. China
    a. Sexuality in China is discussed as a natural procreative process of joining the **yin** and **yang**, the masculine and feminine principles, with the goal of Taoist life being harmony or the blending of yin and yang.
    b. Early Chinese society had numerous sex manuals with explicit pictures and instructions.
    c. The Chinese were unique in stressing the importance of female orgasm due to the belief that a man can maximize his contact with yin energy during a woman's orgasm, which helps him to feed his yang.
    d. Same-sex relations were not discouraged but viewed as a wasteful use of sperm that is needed primarily for impregnation.
    e. The Chinese developed aphrodisiacs and sexual devices such as Ben-wa balls that are inserted into the vagina for sexual pleasure.
    f. **Polygamy** was practiced until very late in Chinese history, with the average middle class male having between three and twelve wives and concubines.
2. India
    a. Hinduism, the religion of India for most of its history, concentrates on **karma**, an individual's cycle of birth and rebirth.
        - Karma involves a belief that suffering in a future life punishes a person's unjust deeds in this life.

- Sex in India was generally viewed as a positive pursuit and even a source of power and magic because marriage and procreation were seen as an important responsibility and part of living a just life.

b. India's social system was basically **patriarchal**, and being born a woman was seen as a punishment for sins committed in previous lives. **Female infanticide** was not uncommon.

c. *The Kamasutra*

- During the third or fourth century B.C.E., the first and most famous of India's sex manuals appeared. Entitled *The Kamasutra,* it talks about sex, love, moral guidance, and making a good home and family.
- It describes eight kinds of love biting, eight stages of oral sex, and nine ways to move the penis in the vagina.
- It proposed that intercourse should be a passionate activity that included scratching, biting, and blows to the back, accompanied by a variety of animal noises.

d. In India, marriage was an economic and religious obligation with families arranging marriages.

e. Widows in India were forbidden to remarry and had to live simply and devote their lives to prayer, leading many widows to the ritual act of sati, throwing themselves on their husband's burning funeral pyre to die.

III. SEXUALITY FROM ST. PAUL TO QUEEN VICTORIA

A. Early Christianity—Chastity Becomes a Virtue

1. Christianity has had much impact on the Western views of sexuality.
2. Jesus was mostly silent on sexual issues but did say that men should be held to the same standards of adultery, divorce, and remarriage as women.
3. Saint Paul was influential in Christian views of sexuality, suggesting that the highest love was love of God and that the ideal was not to allow sexual or human love to compete with love for God.

   a. The ideal situation was **celibacy** although sexuality was not sinful when performed as part of the marital union.

   b. Abstaining from sexual intercourse, or **chastity**, became a sign of holiness.
4. Saint Jerome and Saint Augustine, along with other theologians, were very strong in condemning sexual activity—even renouncing marriage.
5. The legacy of early Christianity was a general association of sexuality with sin. All nonprocreative sex was strictly forbidden, along with contraception, masturbation, and sex for pleasure's sake.

B. The Middle Ages: Eve the Temptress, Mary the Virgin

1. In the early Middle Ages, the influence of the Christian church slowly began to increase. However, the views of sexuality fluctuated during this time.
2. Between the years of about 1050 and 1150, sexuality once again became liberalized, as exemplified by a gay subculture established in Europe.
3. In the thirteenth century, the church cracked down on a variety of groups including, Jews, Muslims, and homosexuals.
4. By the year 1215, **confession** was instituted by the church and priests were taught about the sins **penitents** might have committed.

   a. Sexual sins were most commonly emphasized.

b. All sex outside of marriage was considered sinful, but penance also had to be done for such things as nocturnal emissions, certain marital acts, and violations of modesty such as looks, desires, touches, and kisses.

5. Women were elevated to a place of purity and were considered almost perfect, with the symbol of ideal womanhood embodied by Mary, the mother of God.
6. As women were no longer viewed as temptresses but models of virtue, the idea of romantic love spread throughout popular culture.
7. Women were also said to be the holder of the secrets of sexuality, leading men to employ the services of an **entremetteuse,** an old woman who found prostitutes who could teach them the ways of love and restore potency and virginity before marriage.
8. By the late fifteenth century, the church began a campaign against witchcraft, which they said was inspired by women's insatiable "canal lust," leading to the death of thousands of women.
9. Thomas Aquinas (1225-1274)
    a. theologian and later saint from the Middle Ages, he had the strongest impact on subsequent attitudes towards sexuality.
    b. He drew from the idea of "natural law" to suggest that sex organs were "naturally" intended for procreation while other uses of them were unnatural and immoral.
    c. Aquinas' strong condemnation of sexuality, and especially homosexuality—which he called the worst of all sexual sins—set the tone for Christian attitudes towards sexuality for many centuries.

C. **Islam: A New Religion**

1. Beginning in the sixth century with a man named Muhammad and drawing from Jewish and Christian roots, Islam became a powerful force that spread around the world.
2. Despite examples in the **Koran** of female saints and intellectuals, many women in many Islamic societies were subjugated to men.
3. Sexuality between a man and a woman was only legal when the couple was married or when the woman was a concubine, and sexual intercourse in marriage was a good religious deed for the Muslim male.
4. A man with a strong sex drive was advised to take many wives, leading to **harems**.
    a. Harems involved women who were married to wealthy men who usually lived in secluded areas in their husbands' homes.
    b. Rather than communities of rampant sex and sensuality, harems were self-contained communities where women learned to become self-sufficient in the absence of men.
5. Islamic society has had a freer, more open attitude to sexuality than Christian societies, with erotic writings from medieval times and the presence of sexual contact between men and boys.
6. The Sultans of the Ottoman Empire, who ruled most of the Islamic world from the fifteenth to the twentieth century, had between 300 and 1200 concubines.
    a. Because these women had sexual contact with the sultan once or twice a year, concerns that they would seek sexual satisfaction elsewhere lead to the use of **eunuchs**.
    b. Eunuchs had their testicles, penises, or both removed.

D. **The Renaissance: The Pursuit of Knowledge**

1. During the Renaissance, which began in Italy in the late 1300s, intellectual and artistic thought turned from a focus on God to a focus on human beings; from **asceticism** to sensuality.
2. Women began to make great strides in education and politics, with debates on the value of women.

E. **The Reformation: The Protestant Marital Partnership**

1. In the early sixteenth century, Martin Luther founded a movement known as Protestantism, which differed from previous Christian views in that it allowed divorce, saw marital love as blessed, and considered sexuality not just for procreation but as a natural function.
2. John Calvin, another Protestant reformer, emphasized that women were not just reproductive vessels but men's partners in all things, and he saw the marital union as primarily a social and sexual relationship with companionship as the main goal of marriage.
3. Luther continued to accept the subjugation of women to men in household affairs and excluded women from the clergy due to what he considered their inferior aptitudes for ministry.

IV. THE ENLIGHTENMENT AND THE VICTORIAN ERA

A. The Enlightenment

1. The Enlightenment, an intellectual movement of the eighteenth century, considered sexual pleasure as natural and desirable.
2. Sexuality became so free that there was an unprecedented rise in nonmarital pregnancy although homosexuality was condemned and persecuted.

B. The Victorian Era

1. The Victorian Era, which lasted into the twentieth century, was a time of great prosperity in England when propriety and public behavior became more important and sexual attitudes became more conservative.
2. Conservative values around sexuality were often preached though not always practiced, and pornography, extramarital affairs, and prostitution were common.
3. Women were considered to be virtuous, refined, and fragile and would never admit to a sexual urge.
4. Sylvester Graham recommended sex only twelve times a year to avoid sexual indulgence that could lead to various ailments.
5. The Victorian Era had great influence on the sexuality of England and the United States and can be linked to many of the conservative attitudes held today.

V. SEX IN AMERICAN HISTORY

A. **The Colonies: The Puritan Ethic**

1. The **Puritans** were a religious group who fled England and tried to set up a biblically based society in the New World.
2. They had severe sanctions for sexual transgressions including the death penalty for sodomy, bestiality, adultery, and rape, although they did believe that sexuality was good and proper with marriage.
3. **Bundling** referred to young couples who were allowed to share a bed as long as they were clothed and wrapped in sheets or had a wooden "bundling board" between them.

B. **The United States: Freedom—and Slavery—in the New World**

1. The Liberalization of Sex
    a. After the Revolutionary War, the church's power began to diminish in the United States, and the country entered a period of practical, utilitarian philosophy, which stressed the individual's right to pursue happiness.
    b. People began to speak more openly about sexuality and romantic love.
    c. Prostitution, contraception, and abortion were more visible, and the birth rate decreased.

2. Slavery
   a. Before slavery, the southern colonies had made use of indentured servants, where rape was fairly common.
   b. After slavery, many states passed anti-miscegenation laws prohibiting sexual relations between whites and blacks.
   c. African slaves were accused by whites of having loose sexual morals. Black men were seen as a threat to white women, which served as an excuse for continued abuse.
   d. Despite harsh conditions, there was a strong tendency to regulate sexual behavior among slaves.
   e. Settlers throughout early American history used the sexuality of minorities, Native Americans and Mexicans, as an excuse to oppress them.

C. **The 19th Century: Polygamy, Celibacy, and Comstock Laws**

1. The free love movement began in 1820s and preached that love, not marriage, should be the prerequisite to sexual relations.
2. At the end of the nineteenth century, the medical model of sexuality began to emerge. Americans became obsessed with sexual health emphasizing sexual restraint.
3. Physicians began to argue that homosexuality was an illness rather than a sin despite a number of recorded expressions of same-sex relationships.
4. In the 1870s, Anthony Comstock, a dry-goods salesman, lobbied the legislature to outlaw obscenity, resulting in the Comstock Act of 1873, which prohibited the mailing of obscene and indecent writing or advertisements, including articles about contraception and abortion.

G. **The 20th Century: Sexual Crusaders and Sexologists**

1. In 1905 the **Social Hygiene Movement** convinced legislators that "virtuous" women were catching venereal diseases (STIs) from husbands who frequented prostitutes, leading to blood tests before marriage and a crackdown on prostitution.
2. In the early part of the century, **sexologists** began conducting sexuality research, helping to demystify sex and to make it more respectable to publicly discuss the sexual behaviors and problems of real people.
3. Feminism
   a. The **women's suffrage** movement of the early twentieth century first put women's agendas on the national scene.
   b. Despite being arrested for violating the Comstock laws, Margaret Sanger fought for the rights of women to use birth control, opening a birth control clinic in Brooklyn, which evolved into the contemporary Planned Parenthood.
   c. Feminists of the 1960s argued that they were entitled to sexual satisfaction and that they had the right to control their bodies.
4. Gay Liberation
   a. After World War II, homosexuals were portrayed as perverts and thrown in jails and mental hospitals.
   b. The Mattachine Society and the Daughters of Bilitis were two early organizations for lesbian and gay rights founded in the 1950s.
   c. In 1969, the Stonewall riot became a symbol to the gay community when police raided a gay bar in New York City but were greeted by resistance.
   d. In 1973, after a strong lobbying campaign, the American Psychiatric Association removed homosexuality from the Diagnostic and Statistical Manual (DSM), emphasizing that homosexuality was not a mental disorder.
   e. In more recent years, the AIDS epidemic has mobilized the gay community as it has established clinics, education programs, and political lobbies to fight AIDS.

TEST YOURSELF

Below you will find fill-in-the-blank and short answer essay questions for topics covered in chapter 1. Check your answers at the end of this chapter.

Human Sexuality in a Diverse World

1. While many of our fellow creatures also display complex sexual behaviors, only human beings have gone beyond instinctual mating rituals to create ideas, ________________________, customs, ________________________, and art around the sexual act.

2. Sexuality is studied by ________________________, who specialize in understanding our sexuality.

3. Political scientists may study how sexuality reflects social ________________________.

4. We come from a society that is often called sexually ________________________, yet images of sexuality are all around us.

5. While parents teach their children about safe driving, safe use of fire, and safe hygiene, many are profoundly ________________________ instructing their children on safe sexual practices.

6. Goals for this human sexuality course: To understand the various influences on the development of your sexual ________________________, ________________________, ________________________, and ________________________.

7. Goals for this human sexuality course: To be more ________________________ when talking about sex.

8. We live in a visual culture whose ________________________ we simply cannot escape.

9. The humor in television situation comedies has become more and more ________________________, and ________________________ has begun to appear on prime-time network television shows.

10. Many television shows use ________________________ to increase their ratings.

11. ________________________ ________________________, ________________________, and ________________________ hold us back from expressing many of our sexual needs and desires.

12. The ________________________ is changing patterns of social communication and relationships.

13. ________________________ is the most frequently searched-for topic on the Internet.

14. The Internet allows a person to be ________________________ if they wish and provides the freedom to ask questions, seek answers, and talk to others about sexual issues.

15. Very few of the sexual references from the media watched by adolescents have anything to do with contraception, ________________________ ________________________ ________________________, or pregnancy risk.

16. What makes sexuality a uniquely human trait?

17. In addition to sexologists, what are some of the other professions that study sexuality?

18. How have television and movies changed in the last few decades that relates to sexuality?

19. What are some examples of sexuality related websites on the Internet?

20. What are some of the things that most of the media references to sexuality are lacking?

The Early Evolution of Human Sexuality

21. The evolution of a(n) ___________________________ posture changed forever the way the human species engaged in sexual intercourse.

22. Since a great value was put on having as many children as possible in ancient Egypt—especially sons, for inheritance___________________________ was usually forbidden.

23. Though the Egyptians condemned ___________________________, especially among women, there is ample evidence that it was still fairly common and was the subject much sexual joking.

24. Of all the ancient civilizations, modern Western society owes the most to the interaction of three ancient cultures: the ___________________________, ___________________________, and ___________________________.

25. The Hebrew Bible also tells tales of marital ___________________________ and acknowledges the importance of sexuality in marital relations.

26. In Greek ________________________, an older man would befriend a post-pubescent boy who had finished his orthodox education and aid the boy's continuing intellectual, physical, and sexual development.

27. When the ancient Greek philosophers spoke of love, they did so almost exclusively in ________________________ terms.

28. Plato discussed such an ideal love as man's nonsexual love for another man, and so we have come to call friendships without a sexual element ________________________.

29. In Rome, marriage and sexual relations were viewed as a means to improve one's ________________________ and ________________________ standing.

30. The Tao itself is made up of two principles, ________________________ and ________________________.

31. Sexual instruction and sex manuals were common and openly available in early ________________________ society.

32. In ancient Chinese society, same-sex relations were not discouraged, though male homosexuality was viewed as a wasteful use of ________________________.

33. ________________________ was practiced until very late in Chinese history, and the average middle class male had between three and a dozen wives and concubines.

34. ________________________, the religion of India for most of its history, concentrates on an individual's cycle of birth and rebirth, or ________________________.

35. By about the third or fourth century B.C.E., the first and most famous of India's sex manuals appeared, entitled ________________________.

36. How did the evolution of humans walking upright change sexual relations?

37. Why do we know more about the sexual attitudes and practices of men in ancient times than women?

38. Describe some of the explicit rules about sexuality adopted by Hebrews.

39. Describe the attitudes towards same-sex relations between men in ancient Rome.

40. Where was *The Kamasutra* developed and what does it contain?

Sexuality From St. Paul to Queen Victoria and The Enlightenment

41. Perhaps no single system of thought had as much impact on the Western world as __________________________, and nowhere more so than in its views on sexuality.

42. St. Paul condemned sexuality in a way not found in either Hebrew or Greek thought and also not found anywhere in the teachings of __________________________.

43. Though sexuality itself was not sinful when performed as part of the marital union, the ideal situation was __________________________.

44. The legacy of early Christianity was a general association of sexuality with __________________________.

45. European women in the early Middle Ages were elevated to a place of __________________________ and were considered almost __________________________.

46. Perhaps no person from the Middle Ages had a stronger impact on subsequent attitudes toward sexuality than the theologian (and later saint) __________________________ __________________________.

47. In the sixth century, a man named __________________________ began to preach a religion that drew from Jewish and Christian roots and added Arab tribal beliefs.

48. Many Muslim societies have strong rules of __________________________, which involve covering the private parts of the body (which in women means almost the entire body).

49. __________________________ were not the dens of sex and sensuality that are sometimes portrayed.

50. Since each woman might sleep with the Sultan once or twice a year at best __________________________ were employed to guard against the women finding sexual satisfaction elsewhere.

51. The __________________________ may be summed up as a time when intellectual and artistic thought turned from a focus on God to a focus on human beings.

52. ________________________ ________________________ ________________________ published a tract in 1532 arguing that each of God's creations in Genesis is superior to the one before, and since the human female is the last thing God creates, she must be his most perfect creation.

53. In the early sixteenth century, ________________________ ________________________ challenged papal power and founded a movement known as Protestantism.

54. As liberal as the Enlightenment was, many sexual activities, such as ________________________, were condemned and persecuted.

55. Victorian women were too embarrassed to talk to a ________________________ about their "female problems" and so would point out areas of discomfort on dolls.

56. How did chastity come be seen as a sign of holiness?

57. During the Middle Ages, what groups did the Church begin to crack down upon?

58. In addition to ancient Greeks, what other society celebrated young boys as the epitome of beauty and allowed sexual contact between men and boys?

59. What are the names of the two major Protestant reformers?

60. During what time period did writers argue that human drives and instincts are part of nature's design so one must realize the basic wisdom of human urges and not fight them?

Sex in American History

61. The ________________________ were a religious group who fled England and tried to set up a biblically based society in the New World.

62. As the New World began to grow, it suffered from a lack of ________________________.

63. There was a custom called ________________________, where young couples were allowed to share a bed as long as they were clothed or had a board between them.

64. By the late eighteenth century, as many as ________________________ of all brides in some parts of New England were pregnant.

65. By the late eighteenth century, the birth rate dropped and ________________________ rates rose through the use of patent medicines and folk remedies.

66. After 1670, African slaves became common in the South, and many states passed ________________________ ________________________.

67. The myth of slave sexual looseness is disproved by the lack of ________________________ and very low sexually transmitted disease rates among slaves.

68. The ________________________ ________________________ movement, which began in the 1820s, preached that love, not marriage, should be the prerequisite to sexual relations.

69. The ________________________ community preached group marriage while the ________________________ who, frustrated with all the arguments over sexuality, practiced strict celibacy.

70. During the nineteenth century, an influential group of physicians even argued that women were biologically designed ________________________ and should stay in the marital union.

71. The great poet ________________________ ________________________, now recognized as having been homosexual, at times confirmed his attraction to men, while at other times he denied it.

72. The ________________________ ________________________ of 1873 prohibited the mailing of obscene, lewd, lascivious, and indecent writing or advertisements, including articles about contraception or abortion.

73. After publishing information about birth control, ________________________ ________________________ opened a birth control clinic in Brooklyn.

74. ________________________ were at the forefront of the abortion debate and hailed the legalization of abortion as a great step in achieving women's rights over their own bodies.

75. Four lesbian couples in San Francisco founded the ________________________ of ________________________, the first postwar lesbian organization, in 1955.

76. What minorities, as discussed in the textbook, did early American settlers oppress, using their sexual practices or attitudes as an excuse?

77. What was the name of the law passed in the 1870s that lead to thousands of books, sexual objects, and contraceptive devices being destroyed?

78. What was the name of the movement that was a combination of liberal and traditional attitudes, leading to arrests of prostitutes and the institution of sex education in schools

79. What important book did Betty Friedan write in 1963, which became part of the modern feminist movement?

80. What was the name of the New York gay bar that was the sight of active resistance against police in 1969 and became a symbol to the gay and lesbian community?

POST TEST

Below you will find true/false, multiple-choice and matching quiz items covering the entire chapter. Check your answers at the end of this chapter.

True/False

1. Some people believe that providing sexuality information can increase teenage sexual activity.
2. One of the goals of a sexuality course, according to the authors, is to encourage students to change their sexual attitudes and values.
3. Ancient Egyptian women inserted sponges in their vaginas to prevent pregnancy.
4. The ancient Indian book, *The Kamasutra*, was known for its emphasis on chastity and modesty.
5. Jesus himself was mostly silent on sexual issues such as homosexuality or premarital sex.
6. Same-sex sexual relations had been legal for the first two hundred years that Christianity was the state religion of Rome.
7. During the Enlightenment, same-sex sexual behaviors flourished and were supported by society.
8. The Free Love Movement preached that love, not marriage, should be the prerequisite to sexual relations.
9. In the late nineteenth century, doctors began recognizing the importance of sexual passion and masturbation to the mental health of women.
10. The women's suffrage movement advocated women leaving college to return to marriage.

Multiple-Choice

11. According to the textbook, what is one area of sexuality that is mentioned regularly on television shows such as sitcoms and talk shows?
    a. pregnancy risk
    b. sexual intimacy
    c. sexually transmitted infections
    d. condom use
    e. contraception

12. How did the evolutionary process of humans evolving from quadrupeds to walking upright change sexual intercourse?
    a. It decreased the manipulation of the breasts.
    b. Sense of smell became more important than eyesight.
    c. It decreased the connection between the penis and aggression.
    d. It lead to less body contact.
    e. It lead to easier stimulation of the clitoris.

13. The rules about sexual behavior contained in the Hebrew Bible include ____________________.
    a. acceptance of sexual relations between siblings but not cousins
    b. permissive attitude regarding sex outside of marriage
    c. prohibition against homosexuality
    d. marriage as an economic union rather than a sexual union
    e. All of the above

14. Which ancient culture spoke of love in almost exclusively homoerotic terms?
    a. Chinese
    b. Greek
    c. Hebrew
    d. Muslim
    e. Roman

15. Friendships without a sexual element are called ____________________.
    a. homoerotic
    b. Taoist
    c. platonic
    d. penitent
    e. Puritan

16. Which ancient civilization emphasized the importance of female orgasm?
    a. Chinese
    b. Egyptian
    c. Roman
    d. Greek
    e. None of the above

17. Which of the following is NOT true regarding ancient Indian marriage traditions?
    a. Cohabitation and sex usually began around the age of eight.
    b. Women were forbidden to remarry if their husbands died.
    c. Marriage was an economic and religious obligation.
    d. Families typically arranged marriages for their young children.
    e. Widows had to wear plain clothes and sleep on the ground.

18. Abstaining from sexual intercourse is referred to as ____________________.
    a. bundling
    b. chastity
    c. suffrage
    d. libidism
    e. pederasty

19. Who was directly responsible for the development of chastity as an important Christian virtue?
    a. Thomas Aquinas
    b. Henricus Cornelius Agrippa
    c. St. Paul
    d. Jesus
    e. None of the above

20. The legacy of early Christianity was a general association of sexuality with ____________________.
    a. pleasure
    b. masturbation
    c. homosexuality
    d. sin
    e. witchcraft

21. During the Middle Ages, the church instituted confession and became strict with what types of sexual transgressions?
    a. adultery
    b. nocturnal emissions
    c. homosexuality
    d. violations of modesty
    e. All of the above

22. How did the views of women change during the Middle Ages?
    a. The symbol of womanhood was Eve, who led to Adam's downfall.
    b. They were seen as possible witches with dangerous sexuality.
    c. Woman was no longer a temptress but a model of virtue.
    d. Women were considered ignorant of all sexual matters.
    e. Woman was a field to be cultivated.

23. What theologian during the Middle Ages argued that the sex organs should be used only for procreation?
    a. Martin Luther
    b. Henry Hay
    c. Anthony Comstock
    d. Thomas Aquinas
    e. John Calvin

24. What are self-contained communities where women learned to become self-sufficient in the absence of men?
    a. eunuchs
    b. reform homes
    c. concubines
    d. entremetteuse
    e. harems

25. Why did the Sultans of the Ottoman Empire use eunuchs to guard the large number of female slaves and concubines?
    a. To prevent the women from seeking sexual satisfaction with other men.
    b. Eunuchs were very tall and muscular, which made them excellent guards.
    c. Harems were wild dens of sex and sensuality.
    d. To encourage women to have sex with the men since the Sultan was busy.
    e. All of the above

26. Which of the following is NOT true regarding the Renaissance?
    a. Women made great strides in education and political affairs.
    b. Religious symbolism became more predominant across Europe.
    c. Sober and serious theology became less prominent.
    d. More emphasis was placed on human beings rather than God.
    e. There was a renewed sense of joy and sensuality.

27. Who founded the movement known as Protestantism?
    a. St. Augustine
    b. Thomas Aquinas
    c. Muhammad
    d. Martin Luther
    e. Henricus Cornelius Agrippa

28. During what period in history was sexuality viewed as natural and desirable and erotic literature become popular?
    a. The Middle Ages
    b. The Victorian Era
    c. The Puritan Era
    d. The Enlightenment
    e. The Reformation

29. How were women viewed during the Victorian Era?
    a. as dangerous witches
    b. as knowing secrets of sexual intercourse and love
    c. as sexual temptresses
    d. as fields that men should cultivate as frequently as they wanted
    e. as delicate, fragile, and vulnerable, with no sexual urge

30. Which of the following is NOT true regarding the Puritans?
    a. They practiced a custom called bundling that involved unmarried couples sharing a bed.
    b. They applied the death penalty for adultery and rape.
    c. Puritan men were obligated to have intercourse with their wives.
    d. They strongly believed that sex was for procreation only and not for pleasure
    e. The entire Puritan community was responsible for upholding morality.

31. What statement represents the situation with abortion in the United States after the Revolutionary War?
    a. Women were spending time in jail after inducing abortion.
    b. Newspapers were prohibited from distributing information on abortion.
    c. There was no concept of abortion during this time.
    d. Abortion rates decreased dramatically.
    e. Abortion rates rose through the use of folk remedies.

32. How was sexuality used has an excuse for early white settlers in the United States to oppress and abuse minorities?
    a. Black male slaves were seen as a threat to white women.
    b. Native Americans were viewed as savages due to their acceptance of polygamy.
    c. Mexicans were considered promiscuous because they danced in public.
    d. Whites accused slaves as having loose sexual morals.
    e. All of the above

33. How did physicians respond to sexuality during the latter part of the nineteenth century?
    a. They emphasized the illness aspect of homosexuality and women's sex urges.
    b. They argued the importance of masturbation to a healthy mind.
    c. They stopped performing unnecessary surgery on women.
    d. They declared that women were biologically designed for college.
    e. They claimed that homosexuality was healthy and normal.

34. Who was Anthony Comstock?
    a. He was an important Protestant reformer.
    b. He was a pro-prostitution activist during the Victorian Era.
    c. He lobbied the legislature to outlaw obscenity in the 1870s.
    d. He was a Puritan leader who advocated against adultery.
    e. He was a prominent physician who distributed contraception during the turn of the century.

35. At the turn of the century, what did moral crusaders fault for the spread of prostitution and high rates of sexually transmitted infections?
    a. delaying marriage
    b. nightclubs
    c. new sexual freedoms
    d. youth who rejected traditional morality
    e. all of the above

36. What lead to laws mandating blood tests before marriage?
    a. The Social Hygiene Movement
    b. The Comstock Act
    c. The Mattachine Society
    d. The Free Love Movement
    e. Stonewall

37. Who was an important advocate for birth control during the start of the 20th century?
    a. Anthony Comstock
    b. Margaret Sanger
    c. Alfred Kinsey
    d. Prince Morrow
    e. John Kellog

38. Who wrote *The Feminine Mystique*, which found that a group of educated women felt trapped in the role of housewife and wanted careers in order to have happier, more fulfilled lives?
    a. Margaret Sanger
    b. Simone de Beauvoir
    c. Annie Comstock
    d. Betty Friedan
    e. Leslie Stonewall

39. During the 1950s, what were the Mattachine Society and the Daughters of Bilitis?
    a. Organizations that worked for gay and lesbian rights.
    b. Groups that worked to end the spread of sexually transmitted infections in the United States.
    c. Organizations that fought for birth control rights.
    d. Popular New York nightclubs that were frequented by homosexuals.
    e. Medical organizations for physicians who were trying to curb prostitution.

40. After the Stonewall rebellion, gay and lesbian activism lead to what significant civil rights victory in 1973?
    a. Gays and lesbians were granted the right to marry in New York State.
    b. Gay men could serve openly in the military without fears of dismissal.
    c. Homosexuals could no longer be fired from a job for being gay or lesbian.
    d. Homosexual couples were granted the right to become "domestic partners."
    e. Homosexuality was removed from a list of official psychiatric disorders.

Matching

Column 1		Column 2
A. The Renaissance	_____	41. salesman who fought against obscenity
B. Chinese	_____	42. propriety and public behavior became more important, and sex was not to be spoken of in polite company
C. The Enlightenment	_____	43. ancient civilization that practiced pederasty
D. Greek	_____	44. period of time marked by an emphasis on chastity and penance for sexual transgressions
E. Comstock	_____	45. ancient civilization that viewed marriage as a means to improve economic and social standing
F. Sanger	_____	46. sexual pleasure was considered natural and desirable
G. Roman	_____	47. birth control advocate
H. The Victorian Era	_____	48. period of time marked by women's strides in education and politics
I. McCarthy	_____	49. politician who uncovered homosexuals
J. The Middle Ages	_____	50. ancient civilization that emphasized the joining of the yin and yang

Test Yourself Answer Key

Human Sexuality in a Diverse World

1. laws, fantasies (p. 3)
2. sexologists (p. 3)
3. power (p. 3)
4. repressed (p. 3)
5. uncomfortable (p. 3)
6. knowledge, attitudes, relationships, behaviors (p. 3)
7. comfortable (p. 3)
8. images (p. 4)
9. sexual, nudity (p. 4)
10. sexuality (p. 5)
11. social norms, embarrassment, fear (p. 5)
12. Internet (p. 5)
13. Sex (p. 5)
14. invisible (p. 5)
15. sexually transmitted infections (p. 5)
16. The creation of ideas, laws, customs, fantasies, and art around the sexual act. (p. 3)
17. Biologists, psychologists, physicians, anthropologists, historians, sociologists, political scientists, and those concerned with public health (p. 2)
18. Many of these images have sexual scenes and nudity (pp. 3-5)
19. sexual paraphernalia, pornographic picture libraries, videos, video clips, web-cam sites, chat rooms (p. 5)
20. contraception, sexually transmitted infections, pregnancy risk (p. 5)

The Early Evolution of Human Sexuality

21. upright (p. 6)
22. abortion (p. 7)
23. adultery (p. 7)
24. Hebraic, Hellenistic or Greek, Roman (p. 7)
25. love (p. 8)
26. pederasty (p. 8)
27. homoerotic (p. 8)
28. platonic (p. 8)
29. economic, social (p. 9)
30. yin, yang (p. 9)
31. Chinese (p. 9)
32. sperm (p. 10)
33. Polygamy (p. 10)
34. Hinduism, karma (p. 10)
35. *The Kamasutra* (p. 11)
36. The upright posture emphasized female breasts and the clitoris. It also resulted in face-to-face intercourse with more body contact. (p. 6)
37. Men had more power and are more prevalent in historical texts and literature. (p. 7)
38. Adultery, same-sex sexual relations between men, and incest were forbidden, whereas an emphasis was placed on sexuality in marriage. (pp. 7-8)
39. Early Romans had permissive attitudes toward same-sex sexual behaviors. However, adults who took a receptive position in sexual encounters were viewed negatively. Long-term same-sex relationships did exist. (p. 9)

40. *The Kamasutra* was a sexual manual developed in ancient India that provided instruction on sexual techniques and positions, in addition to moral guidance regarding love, home, and family. (p. 11)

Sexuality From St. Paul to Queen Victoria and The Enlightenment

41. Christianity (p. 11)
42. Jesus (p. 12)
43. celibacy (p. 12)
44. sin (p. 12)
45. purity, perfect (p. 12)
46. Thomas Aquinas (p. 13)
47. Muhammad (p. 13)
48. modesty (p. 13)
49. Harems (p. 13)
50. eunuchs (p. 14)
51. Renaissance (p. 14)
52. Henricus Cornelius Agrippa (pp. 14-15)
53. Martin Luther (p. 15)
54. homosexuality (p. 16)
55. doctor (p. 17)
56. St. Paul viewed that the highest love was love of God, and that the ideal situation was not to allow sexual or human love to compete with love for God. (p. 12)
57. Jews, Muslims, and homosexuals (p. 12)
58. Ancient Arab societies (p. 14)
59. Martin Luther and John Calvin (p. 15)
60. The Enlightenment (p. 16)

Sex in American History

61. Puritans (p. 18)
62. women (p. 18)
63. bundling (p. 18)
64. one-third (p. 18)
65. abortion (p. 18)
66. anti-miscegenation laws (p. 19)
67. prostitution (p. 19)
68. free love (p. 19)
69. Oneida, Shakers (pp. 19-20)
70. procreation (p. 20)
71. Walt Whitman (pp. 20-21)
72. Comstock Act (p. 21)
73. Margaret Sanger (p. 22)
74. Feminists (p. 22)
75. Daughter, Bilitis (p. 23)
76. African slaves, Native Americans, and Mexicans (p. 19)
77. The Comstock Act of 1873 (p. 21)
78. The Social Hygiene Movement (p. 21)
79. *The Feminine Mystique* (p. 22)
80. Stonewall (p. 23)

Post Test Answer Key

True/False

1. T (p. 2)
2. F (p. 3)
3. T (p. 7)
4. F (p. 11)
5. T (p. 11)
6. T (p. 12)
7. F (p. 16)
8. T (p. 19)
9. F (p. 20)
10. F (p. 22)

Multiple-Choice

11. b (p. 5)
12. e (p. 6)
13. c (p. 8)
14. b (p. 8)
15. c (p. 8)
16. a (p. 10)
17. a (p. 11)
18. b (p. 12)
19. c (p. 12)
20. d (p. 12)
21. e (p. 12)
22. c (p. 12)
23. d (p. 13)
24. e (p. 13)
25. a (p. 14)
26. b (p. 14)
27. d (p. 15)
28. d (p. 16)
29. e (p. 16)
30. d (p. 18)
31. e (p. 18)
32. e (p. 19)
33. a (p. 20)
34. c (p. 21)
35. e (p. 21)
36. a (p. 21)
37. b (p. 22)
38. d (p. 22)
39. a (p. 23)
40. e (p. 23)

Matching

41. E (p. 21)
42. H (p. 16)
43. D (p. 8)
44. J (p. 12)
45. G (p. 9)
46. C (p. 16)
47. F (p. 22)
48. A (p. 14)
49. I (p. 23)
50. B (p. 9)

Chapter 2 – Understanding Human Sexuality: Theory and Research

CHAPTER SUMMARY

The study of sexuality is multidisciplinary, using multiple theoretical perspectives including psychological, biological, sociological, sociobiological, feminist, and queer approaches to understanding human sexuality. Sigmund Freud's psychoanalytic theory has become one of the most influential of the psychological theories. Freud's concepts of personality formation can be broken into personality divisions that include the id, ego, and superego. Psychosexual development involves different stages of development during the first six years of life: oral, anal, phallic, latency, and genital. Behavior theory endorses the perspective that it is necessary to observe and measure behavior in an effort to understand it. Social learning theory incorporates the reward and punishment view of operant conditioning from behaviorism with internal events such as feelings, thoughts, and beliefs in order to understand behavior. The motivation to develop the best of our abilities and become self-actualized emerges from humanistic theory, which emphasizes the importance of unconditional positive regard in this quest. Sociologists examine how society influences behaviors, where sociobiological theories connect sociology with evolution. Two politically situated theories, feminist and queer, examine sexuality from the perspective of power and oppression.

Sexuality research has spanned the centuries from the time of ancient Greeks to contemporary times, facing repeated issues of legitimacy, stigma, and a focus upon "problem-driven" research. The pioneers of early sexuality research included: Iwan Bloch, Albert Moll, Magnus Hirschfeld, Richard von Krafft-Ebing, Havelock Ellis, Clelia Mosher, and Katharine Bement Davis. After concentrating in Germany, sexuality research moved to the U.S. at the start of the 20th century. Alfred Kinsey was probably the most influential sex researcher of the during this time with his large-scale research interviews, publication of best-selling sexuality books, and the establishment of the Institute for Sex Research at Indiana University. Masters and Johnson also forged new ground by being the first contemporary scientists to observe and measure the physiological aspects of sexuality in the laboratory. A number of valuable and historical research studies have been conducted on the sexuality of adolescents, older adults, and gay, lesbian, bisexual, and transgendered individuals. The most recent notable large-scale research studies on Americans were the *Janus Report on Sexual Behavior* published in 1993 and the *National Health and Social Life Study (NHSLS)* published in 1994.

Sexuality research must pass standards of validity, reliability, and generalizability. Sexuality researchers have a variety of methodologies from which to choose including case study, questionnaire, interview, direct observation, participant-observation, experimental methods, and correlations. Regardless of which method, researchers must be aware of the challenges and limitations such as ethical issues, volunteer bias, sampling problems, and reliability. Cross-cultural research on sexuality has provided information on how societal values and culture influence sexuality as illustrated by a number of important cross-cultural studies in France, China, and a comprehensive global study funded by Pfizer Pharmaceuticals in 2002 that surveyed 26,000 people from 28 countries. Future research in sexuality needs to answer some new questions but faces challenges due to funding difficulties.

LEARNING OBJECTIVES

After studying Chapter 2, you should be able to:

1. Compare and contrast the different theories about sexuality, and provide definitions of each theory: psychoanalytic, behavioral, social learning, cognitive, humanistic, biological, sociological, sociobiological, feminist, and queer.
2. Summarize the key elements of Sigmund Freud's theories on personality formation and psychosexual development.
3. Describe how six different societal institutions influence the rules and regulations a society holds about sexual expression from a sociological theoretical perspective.
4. Discuss how both feminist and queer theories differ from the other theories of sexuality.
5. Define sexology, and explain the role sexologists play in sexuality research from both a historical and contemporary perspective.
6. Provide examples of "problem-driven" research and research that is not guided by problematic sexuality.
7. Describe some of the societal controversy and stigma sexuality researchers faced in the last 100 years.
8. Identify the seven sexuality researchers who were active during the turn of the 20th century.
9. Discuss the key aspects of Alfred Kinsey's research including his methodology, sampling techniques, findings, and limitations of his research.
10. Explain how interviewer bias and volunteer bias can limit a research study in sexuality.
11. Discuss the key aspects of Masters and Johnson's research including their methodology, sampling techniques, findings, and limitations of their research.
12. Describe the research methodologies that have been used to learn about adolescent sexuality and some of the major areas of inquiry.
13. Summarize some of the research findings on sexuality in older adults.
14. Compare and contrast the research findings of Evelyn Hooker and that of Bell and Weinberg.
15. Identify some of the research findings, strengths, and limitations of the *Janus Report on Sexual Behavior* and the *National Health and Social Life Study.*
16. Define each of the following research terms: validity, reliability, and generalizability.
17. Define the following research methods, providing examples of each: case study, questionnaire, interview, direct observation, participant-observation, experimental methods, and correlational studies.
18. Identify the key elements of experimental methods.

19. Provide examples of some of the problems and issues in sex research: ethical issues, volunteer bias, sampling problems, and reliability.

20. Summarize some of the research findings from the large-scale sexuality studies in France and China.

21. Discuss some of the challenges sexuality research faces in the future.

CHAPTER OUTLINE

I. THEORIES ABOUT SEXUALITY

- The study of sexuality is multidisciplinary and includes psychologists, sexologists, biologists, theologians, physicians, sociologists, anthropologists, and philosophers.
- A **theory** is a set of assumptions, principles, or methods that help a researcher understand the nature of the phenomenon being studied.

A. **Psychological Theories**

1. Psychoanalytic Theory, Sigmund Freud (1856-1939)

a. Personality Formation

- Human behavior is motivated by instincts and drives.
- One of the two powerful drives is the **libido**, which is the life or sexual motivation.
- **Thanatos** is the other most powerful drive, which is the death or aggressiveness motivation.
- There are three levels in which one of two of the divisions of the personality operates: conscious, preconscious, and unconscious
    - **conscious**—the part of the personality that contains the material of which we are currently aware.
    - **preconscious**—the part of the personality that contains thoughts that can be brought into awareness with little difficulty.
    - **unconscious**—all of the ideas, thoughts, and feelings of which we are not and cannot normally become aware.
- The second division of the personality contains the id, ego, and superego.
    - **id**—the collection of unconscious urges and desires that continually seeks expression.
    - **ego**—the part of the personality that mediates between environmental demands (reality), conscience (superego), and instinctual needs (id).
    - **superego**—the social and parental standards an individual has internalized; the conscience.
- **psychoanalysis**—allows the individual to bring unconscious thoughts into consciousness to help resolve a patient's disorder.

b. Psychosexual Development

- Freud believed that our basic personality is formed by events that happen to us in the first six years of life.
- Libido energy is directed to a different **erogenous zone** during each stage of development.
- The first stage is the **oral stage** where the mouth, lips, and tongue are the erogenous zone. A **fixation** (tying up of psychic energies at one stage, resulting in adult behaviors characteristic of that stage) leads to dependency or aggression.

- The second stage is the **anal stage** where the focus is on the anus. **Anal fixation** may include traits such as stubbornness, orderliness, or cleanliness.
- The third and most important stage is the **phallic stage** where the genitals become the erogenous zone.
  - **Oedipus complex**—a male child's sexual attraction for his mother and the consequent conflicts.
  - **Electra complex**—the incestuous desire of the daughter for sexual relations with the father.
  - Karen Horney, a follower of Freud, challenged the concept of penis envy, and other modern feminists have reframed psychoanalytic theory to be less biased against women.
- During the **latency stage** all libido and sexual interest disappears due to the fear and strength of the previous stage.
- The final stage of psychosexual development, the **genital stage,** begins at puberty. The focus becomes the genitals and the ability to engage in adult sexual behaviors is developed.

c. Freud's theories were viewed as controversial in Victorian Vienna.
- Children were sexual from birth.
- Children lusted for the other-sex parent.
- Modern psychologists claim his theories are unscientific and untestable.
- unflattering views of women
  - vaginal orgasms viewed as more mature than clitoral orgasms
  - women seen as neurotic or deficient

2. Behavioral Theory

a. **Behaviorists** believe that it is necessary to observe and measure behavior in order to understand it.

b. Radical behaviorists such as B.F. Skinner endorse **operant conditioning,** where we learn behaviors through reinforcement and punishment.

c. **behavior modification**—therapy based on operant conditioning and classical conditioning principles, used to change behaviors.

d. **aversion therapy**—technique that reduces the frequency of maladaptive behavior by associating it with real or imagined aversive stimuli during a conditioning procedure.

3. Social Learning Theory

a. Albert Bandura argued that external events, such as rewards and punishments, influence behavior, but so do internal events, such as feelings, thoughts, and beliefs.

b. Imitation and identification are important in the development of sexuality, as illustrated by the example of learning gender identity.

4. Cognitive Theory

a. Thoughts and perceptions are responsible for behaviors.

b. The biggest sexual organ is between the ears.

5. Humanistic Theory (or person centered)

a. Humans all strive to develop to the best of their abilities and to become self-actualized.

b. **Self actualization**—the fulfillment of an individual's personalities; the actualization of aptitudes, talents.

c. Becoming self-actualized is easier when children are raised with **unconditional positive regard**—accepting others unconditionally, without restrictions on their behaviors or thoughts.
d. **conditional love**—accepting others conditionally, placing restrictions on their behaviors or thoughts.

B. **Biological Theory**

- The biological theory of human sexuality emphasizes that sexual behavior is primarily a biological process controlled physiologically through inborn, genetic patterns.

C. **Sociological Theories**

1. Sociologists examine how society influences sexual behavior through social norms and regulations.
2. Many institutions influence the rules a society holds about sexual expression.
    a. family
    b. religion
    c. economy
    d. medicine
    e. law
    f. media

D. **Sociobiological Theories**

1. Sociobiology incorporates both evolution and sociology to understand sexual behavior.
2. Sexuality exists for the purpose of reproducing the species and successfully passing on one's genes.
    a. physical attraction
    b. premarital sex
    c. discrepancies between women and men in sexual desires and behaviors
    d. criticism: sociobiologists tend to ignore the influence of both prior learning and societal influences on sexuality.

E. **Feminist Theory**

1. Many feminist researchers attempt to redefine sexology, which has been constructed from a white, middle class, heterosexist, medical, and biological viewpoint.
2. Construction of sexuality is based on power leading to sexual gender inequality.
3. Support for collaborative or group research is often from a qualitative research perspective.

F. **Queer Theory**

1. Developed in 1990s from lesbian and gay studies.
2. Support for a restructuring of sexuality away from heterosexism and homophobia and toward a deconstruction of categories of sex, gender, and desire.

II. SEXUALITY RESEARCH: PHILOSOPHERS, PHYSICIANS, SEXOLOGISTS

A. Ancient Greeks

- Hippocrates, Aristotle, and Plato—forefathers of sex researchers, were the first to elaborate theories regarding sexual responses and dysfunctions, sex legislation, reproduction, contraception, and sexual ethics.

B. 18th century

- First program of public and private sex education and classifications of sexual behavior were established.

C. 19th century

1. Sex research began to focus on the different, dangerous, and unhealthy aspects of sexuality.
2. Victorian attitudes lead to stigma around conducting sex research.
3. Physicians were primary sexuality researchers.
4. Jewish Europeans conducted most of the early sexuality research connecting anti-sex views with anti-Semitism.

D. 20th century

1. Sexuality research arrived in the United States.
2. Research took off in the 1920s as a response to the hygiene movement. Sexuality was viewed as a threat to children and society.
3. Funding was difficult for sexuality researchers to secure.
4. "Agenda" or "problem-driven" research
    a. has lead to a focus on social concerns such as adolescent sexuality and HIV/AIDS
    b. less attention paid to healthy or "typical" sexuality such as love and relationships
5. Much vocal opposition to sexuality research due to fears that research will cause individuals to act out sexually.
6. Sexuality research is multidisciplinary, which can be problematic due to decreased communication.
7. **Sexologists** are researchers, educators, and clinicians who specialize in sexuality and often work to prove themselves as legitimate scientists.
8. Many university programs in sexuality exist with the possibility that they will come together in the future as a separate discipline of sexual science.

III. SEXUALITY RESEARCHERS

A. Iwan Bloch (1872-1922)

1. Berlin dermatologist who believed that historical and anthropological research could help broaden sexuality research.
2. helped to form a medical society for sexology research in Berlin
3. published the *Journal of Sexology*

B. Albert Moll (1862-1939)

1. conservative man who countered the research of Freud and Hirschfeld
2. formed the International Society for Sex Research

C. Magnus Hirschfeld (1868-1935)

1. His work with patients convinced him that negative attitudes toward homosexuals were inhumane, which inspired his dedication to the field of sexual problems.
2. became an expert on homosexuality and other sexual variations
3. opened the first Institute for Sexology, which contained libraries, laboratories, and lecture halls
4. His research was destroyed during WWII by the Nazis.

D. Richard von Krafft-Ebing (1840-1902)

1. one of the most significant medical writers on sexology in the late 19th century
2. Focused on "deviant" sexual behavior, writing a book, *Psychopathia Sexualis,* which focused on individuals who experienced what he considered sexual pathology.

E. Havelock Ellis (1859-1939)

1. English citizen who grew up in Victorian society
2. published famous six-volume *Studies in the Psychology of Sex*
3. established himself as an objective and nonjudgmental researcher

F. Clelia Mosher (1863-1940)

1. first researcher to ask Americans about their sexual behaviors
2. Her interest in helping married women have more satisfying sex lives lead her to interview upper-middle class women about whether they enjoyed sexual intercourse.
3. Her results, suggesting these women viewed intercourse as both for sexual pleasure and procreation, were never published.

G. Katharine Bement Davis (1861-1935)

1. appointed superintendent of a prison and became interested in prostitution and sexually transmitted infections.
2. She believed that lesbianism was not pathological.
3. Her ideas that women might have sexual appetites equal to men's lead male researchers to try to strengthen the family unit.

H. Alfred Kinsey (1894-1956)

1. most influential sex researcher of the 20th century
2. moved the study of sexuality away from the medical model
3. persevered continually despite controversy and stigma
4. Kinsey's early work was **atheoretical** due to what he thought was the lack of previous knowledge on which to base theories.
5. Large-scale study of sexual life histories of 20,000 people through in-depth interviews
6. Concern about **interviewer bias** lead Kinsey to assure that only four people conducted all the interviews.
7. Rather than using **probability sampling**, he used what he called **100% sampling** by interviewing everyone located in various groups and organizations.
8. Institute for Sex Research
   a. Established in 1947 by Kinsey and his associates at Indiana University.
   b. Kinsey published two bestsellers that helped to break down the myths and confusion surrounding sexuality.
      - *The Sexual Behavior of the Human Male,* 1948
      - *The Sexual Behavior of the Human Female,* 1953
   c. He continued to face controversy that lead to the termination of several research grants
   d. *Sex and Morality in the U.S.:* research study published in 1989 to complement Kinsey's original work by focusing more on thoughts, feelings, and attitudes towards sexuality.
   e. *The Kinsey Institute New Report on Sex:* was part of a sexuality quiz in 1989 meant to determine Americans' knowledge about sexuality
      - displayed the lack of sexuality knowledge around AIDS, homosexuality, birth control, and sexual problems
      - age differences: older people acknowledged their lack of knowledge while younger people did not

I. Morton Hunt

1. In 1974 he published *Sexual Behavior in the 1970s,* commissioned by the Playboy Foundation
2. criticized due to low response rate of 20% and the random selection donc by telephone
3. not **generalizable** due to **volunteer bias** and telephone sample

J. William Masters and Virginia Johnson

1. The gynecologist and psychologist were thc first modern scientists to observe and measure the act of sexual intercourse in the laboratory, beginning in 1954.

2. dual sex-therapy team primarily interested in the anatomy and physiology of the sexual response
3. first study: *Human Sexual Response,* 1966
4. first subjects were prostitutes, but that was problematic due to their **chronic pelvic congestion**
5. began to recruit other volunteers for financial reasons, personal reasons, and for the release of sexual tension
6. They didn't recruit a random sample because they felt that they were studying behaviors that happened to most people
7. Instruments measured heart and muscle changes, including **penile strain gauges** and **photoplethysmographs** that measured penile erection and vaginal lubrication.
8. 1970: *Human Sexual Inadequacy* discussed sexual dysfunction
9. Overall research findings:
    a. multiple orgasms in women
    b. dual sexual dysfunction in couples
    c. refuted Freud's theory of separate female orgasms, instead finding that all women need direct or indirect clitoral stimulation in order to have an orgasm.

K. Age-Specific Studies

1. Adolescent studies
    a. 2000: national longitudinal study of adolescent health completed by the National Institute of Child Health and Human Development
        - 125,000 adolescents surveyed through interviews and questionnaires
        - four areas: emotional health, sexuality, violence, and substance use
    b. 1988-1995: longitudinal study on adolescent males, *National Survey of Adolescent Males* (NSAM)
        - face-to-face interviews with 6,500 adolescent males
        - areas covered included sexual contraceptive histories, attitudes about sexuality, contraception, and fatherhood
        - findings showed that a significant number of adolescent males engage in sexual activities beyond vaginal intercourse
    c. 1973: Robert Sorensen: *Adolescent Sexuality in Contemporary America*
        - examined frequency of masturbation, sexual activity, and homosexual behavior
        - concerns with reliability due to many parents not permitting teenagers to participate
    d. 1970s classic study by Melvin Zelnik and John Kanter examined the sexual and contraceptive behavior of 15-19 year-old females.

1. Older Adult Studies
    a. 1981: Vernard Starr and Marcella Weiner explored the sexuality of 800 adults who were between the ages of 60 and 91 years
        - questionnaire and low response rate may have over-represented seniors interested in sex
        - 50 open-ended questions about sexual experience, changes in sexuality that have occurred with age, sexual satisfaction, sex and widowhood, sexual interest, masturbation, orgasm, sex likes and dislikes, and intimacy
        - results showed that older adults continued to view sexuality as important as they aged
    b. 1983: Edward Brecher: Love, Sex and Aging

- survey included questions on attitudes about sex, behaviors, and sexual concerns
- results suggested that older adults were indeed sexual although society considered them nonsexual

c. Current research on older adults supports previous research in that older adults are interested in sexuality, with a survey of adults over the age of 60 expressing that over half of them were sexually active.

L. GLBT Studies (gay, lesbian, bisexual and transgendered)

1. Much research exists but few wide-scale studies.
2. Evelyn Hooker
   a. studied male homosexuality in the 1950s
   b. demonstrated that there was little fundamental psychological difference between gay and straight men
   c. first study to provide evidence that homosexuality was not a psychological disorder
3. Alan Bell and Martin Weinberg
   a. published *Homosexualities,* 1978
   b. compared 5000 homosexual adults with 5000 heterosexual adults
   c. findings refuted stereotypes: Gay men and lesbians do not push unwanted sexual advances onto people, they do not seduce children, and their intimate relationships were similar to heterosexual individuals

M. Other Sexuality Studies

1. The *Janus Report on Sexual Behavior*
   a. published in 1993 by Samuel and Cynthia Janus
   b. 3,000 questionnaires examined regional differences
   c. criticisms included no random selection and the overestimation of sexual behaviors
   d. Findings:
      - Americans in their sixties and seventies are experiencing greatly heightened levels of sexual activity
      - Married couples reported the highest level of sexual activity and satisfaction
      - Three out of five married people said their sex lives improved after marriage
      - Areas in which people live influence overall sexual attitudes and behaviors
      - people who are ultra-conservative are more likely to be involved in frequent or ongoing extramarital affairs than are those who are ultra-liberal
      - men and women are both initiating sexual activity
2. *The National Health and Social Life Study,* 1987
   a. Due to HIV/AIDS crisis, the U.S. Department of Health and Human Services called for large study of American adults.
   b. After securing funding and plans to interview 20,000 people, funding was withdrawn in 1991 due to conservative politicians.
   c. Private funding was secured, and the sample size reduced to 4,369 with a 79% response rate.
   d. NHSLS was the most comprehensive study of sexual attitudes and behaviors since the Kinsey Report with much better sampling procedures.

e. Results suggested that Americans were more sexually conservative than previously thought.
    - The median number of sexual partners since the age of 18 was six for men and two for women
    - 75% of married men and 80% of married women do not engage in extramarital sexuality
    - 2.8% of men and 1.4% of women describe themselves as homosexual or bisexual
    - 75% of men claimed to have consistent orgasms with their partners, while 29% of women did
    - More than one in five women said they had been forced by a man to do something sexual

IV. SEX RESEARCH METHODS AND CONSIDERATIONS

- **validity:** whether or not a test or question measures what it is designed to measure
- **reliability:** the consistency of the measure with repeated testing of the same group
- **generalizability:** ability for the answers of a few subjects to be applicable to the general population

A. Case Study: exploring individual cases to formulate general hypotheses
1. Freud was known for this methodology
2. lack of generalizability

B. Questionnaire vs. Interview
1. Questionnaire or survey research is generally used to identify the attitudes, knowledge, or behavior of large samples.
2. Alfred Kinsey was known for this research methodology.
3. Interviews allow the researcher to establish a rapport with each subject and emphasize the importance of honesty in their study.
4. Interviews can be more time-consuming and expensive than questionnaires.
5. Questionnaires can provide more honesty due to the anonymity of a questionnaire.

C. Direct Observation
1. Masters and Johnson used this type of methodology.
2. People were invited to engage in sexual activity in a laboratory while physiological changes were monitored.
3. Challenges include the difficulty of finding subjects, cost, and generalizability due to lack of a random sample.
4. The limitations include the lack of information on feelings, attitudes, or personal history.

D. Participant-Observation
1. This method involves researchers going into an environment and monitoring what is happening naturally.
2. Limitations include a lack of generalizability and access to private settings.

E. Experimental methods
1. This is the only method that allows researchers to isolate cause and effect.
2. Requires **random assignment**, where subjects are assigned to groups with each subject having an equal chance of being assigned to each group.
3. The **independent variable** is manipulated by the researcher and applied to the subject to determine the effect on the **dependent variable**.
4. Limitations:
    a. often expensive and time-consuming
    b. may not replicate real-world situations

c. especially in some sexuality situations, may be unethical or impossible to replicate in the laboratory

E. Correlations

1. often used when it's not possible to conduct an experiment
2. Example: study a given population to see if there is any correlation between past sexual abuse and later difficulties with intimate relationships
3. The major limitation is that it doesn't provide any information about cause

V. PROBLEMS AND ISSUES IN SEX RESEARCH

A. Ethical Issues

1. Prior to a person's participation in a study of sexuality, it is necessary to obtain **informed consent**, which means that the person knows what to expect from the questions and procedures, how the information will be used, and how confidentiality will be assured.
2. The importance of keeping all information and materials gathered during a research study reflects the concept of confidentiality.

B. Volunteer Bias

1. College student volunteers for sexuality research were found to be more sexually liberal, more sexually experienced, more interested in sexual variety, more likely to have had sexual intercourse and performed oral sex, and report less traditional attitudes than nonvolunteers.
2. difficult to generalize

C. Sampling Problems

1. Samples of convenience are typically college age samples that are convenient for researchers who work at universities.
2. difficult to generalize

D. Reliability

1. difficult to determine reliability with such a sensitive subject as sexuality
2. studies designed specifically for determining reliability of sexuality research
3. Reporting of behavior may be affected by time and location

VI. SEXUALITY RESEARCH IN OTHER COUNTRIES

A. Societies' values and culture influence on sexuality has been one of the topics sexuality researchers have studied the most.

B. *Human Sexual Behavior,*1971: Donald Marshall and Robert Suggs' classic anthropological study examined how sexuality was expressed in several different cultures entitled

C. 1991 & 1992: Largest study done in France in 20 years examined the sexual practices of over 20,000 people between the ages of 18 and 69.

1. telephone interviews with a response rate of 76.5%
2. many teenagers do not use condoms during sexual intercourse because they are too expensive
3. rates of extramarital sexual behavior are decreasing
4. the average French heterosexual engages in sex approximately two times per week.

D. Similar study to the one in France was conducted in Britain and yielded similar findings.

E. 1992: nation-wide survey of sexuality behavior done in China with 23,000 individuals completing the questionnaires.

1. 50% of young people engaged in sex outside of marriage
2. 70% approved of extramarital sex
3. many married couples considered sex as nothing more than a duty or a means of producing offspring
4. the majority of Chinese couples use little or no foreplay before sexual intercourse.

5. adolescent sexuality education does not take place in the Chinese school curriculum

VI. SEX RESEARCH IN THE FUTURE

A. Several nationwide sexuality studies have been cancelled in recent years due to pressure from conservative politicians.
   1. *Survey of Health and AIDS Risk Prevalence*
   2. *American Teen Study* (ATS), 1991

B. All studies must be peer-reviewed by the Institutional Review Board

C. Future research is needed to find:
   1. how health crises such as HIV/AIDS are related to human sexuality
   2. how sexual beliefs, attitudes, and values may contribute to some of these problems
   3. how sexual behaviors and attitudes are shaped by society, culture, and familial contexts

TEST YOURSELF

Below you will find fill-in-the-blank and short answer essay questions for topics covered in chapter 2. Check your answers at the end of this chapter.

Theories about Sexuality

1. A ____________________ is a set of assumptions, principles, or methods that help a researcher understand the nature of the phenomenon being studied.

2. Of the psychological theories of sexuality, the most influential has been the ____________________ theory of Sigmund Freud.

3. According to Freud the two most powerful drives of human behavior are ____________________ and ____________________.

4. Freud believed that the ____________________ was the most important level of personality, containing information that we have no conscious access to such as conflicts or anxiety-producing memories.

5. By five years of age, the last portion of the personality, the ____________________, develops, containing both societal and parental values.

6. During each stage of development, Freud identified a different ____________________ ____________________ where libido energy is directed.

7. Freud reported that the ____________________ stage begins at puberty when the erogenous zone becomes the genitals.

8. To help change unwanted behavior, behaviorists use ____________________ ____________________.

9. Social learning theories believe that ____________________ and ____________________ are important in the development of sexuality.

10. Humanistic psychologists believe that it easier to become self-actualized if we are raised with ____________________ ____________________ ____________________.

11. The biological theory of human sexuality emphasizes that sexual behavior is primarily due to inborn ____________________ patterns.

12. Sociologists are interested in how the ____________________ influences sexual behavior.

13. Sociobiology incorporates both ____________________ and ____________________ to understand sexual behavior.

14. Feminist researchers have made efforts to redefine the biological and over-____________________ of sexuality.

15. The feminist and queer theories share a common political interest — a concern for ____________________ and ____________________ rights.

16. What were two of Freud's most controversial concepts?

17. What are the five psychological theories of sexuality discussed in the textbook?

18. What are three of the many institutions that can influence the rules a society holds about sexual expression?

19. Which theories about sexuality incorporate some aspect of biology?

20. Why would feminist researchers use collaborative research methods such as interviewing?

Sexuality Research

21. The ancient ____________________, including physicians like Hippocrates, and philosophers such as Aristotle and Plato, may be the forefathers of sexuality research.

22. It wasn't until the 18th century that there was increased discussion of sexual ____________________.

23. During the 19th century, sex research was almost exclusively focused on people believed to be ____________________.

24. ____________________ were the primary sexuality researchers in the late 19th century.

25. The majority of the early sexuality studies were done in Europe, primarily in ____________________.

26. Due to the stigma around conducting sexuality research, some researchers used ____________________ to publish their work.

27. Many of the early researchers of sexology were of ____________________ background.

28. Systematic research into sexuality in the U.S. began in the early 1920s, motivated by pressures from the social ____________________ movement, which was concerned about venereal disease and its impact on marriage and children.

29. Although various funding sources for sex research exist, many of these sources are ____________________ ____________________ and do not allow researchers to answer questions that they might have about sexuality.

30. In the late 1980s and early 1990s there was a surge in sexuality research, predominantly driven by concerns over ____________________.

31. Conservative groups believe that research done on topics such as ____________________ ____________________ would encourage young people to have more sex.

32. Sexuality research in the U.S. has become very ____________________, with researchers coming from several different disciplines.

33. Researchers, educators, and clinicians who specialize in sexuality are called ____________________.

34. A unified department of sexual science would enable the field to acquire appropriate research ____________________.

35. Although sexuality research is still in its early stages, it has begun to help remove the ____________________ and ____________________ associated with discussing human sexual behavior.

36. What challenges did sex researchers face during the Victorian period in the late 19th century?

37. Who were the wealthy individuals whose philanthropy provided sexuality researchers with large-scale funding sources during the middle of the 20th century?

38. According to the authors, why are there many people opposed to sexuality research today?

39. What are examples of sexuality research that are not addressed by the focus on problem-driven research?

40. What type of discipline might universities form to help united the different areas studying sexuality?

Sexuality Researchers

41. Iwan Bloch hoped that sexual science would one day have the same structure and ____________________ as other sciences.

42. ____________________ ____________________ was a very conservative man who disliked both Freud and Hirschfeld and tried to counter their research at every opportunity.

43. In 1919 Magnus Hirschfeld opened the first ____________________ ____________________ ____________________.

44. Richard von Krafft-Ebing's primary interest was what he considered ____________________ sexual behavior.

45. Two important female researchers active around the turn of the 20th century were ____________________ ____________________ and ____________________ ____________________ ____________________.

46. Alfred Kinsey was the first researcher to take the study of sexuality away from the ____________________ model and move it towards other scientific disciplines.

47. Kinsey claimed to be ____________________ since he felt that sexuality research was so new, it was impossible to construct theories and hypotheses without first having a large body of information to base them on.

48. In an effort to counter ____________________ ____________________, Kinsey had only himself and three colleagues conduct 20,000 interviews.

49. Kinsey's books *The Sexual Behavior of the Human Male* (1948) and *The Sexual Behavior of the Human Female* (1953) were overnight ____________________, providing the institute with the financial ability to continue its work.

50. Masters and Johnson were the first modern scientists to observe and measure the act of sexual intercourse in the ____________________.

51. Masters and Johnson were studying behaviors they felt happened to most people so they did not feel that they needed to recruit a ____________________ ____________________.

52. Masters and Johnson research results noted the fear of ____________________ and ____________________ anxiety on sexual functioning.

53. Current research into elderly sexuality supports earlier studies, finding that over ____________________ reported that they were sexually active.

54. Evelyn Hooker's 1950s research study was the first to provide evidence that ____________________ was not a psychological disorder.

55. One of the findings from the Janus report released in 1993 was the three out of five married people said their lives improved after ____________________.

56. According to the authors, who was the most influential sex researcher of the 20th century?

57. How many different areas did Kinsey cover in his in-depth interviews for his groundbreaking research study?

58. How did Masters and Johnson obtain their first research subjects for their laboratory study?

59. What is one major societal myth that the findings from the research studies on older adults disproved?

60. What major research study lost federal funding in 1991 due to pressure from conservative politicians?

Sex Research Methods and Considerations

61. Tests of ____________________ determine whether or not a question is measuring what it is designed to measure.

62. Interviews allow the researcher to establish a ____________________ with each subject.

63. ____________________ involves researchers going into an environment and observing what happens naturally.

64. The experimental method is the only research method that enables researchers to isolate ____________________ and ____________________.

65. Experiments are usually much more costly than any of the other methods discussed, both in terms of ____________________ and ____________________ ____________________.

66. The major limitation of a correlation study is that it doesn't provide any information about ____________________.

67. Prior to a person's participation in a research study, it is necessary to obtain ____________________ ____________________.

68. Studies that have examined ____________________ ____________________ in sexuality research conducted with college students generally support the finding that volunteers differ from nonvolunteers.

69. Some studies have found that couples who are sexually satisfied ____________________ their frequency of sexual behavior, while those who are unsatisfied ____________________ them.

70. It is necessary to take the ____________________ and ____________________ of a study into account while evaluating the results.

71. A comprehensive study of sexual behavior in France in 1991 and 1992 found that rates of __extramarital__ sexual behavior were decreasing.

72. The nation-wide sexuality survey conducted in 1991 in China found that ____50____ % of young people engaged in sex outside of marriage.

73. The comprehensive global study of sexuality done by Pfizer Pharmaceuticals was the first global survey to assess ____________________, ____________________, ____________________, and ____________________ ____________________.

74. The ____________________ and sense of ____________________ of the Internet has given sexuality researchers access to a wider group of diverse subjects.

75. In order to be federally funded, all sexuality studies must be ____________________ ____________________, and approved by the investigator's ____________________ ____________________ ____________________.

76. What research methodology what Freud famous for using?

77. What are the challenges and limitations of direct observation as a methodology for sexuality research?

78. What are the problems that are more of a challenge when it comes to conducting sexuality research compared to other types of research?

79. What types of subjects does the term "samples of convenience" typically refer to?

80. What has been learned the most from cross-cultural studies of sexuality?

Post Test

Below you will find true/false, multiple-choice, and matching quiz items covering the entire chapter. Check your answers at the end of this chapter.

True/False

1. Sigmund Freud was a major proponent of the social learning theory of sexuality.

2. Sociologists believe that many societal institutions influence the rules a society holds about sexual expression.
3. Sociobiology incorporates both evolution and sociology to understand sexual behavior.

4. In the late 19th century, the primary sexuality researchers were philosophers and religious leaders.

5. Sexologists are sometimes ridiculed for studying a field of research that is not considered legitimate science.

6. Alfred Kinsey was noted as the first researcher to observe the physiological changes of sexuality in the laboratory.

7. Masters and Johnson supported Sigmund Freud's earlier views that clitoral and vaginal orgasms were physiologically different in women.

8. Experimental methods provide no information on cause.

9. Informed consent is not necessary in sexuality research.

10. Volunteers for sexuality research have been found to be different from nonvolunteers.

Multiple-Choice

11. A theory is ____________________.
    a. a method that explores individual cases to formulate general hypotheses
    b. manipulated by the researcher
    c. a study that is done over a certain period of time, wherein subjects are studied at various intervals
    d. a set of assumptions, principles, or methods that help a researcher understand the nature of the phenomenon being studied
    e. None of the above

12. Freud believed that human behavior is motivated by the ____________________.
    a. libido and thanatos
    b. conscious and unconscious
    c. id and superego
    d. ego and Oedipus complex
    e. erogenous zone and phallic stage

13. According to Sigmund Freud's theory of personality development, which part of the personality functions as the pleasure center?
    a. ego
    b. unconscious
    c. id
    d. superego
    e. anal fixation

14. According to Freud, during the oral stage ____________________.
    a. children become fixated on the genitals
    b. the erogenous zone is the mouth
    c. verbal skills develop
    d. the unconscious undergoes rapid development
    e. the libido experiences latency

15. During his lifetime, Sigmund Freud's theories on sexuality were ____________________.
    a. praised by politicians and religious leaders
    b. a reaction to the views of Alfred Kinsey
    c. were considered controversial
    d. developed from his experimental work in the laboratory
    e. All of the above

16. To help change unwanted behavior, behaviorists would use ____________________.
    a. aversion therapy
    b. a photoplethysmograph
    c. psychoanalysis
    d. conditional love
    e. unconditional positive regard

17. Which theoretical view of sexuality grew out of behaviorism but emphasized that thoughts and feelings had more influence on behaviors than the behaviorists claimed?
    a. humanistic theory
    b. sociobiological theory
    c. social learning theory
    d. psychoanalytic theory
    e. biological theory

18. Which is true of cognitive theorists?
    a. They believe that we all strive to become self-actualized.
    b. They believe that the biggest sexual organ is the brain.
    c. They believe that all human sexual behavior is primarily due to inborn, genetic patterns.
    d. They believe that sexology is dominated by white, middle class, heterosexist attitudes.
    e. They believe that the basic personality is formed by events that happen in the first six years of life.

19. Self-actualization is a concept related to what theoretical perspective?
    a. Social Learning Theory
    b. Behaviorist Theory
    c. Psychoanalytic Theory
    d. Sociobiological Theory
    e. None of the above

20. Feminist scholars believe that the social construction of sexuality is based on ____________________.
    a. the psychosexual stages of development
    b. operant conditioning
    c. evolution and sociology
    d. the id
    e. power

21. What is true about queer theory of sexuality ____________________.
    a. It seeks to abandon the category of "homosexual."
    b. It emerged from lesbian and gay studies.
    c. It resists the model of heterosexuality as a focus.
    d. It centers around a political focus.
    e. All of the above

22. Which of the following is NOT true about sexuality research conducted around the turn of the 20th century?
    a. Research focused on problematic sexual issues such as diseases.
    b. Some sexuality researchers used pseudonyms to publish their research.
    c. The majority of early sexuality studies were conducted in Europe.
    d. Sexuality researchers were credited with helping society deal with sexual disorders.
    e. Sexuality organizations faced low membership rates.

23. Which sexuality researcher from the turn of the 20th century established himself as an objective and nonjudgmental researcher who reported that masturbation was not abnormal?
    a. Magnus Hirschfeld
    b. Clelia Mosher
    c. Richard von Krafft-Ebing
    d. Albert Moll
    e. Havelock Ellis

24. Which research area was of major importance to Katharine Bement Davis?
    a. children's sexual development
    b. lesbianism
    c. sexual dysfunction in couples
    d. pregnancy and birth
    e. women and sexual intercourse

25. What is TRUE regarding Alfred Kinsey's sexuality research?
    a. He examined sexual development using case studies from his sex therapy practice.
    b. He was the first researcher to observe sexual intercourse in a laboratory setting.
    c. He published best-selling books on sexuality that helped to decrease societal guilt.
    d. His research was based upon psychoanalytic theory of sexuality.
    e. He used experimental methods as his research design.

26. Alfred Kinsey recruited research subjects from all of the following except ____________________.
    a. churches
    b. hospitals
    c. colleges and universities
    d. Internet chat rooms
    e. homosexual groups

27. How was the *Sex and Morality in the U.S.* report published by the Kinsey Institute considered different from Kinsey's original work?
    a. It focused on thoughts, feelings, and attitudes rather than just sexual behaviors.
    b. It used experimental methods that isolated cause and effect.
    c. It was based on psychoanalytic theory rather than remaining atheoretical.
    d. It focused on adolescent sexual behaviors rather than only adult sexuality.
    e. It used a case study type of methodology.

28. Which sexuality researchers used penile strain gauges and photoplethysmographs?
    a. Kinsey and Hooker
    b. Mosher and Krafft-Ebing
    c. Bell and Weinberg
    d. Masters and Johnson
    e. Starr and Weiner

29. Current sexuality research on older adults has found that ____________________.
    a. the majority don't know what masturbation is
    b. over half reported that they were sexually active
    c. 85% feel that they have no sexual interest
    d. a large number of them refuse to participate in sexuality studies
    e. 95% have engaged in premarital sex

30. Evelyn Hooker's research ____________________.
    a. evaluated the sexual satisfaction of lesbians
    b. compared gay and straight men, finding large differences in the psychological functioning
    c. investigated the likelihood that gay men will push unwanted sexual advances onto other people
    d. found that a majority of homosexuals conformed to societal stereotypes
    e. compared gay and straight men, finding little psychological differences

31. What is NOT true about case study research?
    a. It was the method used by Sigmund Freud.
    b. It is generally used to identify the attitudes, knowledge, or behavior, or large samples.
    c. It does not allow researchers to generalize to the general public.
    d. The sample size is small.
    e. It explores individual cases to formulate hypotheses.

32. What is an advantage of interviews as a sexuality research method?
    a. They are inexpensive.
    b. They are quick to develop and administer.
    c. Researchers can establish a rapport with subjects.
    d. Researchers can isolate cause and effect.
    e. Subjects can remain anonymous.

33. The method of direct observation was used most notably by ____________________.
    a. Iwan Bloch
    b. Katharine Bement Davis and Havelock Ellis
    c. William Masters and Virginia Johnson
    d. Sigmund Freud
    e. Evelyn Hooker

34. A major limitation of direct observation is ____________________.
    a. volunteer bias
    b. operant conditioning
    c. self-actualization
    d. honesty
    e. All of the above

35. What is a key element of participant-observation?
    a. a laboratory
    b. a therapist's office
    c. an independent variable
    d. a natural setting
    e. a tape recorder

36. Which of the following is an important ethical issue in sexuality research?
    a. validity
    b. unconditional positive regard
    c. confidentiality
    d. behavior modification
    e. interview methodology

37. When researchers influence a research study because of their own ideas about the research, it is called ____________________.
    a. informed consent
    b. validity
    c. interviewer bias
    d. reliability
    e. fixation

38. Which of the following research methods eliminates volunteer bias?
    a. direct observation
    b. interviews
    c. telephone surveys
    d. samples of convenience
    e. None of the above

39. Cross-cultural studies have provided sexuality researchers with information on how ____________________.
    a. each culture develops its own rules and regulations about sexual behaviors.
    b. societies' values and culture influence sexuality.
    c. Chinese and French societies view sexuality.
    d. people living in different countries view sex and intimacy in relationships.
    e. All of the above

40. Sexuality studies must be peer-reviewed in order ____________________.
    a. to be federally funded
    b. to be considered direct observation
    c. to emerge from a sociological theory of sexuality
    d. to be considered confidential
    e. to achieve validity

Matching

Column 1		Column 2
A. Generalizability	________	41. learning resulting from the reinforcing response a subject receives following a certain behavior
B. Direct Observation	________	42. a research method that involves researchers going into an environment and monitoring what is happening naturally
C. Operant Conditioning	________	43. research method used by Masters and Johnson
D. Experimental Methods	________	44. accepting others without restrictions on their behaviors or thoughts
E. Participant-Observation	________	45. research method used by Alfred Kinsey and Clelia Mosher
F. Validity	________	46. a research method that involves random selection and the manipulation of the independent variable
G. Case Study	________	47. ability for the answers of a few subjects to be applicable to the general population
H. Unconditional Positive Regard	________	48. the consistency of the measure with repeated testing of the same group
I. Reliability	________	49. research method used by Sigmund Freud
J. Interviews	________	50. whether or not a test or question measures what it is designed to measure

TEST YOURSELF ANSWER KEY

Theories about Sexuality

1. theory (p. 29)
2. psychoanalytic (p. 29)
3. libido, thanatos (p. 29)
4. unconscious (p. 30)
5. superego (p. 30)
6. erogenous zone (p. 30)
7. genital (p. 31)
8. behavior modification (OR aversion therapy) (p. 32)
9. imitation, identification (p. 32)
10. unconditional positive regard (p. 33)
11. genetic (p. 33)
12. society (p. 33)
13. evolution, sociology (p. 34)
14. medicalization (p. 35)
15. women, GLBT (p. 35)
16. personality formation and psychosexual development (p. 29)
17. psychoanalytic, behavioral, social learning, cognitive, and humanistic (pp. 29-35)
18. family, religion, economy, medicine, law, and the media (p. 34)
19. biological, sociobiological, and psychological (pp. 29-35)
20. They can provide rich, qualitative data and offer an in-depth understanding of individuals (p. 35)

Sexuality Research

21. Greeks (p. 36)
22. ethics (p. 36)
23. sick (p. 37)
24. physicians (p. 37)
25. Germany (p. 37)
26. pseudonyms (p. 37)
27. Jewish (p. 37)
28. hygiene (p. 37)
29. agenda-driven (p. 37)
30. HIV/AIDS (p. 37)
31. adolescent sexuality (p. 38)
32. fragmented (p. 39)
33. sexologists (p. 39)
34. funds (p. 40)
35. stigma, ignorance (p. 40)
36. They found that they suddenly lost their professional status, were accused of having the very sexual disorders they studies, or were viewed as motivated solely by lust, greed, or fame. (p. 37)
37. John D. Rockefeller and Andrew Carnegie (p. 37)
38. normal child sexual development and sexual expression in loving long-term relationships (p. 37)
39. Some believe that the mystery surrounding sexuality will be taken away and research donc on adolescent sexuality would encourage adolescents to have more sex. (p. 38)
40. sexual science (p. 40)

Sexuality Researchers

41. objectivity (p. 40)
42. Albert Moll (pp. 40-41)

43. Institute for Sexology (p. 41)
44. deviant (p. 41)
45. Clelia Mosher, Katharine Bement Davis (p. 42)
46. medical (p. 43)
47. atheoretical (p. 43)
48. interview bias (p. 44)
49. best-sellers (p. 44)
50. laboratory (p. 46)
51. random sample (p. 46)
52. failure, performance (p. 46)
53. half (p. 48)
54. homosexuality (p. 48)
55. marriage (p. 49)
56. Alfred Kinsey (p. 43)
57. 13 (p. 44)
58. They hired prostitutes. (p. 46)
59. that older adults are not sexually active (pp. 47-48)
60. The National Health and Social Life Study (NHSLS) (p. 49)

Sex Research Methods and Considerations

61. validity (p. 50)
62. rapport (p. 51)
63. Participant-observation (p. 51)
64. cause, effect (p. 51)
65. finances, time commitment (p. 51)
66. cause (p. 52)
67. informed consent (p. 52)
68. volunteer bias (p. 52)
69. overestimate, underreport (p. 53)
70. time, location (p. 53)
71. extramarital (p. 54)
72. 50 (p. 54)
73. behaviors, attitudes, beliefs, sexual satisfaction (p. 54)
74. accessibility, anonymity (p. 56)
75. peer-reviewed, Institutional Review Board (p. 57)
76. case study (p. 50)
77. difficulty of finding subjects, expense, lack of generalizability, and no information on feelings, attitudes, and personal history (p. 51)
78. ethical issues, volunteer bias, sampling problems, and reliability (p. 52)
79. college age populations (p. 53)
80. how societies' values and cultures influence sexuality (p. 54)

Post Test Answer Key

True/False

1. F (p. 29)
2. T (p. 34)
3. T (p. 34)
4. F (p. 37)
5. T (p. 39)
6. F (pp. 43-44)
7. F (p. 46)
8. F (p. 51)
9. F (p. 52)
10. T (p. 52)

Multiple-Choice

11. d (p. 29)
12. a (p. 29)
13. c (p. 30)
14. b (p. 30)
15. c (p. 31)
16. a (p. 32)
17. c (p. 32)
18. b (p. 32)
19. e (p. 33)
20. e (p. 35)
21. e (pp. 35-36)
22. d (p. 37)
23. e (p. 42)
24. b (p. 42)
25. c (p. 43)
26. d (p. 44)
27. a (p. 44)
28. d (p. 46)
29. b (pp. 47-48)
30. e (p. 48)
31. b (p. 50)
32. c (p. 51)
33. c (p. 51)
34. a (p. 51)
35. d (p. 51)
36. c (p. 52)
37. c (p. 52)
38. e (p. 52)
39. e (pp. 53-56)
40. a (p. 57)

Matching

41. C (p. 32)
42. E (p. 51)
43. B (p. 51)
44. H (p. 33)
45. J (p. 51)
46. D (p. 51)
47. A (p. 50)
48. I (p. 53)
49. G (p. 50)
50. F (p. 50)

Chapter 3 – Gender Development, Gender Roles, and Gender Identity

CHAPTER SUMMARY

People tend to use the words "sex" and "gender" synonymously when "sex" actually refers to the biological aspects of being male or female, and "gender" refers to the behavioral, psychological, and social characteristics of men and women. Complex organisms reproduce through sexual reproduction, where two parents each donate a gamete that combine to create a new organism. Most cells in the human body contain forty-six chromosomes arranged in twenty-three pairs. The two sex chromosomes are made up of an X chromosome donated by the woman's ovum and either an X or Y chromosome donated by the man's sperm. If the male contributes an X chromosome, the child will be female (XX), and if it is a Y, the child will be male (XY).

During the first few weeks of development, female and male embryos are identical. Around the fifth to sixth week, they begin to differentiate based upon the presence of hormones that lead to the development of external and internal sex organs. Atypical sexual differentiation may be caused by hormone or genetic irregularities and can result in an infant born with atypical genitals or external genitals that do not match the infant's genetic sex. Atypical sexual differentiation includes such syndromes as Klinefelter's, Turner's, triple-X, adrenogenital syndrome (AGS), and androgen-insensitively syndrome (AIS).

Culture and social structure interact to create sex typing, a way of thinking that splits the world into two basic categories—male and female—and suggests that most behaviors, thoughts, actions, professions, emotions, and so on fit one gender more than the other. Male stereotypes tend to be narrower than female stereotypes, and men who want to conform to society's ideas of masculinity have less flexibility in their behavior than women who want to live up to feminine stereotypes. The messages women receive from modern North American culture are contradictory, such as messages about beauty and careers. In the 1970s, researchers suggested that androgyny, having both masculine and female characteristics, was a desirable state because androgynous individuals have greater flexibility in behavior because they have a greater repertoire of possible reactions to a situation. Individuals who do not conform to society's gender role rules often fall under the categories of transgender, transsexual, transvestite, and drag queen.

A number of theories of gender role development have arisen to explore how girls and boys learn gender roles. Some of the theories include social learning, cognitive development, gender schema, and gender hierarchy. Gender role socialization continues throughout the life cycle. Young children often learn gender roles through modeling and homosocial play. By adolescence, gender roles are usually established, but gender roles can be challenging for young people who don't fit traditional expectations. By adulthood, careers and families are two places where gender role expectations have played out throughout the years. However, stereotypes and expectations for women and men are changing in workplaces and in family settings. Finally, as people age, gender roles relax and become less restrictive. It is the content of gender roles, not their existence, which societies can alter to provide each person an opportunity to live without being judged by stereotypes of gender.

LEARNING OBJECTIVES

After studying Chapter 3, you should be able to:

1. Describe the process of fertilization at the chromosomal level.
2. Summarize the process of sexual differentiation during gestation.
3. Discuss how female and male embryos are similar and different.
4. Explain the role that hormones play in prenatal development.
5. List and define the types of chromosomal irregularities that can occur during fertilization.
6. Compare the hormonal irregularities that can occur during gestation.
7. Compare and contrast the six theories of gender role development.
8. Provide examples of gender role development from the social learning and cognitive development theoretical perspectives.
9. Summarize Sandra Bem's gender schema theory.
10. Explain the concept of gender hierarchy.
11. Discuss three theories of gender role development from the perspective of gender hierarchy.
12. Define masculinity and femininity.
13. Summarize some of the research on the biological and environmental causes of gender roles.
14. Explain the concept of androgyny with examples of research studies.
15. Discuss the concept of transgender.
16. Provide definitions and examples of "Third Gender" categories in different cultures.
17. Summarize some of the research findings on gender roles in childhood and adolescence.
18. Describe how career experiences are related to gender roles.
19. Discuss how the family has been a source for gender role expectations and how those roles are changing.
20. Explain how gender roles influence people's lives as they age.

CHAPTER OUTLINE

- **Sex** – The biological aspects of being male or female.
- **Gender** – The behavioral, psychological, and social characteristics of men and women.

I. PRENATAL DEVELOPMENT: X AND Y MAKE THE DIFFERENCE

- Complex organisms reproduce through **sexual reproduction,** where two parents each donate a gamete or germ cell, which combine to create a new organism.
- The germ cells from the sperm and from the ovum each contains half of the new person's genes, which direct the development of the genitals and reproductive organs, and set the biological clock running to trigger puberty, female **menopause**, and male **andropause**.
- Most cells in the human body contain forty-six **chromosomes** arranged in twenty-three pairs.
- Twenty-two of the pairs look almost identical and are referred to as **autosomes**.
- The two **sex chromosomes** are made up of an X chromosome donated by the woman's ovum and either an X or Y chromosome donated by the man's sperm.
- If the male contributes an X chromosome, the child will be female (XX), and if it is a Y, the child will be male (XY).
- During **fertilization**, a sperm and egg, each containing twenty-three chromosomes (called haploid), join to produce a **zygote** containing forty-six chromosomes (called diploid). The zygote can now undergo **mitosis**, cell division, and reproduce its forty-six chromosomes.
- The forty-six chromosomes are threadlike bodies made up of over 100,000 genes, each of which contains **deoxyribonucleic acid (DNA)** which acts as a blueprint for how every cell in the organism will develop.

A. Sexual Differentiation in the Womb

- A human embryo typically undergoes about nine months of **gestation.**
- At about four to six weeks, the first tissues that become the embryo's **gonads** develop.

1. Internal sex organs

   a. Around the fifth to sixth week of gestation the primitive gonads form, which can potentially develop into **testes** or **ovaries**.

   b. The primitive duct system, the **Müllerian duct** (female) or the **Wolffian duct** (male), develop by the tenth or eleventh week.

   c. In female embryos, hormones lead to the disappearance of the Wolffian ducts, and the Müllerian duct fuses to form the uterus, the inner third of the vagina, and the Fallopian tubes.

   d. In male embryos, the testes begin producing **Müllerian inhibiting factor (MIF)** and **testosterone,** which cause the Müllerian ducts to disappear.

2. External sex organs

   a. Male and female organs that began from the same prenatal tissue are called **homologous**.

   b. In females, hormones cause the mound of skin beneath the umbilical cord (the genital tubercle) to develop into the clitoris, the labia minora, the vestibule, and the labia majora.

   c. In males, **androgen** stimulates the skin beneath the umbilical cord to develop the penis, the urethra, and the scrotum.

3. Hormonal Development and Influences

   a. **Endocrine glands**, such as the gonads, secrete hormones directly into the bloodstream to be carried to the target organs.

   b. The ovaries produce the two major female hormones, **estrogen** and **progesterone**.

c. Estrogen is an important influence in the development of female sexual characteristics throughout life, while progesterone regulates the menstrual cycle and prepares the uterus for pregnancy.
d. The testes produce androgen, which influences the development of male sexual characteristics throughout life.

4. Brain Differentiation
a. Most hormonal secretions are regulated by the brain, yet hormones also affect the development of the brain.
b. Female brains control menstruation and must signal the release of hormones in a monthly cycle, whereas male brains signal release of hormones continuously.

B. **Atypical Sexual Differentiation**

- Atypical sexual differentiation may be caused by hormone or genetic irregularities and can result in an infant born with ambiguous genitals or external genitals that do not match the infant's genetic sex.

1. Sex chromosome disorders
a. **Klinefelter's syndrome**
- occurs in about one in 700 live male births
- an ovum containing an extra X chromosome is fertilized by a Y sperm, resulting in XXY.
- Men with XYY are infertile, have small testicles, low levels of testosterone, can have **gynecomastia**, and are tall with feminized boy contours.
- **Testosterone therapy** can enhance the development of **secondary sexual characteristics**.

c. **Turner's syndrome**
- occurs in about one in 2,500 live female births.
- an ovum without any sex chromosome is fertilized by an X sperm, resulting in XO.
- External genitalia look typical, but ovaries do not develop fully, causing **amenorrhea** and infertility.
- Turner's syndrome can cause short stature, immature breast development, abnormalities of certain internal organs, and mental retardation.
- Administration of estrogen and progesterone can enhance the development of secondary sexual characteristics.

d. **XYY syndrome** and **triple-X syndrome**
- occur rarely
- may cause slight mental retardation and fertility problems.

2. Hormonal irregularities
a. A **hermaphrodite** is born with fully formed ovaries and fully formed testes, which is very, very rare.
b. Most individuals with intersex conditions are **pseudohermaphrodites**.
c. **Adrenogenital syndrome (AGS)**
- A genetic female (XX) is exposed to large amounts of androgens during prenatal development.
- The internal organs remain female, but the external organs may appear masculinized.

d. **Androgen-insensitivity syndrome (AIS)**
- A genetic male (XY) does not respond to testosterone and develops internal testes with female external genitalia.

- AIS is typically determined in adolescence when girls do not menstruate

II. GENDER ROLES AND GENDER TRAITS

- **gender roles**—culturally defined behaviors seen as appropriate for males and females
- **gender traits**—innate or biologically determined gender-specific behaviors

A. **Girls Act Like Girls, Boys Act Like Boys**

1. **Masculinity and Femininity**

a. Masculinity and femininity refer to the ideal cluster of traits that society attributes to each gender.

b. Most societies have cultural figures who are supposed to embody the traits of masculinity and femininity and serve as models for socializing youths into their gender roles.

c. Many studies have documented less gender role stereotyping in African American populations than in Caucasian populations.

d. Changing gender roles can result in confusion, fear, and even hostility in society because gender roles can allow for comfortable interaction between the sexes, since people can assume how they are supposed to behave in relation to the other sex.

B. **Are Gender Roles Innate?**

1. Behaviors are complex and are almost always interactions between one's innate biological capacities and the environment in which one lives and acts.

2. Research has explored various physical traits, such as size, and emotional traits, such as mothering, to determine what traits might be innate, but no differences between the sexes are universally accepted by researchers.

C. **Studying Gender**

1. During the 1970s and 1980s, the focus of gender research was on girls, whereas increasing research is now being conducted on boys in addition to girls.

2. Researchers have also been interested in whether people overestimate or underestimate gender characteristics.

III. GENDER ROLE THEORY

- **Socialization**—the process whereby an infant is taught the basic skills for functioning in society

A. **Evolutionary Biology: Adapting to Our Environment**

B. **Social Learning Theory**

1. We learn our gender roles from our environment, and from the same system of rewards and punishments that we learn our other social roles.

2. Children learn to model their behavior after the same-gender parent to win parental approval.

3. Children also see models of the "appropriate" ways for their genders to behave in their books, on television, and when interacting with others.

C. **Cognitive Development Theory: Age-State Learning**

1. All children go through a universal pattern of development following the ideas of Piaget (1951), who suggested that social attitudes in children are mediated through their cognitive developmental level.

2. Children begin to categorize themselves based on the information they receive about their genders.

3. Rigid gender role behavior should decrease after the age of seven or eight, once children realize that gender roles are social and arbitrary.

D. **Gender Schema Theory: Our Cultural Maps**

1. According to Sandra Bem (1981; 1983; 1987), children think according to **schemas**, which are cognitive mechanisms that organize our world.
2. **Gender schemas** organize our thinking about gender as determined by culture.
3. The gender schema becomes so ingrained that people do not even realize that they are created by stereotypes.

E. **Gender Hierarchy Theories: Power and Subordination**

- Men tend to be more frequently assigned to formal positions of power and authority in society, creating a **gender hierarchy**.
- Theories have been developed to explain why societies value traits defined as male more than traits defined as female.

1. Chodorow's Development Theory
   a. Chodorow (1987) suggests that boys come to devalue the female role due to the separation process they must endure with their mothers.
   b. To develop heterosexually, girls must separate from their mothers, which results in idealizing the qualities of their father the male role.
2. Ortner's Culture/Nature Theory
   a. Since culture, in the broad sense, sets human beings apart from animals, while childbirth and child-rearing are traits of all animals, men's cultural roles are valued over women's more biological roles.
   b. Women are concerned with the "domestic domain" and men with the "public domain," but since the public domain includes the domestic domain, women's sphere of influence—the family—is subordinate to men's.
3. MacKinnon's Dominance Theory
   a. "Male" and "female" are not biological categories but social and political categories.
   b. Men define what is male and what is female, thus dominating social life.

IV. VARIETIES OF GENDER

- Culture and social structure interact to create **sex typing**, a way of thinking that splits the world into two basic categories—male and female—and suggests that most behaviors, thoughts, actions, professions, emotions, and so on, fit one gender more than the other.

A. **Masculinity**

1. Williams & Best (1982) collected data about masculinity and femininity in thirty different countries and found that all had a general view of men as stronger, more active, and higher in achievement, autonomy, and aggression.
2. It is difficult for men to live up to the strong social demands of being male due to the contradictions inherent in masculinity.
3. Male stereotypes tend to be narrower than female stereotypes, and men who want to conform to society's ideas of masculinity have less flexibility in their behavior than women who want to live up to feminine stereotypes.

B. **Femininity**

1. Sheila Rothman (1978) has argued that American society has gone through a number of basic conceptions of what "womanhood" should be, such as "virtuous" or "wife-companion."
2. The messages women receive from modern North American culture are contradictory, such as messages about beauty and careers.

C. **Androgyny: Feminine and Masculine**

1. Androgyny refers to having both masculine and female characteristics.
2. In the 1970s, researchers suggested that androgyny was a desirable state, since androgynous individuals have greater flexibility in behavior because they have a greater repertoire of possible reactions to a situation.

D. **Transgenderism**

1. **transgender**—people who live full or part-time in the other gender's role and derive psychosocial comfort in doing so.
2. **transsexuals**—a person who feels that he or she is trapped in the body of the wrong gender.
3. **transvestite**—a person who dresses in the clothing of the other gender and derives pleasure from doing so.
4. **drag queens/female impersonators**—professional actors, typically gay men, who dress in flamboyant women's clothing to perform for a variety of reasons.

E. **Transsexualism: When Gender and Biology Don't Agree**

1. A male transsexual feels that he is a female "trapped" in a man's body; his gender identity is inconsistent with his or her biological sex which is called **gender dysphoria**.
2. **Sex reassignment surgery (SRS)** refers to the procedure where transsexuals have their bodies surgically altered to resemble those of the other sex.
3. More males than females report experiencing gender dysphoria.
4. Most transsexuals report a life-long desire to be a member of the other sex, which is typically satisfied only through sex reassignment surgery.
5. The process of seeking gender reassignment is long and complicated, involving a process of counseling, hormone therapy, and surgery.

F. **Third Genders: Other Cultures, Other Options**

1. Some cultures have a gender category that is neither male nor female—a third gender.
2. A category of individuals in many Native American societies called a **berdache** was usually a biological male who was effeminate or androgynous in behavior and who took on the social role of females.
3. Third gender categories with varying roles and positions in society have existed in a variety of cultures including Oman, India, Thailand, Hawaii, and Tahiti.

G. **Asexualism**

- An individual born either XX or XY but without internal or external sexual organs.

V. GENDER ROLE SOCIALIZATION FROM BABYHOOD THROUGH OLD AGE

- **gender-identity disorder**—diagnosis given to children who have a strong and persistent identification with the other sex or the gender role of the other sex and may be uncomfortable with their own biological sex or gender role.

A. **Childhood: Learning by Playing**

1. As early as age two, **modeling behavior** begins to emerge, and children begin to realize that objects and activities are appropriate to specific genders, and they believe that these gender rules are fixed rather than flexible.
2. Early in childhood, gender segregation or **homosocial play** begins, which may be due to different playing styles, attraction to similarities, or learned social roles.
3. During the school years, children become judged by peers and teachers based on gender roles, with boys experiencing more rejection from peers for violating gender stereotypes.

B. **Adolescence: Practice Being Female or Male**

1. By adolescence, gender roles are firmly established, and they guide adolescents through their exploration of peer and romantic relationships.
2. Gender role expectations can be challenging in adolescence for gay, lesbian, bisexual, and transgender adolescents.
3. Gender roles in adolescence are changing, particularly in regards to heterosexual relationships.

C. **Adulthood: Careers and Families**

1. Gender socialization pressures shape career choices for women and men.

2. In recent years, men's and women's roles have been changing in the workplace.
3. Women and family life
    a. Most women are still taught that their primary sense of satisfaction and identity should be derived from their roles as wives and mothers, yet studies show that women whose sole identities are as wives and mothers have higher rates of mental illness and suicide than single or married women who work.
    b. As more women enter the workforce, women are struggling with guilt feelings for being away from their homes and children, with a societal dispute forcing mothers to take sides on the issue.
4. Men and family life
    a. Recent research suggests that many men are taking more responsibility for childcare and domestic chores.
    b. Because of the changing workforce, the numbers of two-parent, heterosexual families with wives as the primary wage earners are increasing.

D. **The Senior Years**
1. "Empty nest syndrome" and retirement can be two issues that adults face as they age, affecting women and men in different ways.
2. As people age, gender roles relax and become less restrictive.

VI. DIFFERENT, BUT NOT LESS THAN: TOWARD GENDER EQUALITY

A. Epstein (1986; 1988) believes that gender distinctions begin with basic, human, dichotomous thinking like good-bad, soft-hard, and male-female.
B. Rogers (1978) argues that we cannot apply Western notions of gender equality to countries with fundamentally different systems.
C. It is the content of gender roles, not their existence, that societies can alter to provide each person an opportunity to live without being judged by stereotypes of gender.

TEST YOURSELF

Below you will find fill-in-the-blank and short answer essay questions for topics covered in chapter 3. Check your answers at the end of this chapter.

Prenatal Development

1. People often use the words ____________________ and ____________________ synonymously, even though they have different meanings.

2. Most cells in the human body contain forty-six ____________________

3. During ____________________, a haploid sperm and haploid egg join to produce a diploid ____________________.

4. A human embryo normally undergoes about nine months of ____________________.

5. In the first few weeks of development, XX (female) and XY (male) embryos are ____________________.

6. Around the fifth to sixth week of gestation, the primitive gonads form, and at this point they can potentially develop into either ____________________ or ____________________.

7. In female embryos, the lack of male hormones results in the regression and disappearance of the Wolffian ducts, and the ____________________ duct fuses to form the uterus and inner third of the ____________________.

8. Male and female organs that began from the same prenatal tissue are called ____________________.

9. In females, the genital tubercle develops into the ____________________, the labia minora, the vestibule, and the ____________________ ____________________.

10. In males, the ____________________ begin androgen secretion by the eighth or ninth week, which begins to stimulate the development of ____________________ ____________________.

11. ____________________ glands, such as the gonads, secrete hormones directly into the bloodstream to be carried to the target organs.

12. The ____________________ produce the two major female hormones, estrogen and progesterone.

13. Most hormonal secretions are regulated by the ____________________, in particular by the ____________________, which is the body's single most important control center.

14. ____________________ syndrome occurs when an ovum containing an extra X chromosome is fertilized by a Y sperm, giving a child forty-seven chromosomes altogether.

15. ____________________ syndrome is very uncommon, occurring when an ovum without any sex chromosome is fertilized by an X sperm, which gives the child forty-five chromosomes altogether.

16. What is the name for the chromosomes that determine whether a person is male or female?

17. What does dihydrotestosterone (DHT) do during prenatal development?

18. What is the role of Müllerian inhibiting factor?

19. Describe the chromosomal makeup of Turner's syndrome.

20. What does AIS stand for?

Gender Roles/Gender Role Theory

21. Even gender researchers are socialized into accepted ____________________ ____________________ from birth, which may make it very difficult for them to avoid projecting their own gender biases onto the research.

22. ____________________ and ____________________ refer to the ideal cluster of traits that society attributes to each gender.

23. Changing gender roles can result in ____________________, fear, and even ____________________ in society.

24. The process by which an infant who knows nothing becomes a toddler who has the basic skills for functioning in society is called ____________________.

25. Social learning theory suggests that we learn our gender roles from our ____________________.

26. Children may learn to model their behavior after the ____________________ parent to win parental approval.

27. ____________________ ____________________ theory assumes that all children go through a universal pattern of development, and there really is not much parents can do to alter it.

28. According to cognitive development theory, as a child matures, he or she becomes more aware that gender roles are ____________________ and ____________________.

29. According to Sandra Bem, children think according to ____________________, which are cognitive mechanisms that organize our world.

30. Bem suggests that we all have a ____________________ ____________________, which organizes our thinking about gender.

31. Sandra Bem argues that our gender schema is more powerful than other schemas and is used more often because our ____________________ puts so much emphasis on gender and gender differences.

32. Men tend to be more frequently assigned to formal positions of power and authority in society, creating what we might call a ____________________ ____________________.

33. Nancy Chodorow draws from ____________________ theory to argue that girls and boys undergo fundamentally different psychological developmental processes.

34. While Chodorow draws from the experiences of individuals, Ortner's theory looks at ____________________ as a whole.

35. Catharine MacKinnon dismisses ____________________ arguments about gender and argues that gender itself is fundamentally a system of ____________________.

36. What are gender roles?

37. Which gender role theory relies upon the idea of modeling to gain parental approval?

38. According to cognitive development theory, at what age should rigid gender role behavior decrease?

39. Who developed the theory of gender schemas?

40. What is a gender hierarchy?

Varieties of Gender

41. ____________________ and ____________________ ____________________ interact to create sex typing, a way of thinking that splits the world into two basic categories—male and female.

42. Gender is socially constructed; that is, ____________________ decide how gender will be defined and what it will mean.

43. The more one examines the categories of gender that really exist in the social world, the clearer it becomes that gender is more ____________________ than just splitting the world into male and female.

44. From the moment of a baby's birth, almost every society has ____________________ expectations of its males and females.

45. ____________________ stereotypes tend to be narrower than ____________________ stereotypes.

46. The messages a woman receives from modern North American culture are ____________________.

47. Sandra Bem considers those who have a high score on both masculinity and femininity to be ____________________.

48. ____________________, according to Bem, allows greater flexibility in behavior because people have a greater repertoire of possible ____________________ to a situation.

49. Bem saw androgyny as a potential way to overcome gender ____________________.

50. It is estimated that ____________________ to ____________________ percent of the population fails to conform to prescribed gender roles.

51. A transsexual's gender identity is inconsistent with his or her ____________________ ____________________.

52. Overall, more males than females experience gender ____________________, though the exact degree of differences is in dispute.

53. Gender ____________________ surgery was developed to help bring transsexuals' biology into line with their inner lives.

54. Some cultures challenge our notions and even have a gender category that is neither ____________________ nor ____________________.

55. Many traditional Native American societies had a category of not-men/not-women known as ____________________.

56. What is sex typing?

57. How are male stereotypes harmful to men?

58. How are female stereotypes harmful to women?

59. What is a berdache?

60. What is the name of individuals in India who undergo ritual castration and live in another gender class?

Gender Role Socialization/Toward Gender Equality

61. ____________________ are treated more harshly when they adopt cross-gender characteristics than ____________________ are.

62. Children who have a strong and persistent identification with the other sex or the gender role of the other sex and are uncomfortable with their own biological sex or gender role may be diagnosed with a ____________________ ____________________ ____________________.

63. Research shows that parents tend to be more restrictive of ____________________ babies and allow ____________________ more freedom and less intervention.

64. As early as age two, ___modeling___ behavior begins to emerge.

65. As the child begins to show more complex behaviors, he or she is usually ____________________ for displaying gender-stereotyped behaviors.

66. Early in childhood, gender segregation in play begins, which is also known as ____________________ ____________________.

67. During the school years, gender roles become the measure by which children are judged by their ____________________.

68. By ____________________, gender roles are firmly established, and they guide adolescents through their exploration of peer relationships.

69. Adolescence can be a particularly difficult time for those who are ____________________, ____________________, or ____________________.

70. As men and women grow into adulthood, they tend to derive their gender identity primarily in two realms, their ____________________ and their ____________________ lives.

71. Research suggests that characters in television commercials are portrayed as having more authority if they are ____________________ or ____________________.

72. Society teaches men that ____________________ achievement is, in large part, the measure of their worth.

73. As long as society portrays a woman's "real" job as that of mother, women will feel guilty when they choose to be ____________________ outside the family.

74. As people age, gender roles ____________________ and become less ____________________.

75. The basic human process of splitting the world into opposites like good—bad tends to ____________________ differences between things, including the sexes.

76. What type of rules do children develop around gender at the age of two?

77. How have gender roles changed for adolescents over the last few years?

78. How have gender roles changed for men with respect to career choices in the last few years?

79. In recent years, how has family life changed for heterosexual men with partners and children?

80. How do gender roles change as people age?

POST TEST

Below you will find true/false, multiple-choice, and matching quiz items covering the entire chapter. Check your answers at the end of this chapter.

True/False

1. Most cells in the human body contain 23 chromosomes.

2. Both male and female parents can donate either an X or a Y chromosome.

3. During prenatal development, gonads appear only in male embryos.

4. The majority of traits that were considered to be masculine traits 30 years ago are still considered to be masculine traits today.

5. Social learning theory suggests that children see models for "appropriate" gender on television and in books.

6. Gender schemas are genetic patterns that cause children to act in stereotypical patterns.

7. Chodorow suggest that psychological needs of boys and girls result in a devaluation of the female and an overvaluation of the male.

8. Male stereotypes tend to be narrower than female stereotypes.

9. Female impersonators are always transgendered.

10. Many religious and cultural systems clearly differentiate gender roles.

Multiple-Choice

11. The chromosomal makeup of the typical female is ____________________.
    a. XO
    b. YX
    c. XX
    d. XXY
    e. XY

12. During fertilization ____________________.
    a. a haploid sperm and a diploid egg join to produce a tetraploid zygote.
    b. a diploid sperm and a diploid egg join to form a gamete
    c. autosomes combine to create a gonad
    d. a diploid zygote is produced
    e. two haploid gonads combine to produce a germ cell

13. The clitoris develops from the ____________________.
    a. Wolffian duct
    b. genital tubercle
    c. gonads
    d. labia majora
    e. testicles

14. Progesterone is a female hormone that ____________________.
    a. helps to prepare the uterus for the implantation of the fertilized ovum
    b. is secreted by the vagina
    c. stimulates a deepening voice in both men and women
    d. is not necessary for pregnancy
    e. All of the above

15. What is the difference between how hormones are released in males and females?
    a. via the Müllerian duct for males and the ovum for females
    b. via the autosomes for females and the chromosomes for males
    c. once a week for males and once a day for females
    d. monthly for females and continuously for males
    e. continuously for females and once every 14 days for males

16. Hormone and/or genetic irregularities during gestation may lead to ___________________.
    a. andropause
    b. atypical sexual differentiation
    c. gamete reproduction
    d. the development of deoxyribonucleic acid
    e. mitosis

17. Which of the following sex chromosome syndromes occur when a Y sperm fertilizes an ovum containing any extra X chromosome?
    a. triple-X syndrome
    b. Müllerian syndrome
    c. XYY syndrome
    d. Turner's syndrome
    e. Klinefelter's syndrome

18. What is the name of the hormonal irregularity that occurs when a genetic female (XX) experiences an overproduction of androgen?
    a. triple-X syndrome
    b. androgen-insensitivity syndrome (AIS)
    c. berdachism
    d. Klinefelter's syndrome
    e. Adrenogenital syndrome (AGS)

19. Which of the following is NOT true of Androgen-insensitivity syndrome (AIS)?
    a. An individual with AIS is a genetic male (XY).
    b. An individual with AIS has an extra X chromosome (XXY).
    c. It is often first noticed when a teenage girl fails to menstruate.
    d. An individual with AIS has internal testes.
    e. An individual with AIS has not female internal reproductive organs.

20. What are culturally defined behaviors seen as appropriate for males and females?
    a. gender hierarchy
    b. gender dysphoria
    c. homosocial play
    d. androgyny
    e. gender roles

21. Femininity refers to ___________________.
    a. the ideal cluster of traits that society attributes to females
    b. the innate traits the are expressed when one is biologically female
    c. the guidelines set forth in the 1960s by the researcher Sandra Bem
    d. a set of behaviors that are generally thought to be the result of exposure to estrogen
    e. All of the above

22. Who studied the gender roles of the Tchambulis tribe of New Guinea?
    a. Charles Klinefelter
    b. Margaret Mead
    c. Charles Darwin
    d. Margaret Sanger
    e. Catharine MacKinnon

23. Gender roles are largely ____________________.
    a. socially constructed
    b. biologically determined
    c. unrelated to culture
    d. what occurs when men wear women's clothing
    e. constant across cultures

24. What is the process called where an infant is taught the basic skills for functioning in society?
    a. hierarchy
    b. Ortner's nature theory
    c. socialization
    d. Piaget's psychoanalytic processes
    e. mediation

25. When referring to gender roles development, Social Learning Theory suggests that ____________________.
    a. gender hierarchy is the result of men's attempt to dominate social life
    b. we learn our gender roles from our environment, from the system of rewards and punishments that we learn our other social roles
    c. boys and girls devalue the female and overvalue the male
    d. we all go through a universal pattern of development, and there really is not much parents can do to alter it
    e. All of the above

26. Which theory of gender role development suggests that as the child's brain matures, the child goes through stages where his or her understanding of gender changes in predictable ways?
    a. Klinefelter's Theory of Gender
    b. Chodorow's Developmental Theory
    c. Bem's Androgyny Theory of Gender
    d. Homosocial Learning Theory
    e. Cognitive Development Theory

27. Whose theory of gender role development emphasizes the importance of gender schemas?
    a. Sandra Bem
    b. Catharine MacKinnon
    c. Michael Piaget
    d. Joseph Turner
    e. Nancy Chodorow

28. Gender Hierarchy Theories attempt to explain ____________________.
    a. how gender is programmed through our genes
    b. how gender schemas organize children's thinking and behavior
    c. the reasons that adolescents engage in homosocial play
    d. the basic reasons societies value traits they see as male over traits they see as female
    e. how the system of rewards and punishments that society provides for gender roles are embedded in genetic wiring

29. Which of the following statements is NOT true of Ortner's Culture/Nature Theory?
    a. Men's cultural roles are valued over women's more biological roles.
    b. It looks at society as a whole.
    c. It emphasizes the role hormones play in gender development.
    d. Male and masculine things are aligned with "culture."
    e. Women's concerns are more biological and inward.

30. Which of the following statements reflects MacKinnon's Dominance Theory?
    a. Gender hierarchy is the result of biological and genetic factors.
    b. Gender schemas organize our thinking about gender.
    c. The mother becomes the source of personal identification for both girls and boys.
    d. Gender is a system of dominance rather than a system of biological differences.
    e. None of the above

31. What is the term for cognitive thinking patterns that divide the world into male and female categories and suggest the appropriate behaviors, thoughts, actions, professions, and emotions for each?
    a. sex typing
    b. cognitive imaging
    c. gender fixation
    d. sexual cognitions
    e. sex patterns

32. Androgyny is ____________________.
    a. a class of hormones that promotes the development of male genitalia
    b. a disorder involving an overproduction of androgen in the adrenal glands
    c. a genetic syndrome involving an extra Y chromosome
    d. having both masculine and feminine characteristics
    e. when someone of one sex identifies as being a member of the other sex

33. What is the definition of transvestite?
    a. an individual born either XX or XY but without internal or external sexual organs
    b. a person who feels that he or she is trapped in the body of the wrong gender
    c. professional actors, typically gay men, who dress in flamboyant women's clothing to perform for a variety of reasons
    d. an individual in many Native American societies who took on the social role of females
    e. a person who dresses in the clothing of the other gender and derives pleasure from doing so

34. Having one's gender identity inconsistent with one's biological sex is called ____________________.
    a. socialization
    b. androgyny
    c. drag queen
    d. asexuality
    e. gender dysphoria

35. Which of the following terms is NOT associated with a type of "Third Gender" category?
    a. hijra
    b. xanith
    c. elmlar
    d. kathoey
    e. berdache

36. Modeling behavior begins to emerge as early as age ___________________.
    a. two
    b. three
    c. four
    d. five
    e. six

37. Gender roles are firmly established by ___________________.
    a. 10 weeks of age.
    b. age 3
    c. age 6
    d. adolescence
    e. 6 months

38. Media research suggests that what group of people are viewed as powerful in television commercials?
    a. African-American men
    b. white women
    c. Asian women
    d. African-American women
    e. white men

39. According to research discussed in the textbook, American women do ___________________ percent more housework than American men?
    a. 20
    b. 50
    c. 70
    d. 100
    e. Nowadays women don't tend to do more housework than men.

40. The term "mommy wars" describes the situation in which ___________________.
    a. women compete over who is the best mother
    b. women and their male partners experience a period of hostility after the birth of a child
    c. women experience a psychological struggle over the decision to have children
    d. working mothers and stay-at-home mothers defend their different decisions
    e. working mothers feel guilty and at war with themselves for being away from their children

Matching

Column 1		Column 2
A. Margaret Mead	____	41. famous jazz musician who was biological female but lived as a male
B. Klinefelter's Syndrome	_____	42. a class of hormones that are produced in the testes in men and by the adrenal glands in both women and men.
C. gender	_____	43. a class of hormones that affect the menstrual cycle
D. estrogen	_____	44. the biological aspects of being male or female
E. Billy Tipton	_____	45. XO
F. Bem	_____	46. the behavioral, psychological, and social characteristics of men and women
G. androgen	_____	47. researcher who found that gender roles were not always the same across different cultures.
H. sex	_____	48. XXY
I. Chodorow	_____	49. researcher who developed the theory of gender schemas
J. Turner's Syndrome	_____	50. researcher who developed the psychoanalytic theory of gender role development that explains how children come to devalue females

TEST YOURSELF ANSWER KEY

Prenatal Development

1. sex, gender (p. 62)
2. chromosomes (p. 64)
3. fertilization, zygote (p. 64)
4. gestation (p. 65)
5. identical (p. 65)
6. testes, ovaries (p. 65)
7. Müllerian, vagina (p. 65)
8. homologous (p. 65)
9. clitoris, labia majora (p. 65)
10. testes, male genitalia (p. 65)

11. Endocrine (p. 66)
12. ovaries (p. 66)
13. brain, hypothalamus (p. 67)
14. Klinefelter's (p. 69)
15. Turner's (p. 69)
16. sex chromosomes (p. 64)
17. It stimulates the development of the male external sex organs. (p. 65)
18. It causes the Müllerian ducts to disappear. (p. 65)
19. The ovum provides no sex chromosomes, resulting in XO, 45 chromosomes all together. (p. 69)
20. Androgen-insensitivity syndrome (p. 71)

Gender Roles/Gender Role Theory

21. gender roles (p. 72)
22. masculinity, femininity (p. 73)
23. confusion, hostility (p. 73)
24. socialization (p. 77)
25. environment (p. 77)
26. same-gender (p. 77)
27. Cognitive development (p. 77)
28. social, arbitrary (p. 78)
29. schemas (p. 78)
30. gender schema (p. 78)
31. culture (p. 78)
32. gender hierarchy (p. 78)
33. psychoanalytic (p. 79)
34. society (p. 79)
35. biological, dominance (p. 80)
36. They are culturally defined behaviors seen as appropriate for males or females. (p. 72)
37. social learning theory (p. 77)
38. 7 or 8 (p. 78)
39. Sandra Bem (p. 78)
40. It is the fact that men tend to be more frequently assigned to formal positions of power and authority in society. (p. 78)

Varieties of Gender

41. culture, social structure (p. 81)
42. societies (p. 81)
43. complicated (p. 81)
44. different (p. 81)
45. Male, female (p. 83)
46. contradictory (p. 85)
47. androgynous (p. 85)
48. Androgyny, reactions (p. 85)
49. stereotypes (p. 85)
50. 10, 15 (p. 86)
51. biological sex (p. 87)
52. dysphoria (p. 88)
53. reassignment (p. 88)
54. male, female (p. 88)
55. berdaches (p. 88)

56. It is a way of thinking that splits the world into two basic categories—male and female—and suggests that most behaviors, thoughts, actions, professions, emotions and so on fit one gender more than the other (p. 21)
57. They tend to be narrower than female stereotypes, and men who want to conform to society's ideas of masculinity have less flexibility in their behavior. For example, boys are often taught not to cry in public. (p. 26)
58. Women receive contradictory messages about their bodies, careers, and motherhood. (pp. 84-85)
59. It is a category of not-men/not-women in many traditional Native American societies. They were usually a biological male who was effeminate or androgynous in behavior and who took on the social role of females (pp. 36-37)
60. hijra (p. 89)

Gender Role Socialization /Toward Gender Equality

61. boys, girls (p. 90)
62. gender-identity disorder (p. 90)
63. girl, boys (p. 91)
64. modeling (p. 91)
65. rewarded (p. 92)
66. homosocial play (p. 92)
67. peers (p. 92)
68. adolescence (p. 93)
69. transgendered, homosexual, bisexual (p. 93)
70. career, family (p. 94)
71. white, male (p. 94)
72. career (p. 94)
73. productive (p. 95)
74. relax, restrictive (p. 96)
75. exaggerate (p. 96)
76. Gender for young children is universal rather than flexible. For example, only women can wear skirts or have long hair. (p. 91)
77. Girls are more likely to take initiative when it comes to dating situations. (p. 93)
78. Men have been entering more "female dominated" fields, such as physical therapy and library science, and have also been taking on more childcare responsibilities (p. 94)
79. Research suggests that many men are taking more responsibility for child care and are assuming more domestic chores. (p. 95)
80. Gender roles relax and become less restrictive. For example, older men tend to do more housework than younger men. (p. 96)

POST TEST ANSWER KEY

True/False

1. T (p. 64)
2. F (p. 64)
3. F (p. 65)
4. F (p. 74)
5. T (p. 77)
6. F (p. 78)
7. T (p. 79)
8. T (p. 83)
9. F (p. 85)
10. T (p. 96)

Multiple-Choice

11. c (p. 64)
12. d (p. 64)
13. b (p. 65)
14. a (p. 66)
15. d (p. 67)
16. b (p. 69)
17. e (p. 69)
18. e (p. 71)
19. b (p. 71)
20. e (p. 72)
21. a (p. 73)
22. b (p. 74)
23. a (pp. 72-74)
24. c (p. 77)
25. b (p. 77)
26. e (pp. 77-78)
27. a (p. 78)
28. d (pp. 78-79)
29. c (p. 79)
30. d (pp. 80-81)
31. a (p. 81)
32. c (p. 85)
33. e (p. 85)
34. e (p. 87)
35. c (pp. 88-89)
36. a (p. 91)
37. d (p. 93)
38. e (p. 94)
39. c (p. 95)
40. d (p. 95)

Matching

41. E (p. 86)
42. G (p. 65)
43. D (p. 66)
44. H (p. 62)
45. J (p. 69)
46. C (p. 62
47. A (p. 74)
48. B (p. 69)
49. F (p. 85)
50. I (p. 79)

Chapter 4 – Male Sexual Anatomy and Physiology

CHAPTER SUMMARY

The male sexual and reproductive system is a complex series of glands and ducts. The external sex organs are composed of the penis and the scrotum. The penis is made up of three cylinders of erectile tissue that engorge with blood during erection. The penis also involves a glans penis, or head, and a root that extends far into the body. The scrotum provides a cooling tank for the testicles, keeping them the right temperature for sperm production.

The internal sex organs included the testicles, epididymis, vas deferens, seminal vesicles, ejaculatory duct, prostate gland, and the Cowper's glands (bulbourethral glands). Sperm and testosterone production are the two main functions of the testicles. Sperm production, or spermatogenesis, takes place in the seminiferous tubules, 300 microscopic tubules in the testicles. Sperm formation in men typically takes about 72 days, with men producing about 300 million sperm per day. Testosterone production occurs in the Leydig cells located in the testicles. Once sperm leave the testicles, it matures in each epididymis, adjacent to each testicle. It then travels through the vas deferens. The seminal vesicles contribute a rich fluid to provide nutrients for the sperm, making up 60-70% of the semen. The fluid then travels through the ejaculatory duct and picks up more fluid from the prostate gland, which secretes an alkaline fluid that helps neutralize the acidity of the vagina and provides 25-30% of the semen. The Cowper's glands secrete a small amount of fluid before ejaculation to clean and lubricate the urethra for the semen.

The physiological start of puberty for boys starts at the age of ten with the release of gonadotropin releasing hormone. Although spermatogenesis begins at about 12 years of age, the ejaculation of mature sperm doesn't occur until approximately age 13 or 14. In their 70's or 80's, men experience a process called andropause when their blood testosterone concentrations decrease, leading to some physiological changes such as decreased sperm production and possible fatigue and osteoporosis.

There are a number of important diseases of the male reproductive organs that can occur from birth to old age. Cryptorchidism is the name of the condition where the testicles do not descend into the scrotum during gestation and should be treated by surgery at some point to prevent testicular cancer and infertility. Testicular torsion refers to the twisting of a testicle's spermatic cord that can cause severe pain and swelling and may result from exercise, sexual activity, or even sleeping, and must be treated to prevent the loss of the testicle. Priapism and Peyronie's disease are two painful conditions involving erections, with the first one involving a painful and persistent erection and the latter being a curvature of the penis that may be painful and may make sexual intercourse difficult.

Testicular, penile, and prostate are three types of cancer that can develop in men. Testicular cancer is the most common type of cancer in men aged 20-44. Early detection using monthly testicular exams to feel for lumps on the testicles is important in diagnosing and treating this very curable form of cancer. Penile cancer is quite rare, but a physician should treat penile lesions. Prostate cancer is the most common cause of cancer deaths of men over 60 years in the United States. Regular prostate exams and blood tests help to detect prostate cancer.

LEARNING OBJECTIVES

After studying Chapter 4, you should be able to:

1. List and define the male external sexual and reproductive organs.
2. List and define the male internal sexual and reproductive organs.
3. Summarize the structures of the penis.
4. Compare and contrast spermatogenesis and testosterone production.
5. Explain the pathway of sperm from spermatogenesis to ejaculation.
6. Compare and contrast the role of the seminal vesicles, prostate gland, and Cowper's gland.
7. Describe the physiological process of ejaculation.
8. Describe the components of ejaculate.
9. Explain the changes that boys experience during puberty.
10. Describe the relationship between testosterone and sperm production in the negative feedback loop that takes place in males.
11. Define the term andropause.
12. List at least four diseases or conditions of the male reproductive organs.
13. Define cryptorchidism, testicular torsion, priapism, Peyronie's Disease, and hydrocele.
14. Explain how anabolic-androgenic steroid use affects a human male's body, including the sexual and reproductive effects.
15. Describe an inguinal hernia.
16. Identify the most common types of cancer in young men and older men.
17. Describe the detection and diagnosis procedure for testicular, penile, and prostate cancers.
18. List the symptoms of testicular cancer and prostate cancer.
19. List the treatment options for testicular cancer and prostate cancer.

CHAPTER OUTLINE

I. THE MALE SEXUAL AND REPRODUCTIVE SYSTEM

A. **External Sex Organs**

1. **Penis**

a. The penis contains the urethra which carries urine and semen to the outside of the body.
b. It engorges with blood.
c. The penis is composed of three cylinders which contain sponge-like erectile tissue (no bone or major muscle).

- **corpora cavernosa**—two cylinders lie on the upper sides of the penis
- **corpus spongiosum**—central cylinder lies on the bottom and contains the urethra

d. **Glans Penis**

- composed of the corona, frenulum, and meatus
- sensitive to stimulation
- the prepuce is a fold of skin that can cover all or part of the glans called the foreskin, which is removed during circumcision

e. **Root** (of the penis)

- attaches to internal pelvic muscles
- extends far into the body to the perineum

f. **Erection**

- Nerve fibers swell the arteries of the penis, allowing blood to rush into the corpora cavernosa and corpus spongiosum.
- Erection is a reflex that can occur without tactile stimulation, as in spinal injuries or during sleep.

2. **Scrotum**

a. The scrotum is the loose, wrinkled pouch beneath the penis, which contains the testicles.
b. The testicles are outside the body to keep sperm temperature lower than body temperature.

- The skin on the scrotum contains sweat glands that help regulate testicular temperature.
- The **cremaster muscle** draws the testes closer to the body when it's cold.

B. **Internal Sex Organs**

1. **Testes** (or **Testicles**)

a. These are egg-shaped glands that rest in the scrotum, each about two inches in length and one inch in diameter.
b. The left testicle usually hangs lower than the right, but it might be reversed in some men.
c. **Spermatogenesis** (sperm production) is one of the two main functions of the testicles

- Sperm is produced and stored in 300 microscopic tubes in the testes called **seminiferous tubules**.
- A **spermatogonium** develops in the cells lining the outer wall of the seminiferous tubules and moves toward the center while being nourished by Sertoli cells.

- S spermatogonium develops into **spermatocytes** and eventually divides into **spermatids**.
- The sperm is composed of a head, which contains an enzyme aiding in fertilization of the egg, a midpiece, which generates energy, and a **flagellum**, or tail, which propels it.
- Sperm formation in men typically takes about 72 days. Men produce about 300 million sperm per day.

d. **Testosterone Production** (the 2nd main function of the testicles) occurs in the **Leydig cells**.

e. The **epididymis** is a comma-shaped organ that sits on top of the testicle. Immature sperm mature in the epididymis for about 10-14 days.

2. The Ejaculatory Pathway

a. The **vas deferens** is an 18-inch tube that carries the sperm from the testicles to the urethra.

b. During **ejaculation** (the physiological process whereby the seminal fluid is ejected from the penis), sperm pass through the epididymis, the vas deferens, the ejaculatory duct, and the urethra, picking up fluid along the way from three glands.

c. The **seminal vesicles** are glands adjacent to the ampulla that provide nutrition for the sperm and make up about 60-70% of the volume of the ejaculate.

d. The vas deferens and the duct from the seminal vesicles merge into a common **ejaculatory duct**, a straight tube that passes into the prostate gland and opens into the urethra.

e. The **prostate gland**, a walnut-size gland, sits at the base of the bladder.

- secretes about 25-30% of the ejaculate
- higher pH secretions that help neutralize vaginal acidity for sperm
- Annual prostate exams (during rectal exams) are recommended for men over 35 years old due to possible enlargement or cancer

f. The **bulbourethral** or **Cowper's glands** are two pea-size glands just beneath the prostate gland.

- secrete a fluid that lubricates the urethra and neutralizes acidic urine
- secrete the pre-ejaculatory fluid during arousal that may contain live sperm

g. **Ejaculation**

- Stimulation builds to a threshold leading to the first stage, when the epididymis, seminal vesicles, and prostate empty fluid into the urethral bulb while the bladder is closed off to prevent urine from being expelled.
- The second stage leads to contractions, the expulsion of semen, and usually orgasm.
- After ejaculation, the arteries supplying blood to the penis narrow and the penis becomes limp.
- The average amount of time until another erection is 10-20 minutes and depends upon a variety of factors such as age and physiology.

h. **Ejaculate** (semen)

- averages about 1-2 teaspoons or 2-5 milliliters in quantity
- 50-150 million sperm per milliliter

C. **Other Sexual Organs**

1. The **Breasts**

a. Men's breasts are mostly muscle but do have nipples and areolas.

b. Some men experience sexual pleasure from nipple stimulation.
c. Gynecomastia, or breast growth in men, is common in puberty and old age and may last anywhere from a few months to a few years.
d. Breast cancer does occur in men but is often detected when it's at an advanced stage.

2. Other Erogenous Zones: Many men experience pleasure from stimulation of the scrotum, testicles, and anus.

D. **The Male Maturation Cycle**

1. Male Puberty

a. During a boy's early life, his testicles do not produce sperm or testosterone.
b. Around ten years of age the hormonal cycle begins in the brain.

- The hypothalamus begins releasing gonadotropin releasing hormone (GnRH).
- GnRH stimulates the anterior pituitary gland to send out follicle-stimulating hormone (FSH) and luteinizing hormone (LH).
- FSH and LH flow through the circulatory system to the testes where LH stimulates the production of testosterone.
- Testosterone and LH stimulate sperm production.

c. As puberty progresses, the testicles increase in size, and penis size increases about a year later.
d. The epididymis, prostate, seminal vesicles, and Cowper's glands increase in size over the next few years.
e. Increased testosterone stimulates overall growth spurt and the development body hair, voice deepening, and increased metabolism.
f. Spermatogenesis begins at about 12 years of age but the ejaculation of mature sperm occurs at about 13-14 years of age.

2. Male Andropause

a. Men experience **andropause** in their 70s or 80s when blood testosterone concentrations decrease.
b. Spermatogenesis decreases, ejaculate becomes thinner, and ejaculatory pressure decreases.
c. The reduction in testosterone production may lead to decreased muscle strength and fatigue.
d. Men may also experience osteoporosis and anemia from decreasing hormone levels.

II. MALE REPRODUCTIVE AND SEXUAL HEALTH

A. **Diseases of the Male Reproductive Organs**

1. **Cryptorchidism**

a. During the 8th and 9th month of gestation, the testicles descend into the scrotum through the inguinal canal.
b. Cryptorchidism occurs if the testicles do not descend.
c. Due to the higher temperature of the abdomen, sperm production cannot take place, and infertility can result if the testes are not removed during childhood.
d. Testicles that remain in the abdomen are also at an increased risk of developing cancer.
e. Cryptorchidism in infants can be treated with surgery to relocate the testis to the scrotum.

2. **Testicular Torsion**

a. Testicular torsion refers to the twisting of a testicle's spermatic cord.

b. It typically results from abnormal development of the spermatic cord or membrane covering the testicle.
c. Most common from puberty to age 25, although it can happen at any age after exercise, sexual intercourse, or while sleeping.
d. Severe pain and swelling are the most common symptoms, although abdominal pain, tenderness and aching may also occur.
e. A physician must diagnose and untwist the cord within 24 hours to prevent the loss of the testicle.

3. **Priapism**
a. Priapism is a painful and persistent erection that is not associated with sexual desire or excitement but instead is due to blood becoming trapped in the erectile tissue of the penis.
b. Drug use (cocaine, marijuana, or anticoagulants) is the most common cause, although in many cases the cause is unknown.
c. Draining blood from the penis with a syringe is one treatment, and surgery may be necessary in some cases.
d. Sexual functioning may be affected if effective treatment does not occur.

4. **Peyronie's Disease**
a. Peyronie's Disease refers to abnormal calcifications in the penis, which may cause painful curvature and may make intercourse difficult or impossible.
b. Crystal deposits in the connective tissue, trauma, excessive calcium levels, or calcification may contribute to Peyronie's disease.
c. It typically lasts 2 years and may disappear suddenly but can also be treated with medication or surgery.

B. **Other Conditions that Affect the Male Reproductive Organs**

1. Anabolic-Androgenic Steroid Use
a. Approximately 12% of adolescent males admitted to using anabolic-androgenic steroids (AAS) at some point in their lifetime, while 1-2% of adolescent females used them.
b. AAS use has been associated with damaging effects to the liver, serum lipids, and the reproductive system, including shrinkage of the testicles.
c. Erectile disorder and testicle shrinking are two sexual-related risks of AAS, but other effects include aggressive behavior, mental illness, and masculinization in women.

2. **Inguinal Hernia**
a. An inguinal hernia is caused when the intestine pushes through the opening in the abdominal wall into the inguinal canal.
b. The intestine pushes down on the testicles, causing a bulge in the scrotum which may lead to pain and can be treated by surgery.

3. **Hydrocele**
a. A hydrocele is a condition that causes a scrotal mass due to an excessive accumulation of fluid within the tissue surrounding the testicle.
b. It may be due to an overproduction of fluid or poor reabsorption of the fluid and may lead to pain and swelling but can be treated by removing the fluid.

C. **Cancer of the Male Reproductive Organs**

1. Testicular Cancer
a. Testicular cancer is the most common cancer in men aged 20-44.
b. Few symptoms occur until an advanced stage of the cancer, highlighting the importance of early detection.
c. Symptoms included painless testicular mass or harder consistency of the testes, lower back pain, gynecomastia, shortness of breath, or urethral obstruction.

d. The incidence of testicular cancer has increased, but the cure rates have also increased, leading it to be the most curable form of cancer.
e. Treatment may involve radiation, chemotherapy, or removal of the testicle.
f. Testicular cancer may affect future fertility but usually only temporarily.
g. Monthly testicular exams are very important for the early detection of testicular cancer.

2. Penile Cancer
   a. Cancer of the penis is uncommon.
   b. Any lesions on the penis need to be taken seriously and diagnosed by a doctor so potential penile cancer can be treated.
3. Prostatic Disease
   a. **Benign prostatic hypertrophy (BPH)** is an enlarged prostate, which is a natural occurrence in most men as they age. It becomes a problem when it blocks urine.
   b. Prostate cancer is the most common cause of cancer deaths among men over 60 years in the United States.
   c. 80% of men with prostate cancer are 65 years or older, and African-American men have the highest prostate cancer rates in the world.
   d. Risk factors include age, race, a diet high in fat, and genetics.
   e. Symptoms include pain in the lower back, pelvic region or upper thigh, inability to urinate or decreased urinary pressure, pain or burning during urination, and frequent urination.
   f. Prostate exams during rectal exams can detect prostate cancer at a curable stage, and the prostate-specific antigen (PSA) blood test is an important recent advance because it can measure levels of molecules that are overproduced by prostate cancer cells.
   g. Treatment for prostate cancer varies and may involve watching and waiting the progression of the disease, radical prostatectomy, radiation, cryosurgery, or experimental drugs that attack cancer cells.
   h. Treatment options need to be weighed carefully since they can sometimes lead to erectile dysfunction and incontinence.

TEST YOURSELF

Below you will find fill-in-the-blank and short answer essay questions for topics covered in chapter 4. Check your answers at the end of this chapter.

The Male Sexual and Reproductive System

1. There is no ________________ and little ________________ in the human penis.

2. The ________________ is a continuation of the loose skin that covers the penis as a whole to allow it to grow in size during erection.

3. The ________________ muscle of the scrotum contracts and expands to regulate the temperature of the testicles.
4. Sperm is produced and stored in some 300 microscopic tubes located in the testes and known as ________________ ________________.

5. Human sperm formation requires approximately ________________ days, yet the human male produces about ________________ sperm per day.

6. Once formed, immature sperm enter the seminiferous tubules and migrate to the _______________, where they mature for about ten to fourteen days and where some faulty or old sperm are reabsorbed.

7. _______________ is the physiological process whereby the seminal fluid is forcefully ejected from the penis.

8. The _______________ gland, a walnut-size gland sitting at the base of the bladder, produces several substances that are thought to aid sperm in their attempt to fertilize an ovum.

9. The _______________ glands have ducts that open right into the urethra and produce a fluid that cleans and lubricates the urethra for the passage of sperm, neutralizing any acidic _______________ that may remain in the urethra.

10. Most men have between _______________ and _______________ contractions during orgasm.

11. Once orgasm subsides, the _______________ supplying _______________ to the penis narrow, the veins taking the blood out enlarge, and the penis usually becomes limp.

12. If there are less than twenty million sperm per milliliter, the male is likely to be _______________.

13. _______________ is common both in puberty and old age and usually lasts anywhere from a few months to a few years.

14. The development of _______________ and the sexual fluid glands allows a boy in puberty to experience his first wet orgasms, though, at the beginning, they tend to contain a very low live sperm count.

15. As they age, men's blood _______________ concentrations decrease.

16. What are three parts of the glans penis?

17. Where is testosterone produced?

18. How do the secretions from the seminal vesicles contribute to the sperm?

19. How many sperm are in one milliliter of semen?

20. At what age do men experience andropause?

Male Reproductive and Sexual Health

21. The testicles of a male fetus begin high in the abdomen near the kidneys, and, during gestation, descend into the scrotum through the ________________ ________________.

22. Testicular torsion is most common in men from puberty to the age of ________________.

23. ________________ is a painful and persistent erection that is not associated with sexual desire or excitement, when ________________ becomes trapped in the erectile tissue of the penis and is unable to get out.

24. Every male has individual ________________ to his penis when it becomes erect.

25. ________________ ________________ occurs in the connective tissue of the penis, and although some cases are asymptomatic, others develop penile nodules, which can cause severe erectile pain.

26. The best-documented effects of anabolic-androgenic steroids are to the liver, serum lipids, and the ________________ system, including shrinkage of the ________________.

27. An ________________ ________________ is caused when the intestine pushes through the opening in the abdominal wall into the inguinal canal.

28. A hydrocele is a condition that causes a scrotal mass that occurs when there is an excessive accumulation of fluid within the tissue surrounding the ________________.

29. ________________ cancer is the most common malignancy in men aged twenty to forty-four years.

30. All men should do ________________ self-examinations ________________.

31. Any lesion on the ________________ must be examined by a physician, for benign and malignant conditions can be very similar in appearance, and sexually transmitted infections can appear as lesions.

32. As men age, their prostate glands enlarge, called ________________ ________________ ________________, which may block urination and may require surgery.

33. Although all men can get prostate cancer, it is found most often in men over the age of ________________.

34. In 1986 the U.S. Food and Drug Administration approved the _______________ _______________ _______________ blood test that measures levels of molecules that are overproduced by prostate cancer cells.

35. There are many _______________ for prostatic disease, and almost all are _______________.

36. What is the name of the condition when the testes fail to descend into the scrotum during gestation?

37. What is the most common known cause of priapism?

38. What percent of adolescent males acknowledge using anabolic-androgenic steroids at some point in their lifetime?

39. What is the most common symptom of testicular cancer?

40. What are some of the risk factors that have been linked to prostate cancer?

POST TEST

Below you will find true/false, multiple-choice, and matching quiz items covering the entire chapter. Check your answers at the end of this chapter.

True/False

1. Some researchers suggest that circumcision in men reduces sexual excitability.
2. The cremaster muscle of the scrotum contracts and expands in order to regulate testicular temperature.
3. Sperm can build up in the testicles and cause problems including possible infection and pain.

4. Sperm is stored in 300 microscopic tubes located in the vas deferens.

5. The vas deferens is an 18-inch tube that carries the sperm from the testicles and propels it towards the urethra during ejaculation.

6. Another name for the prostate gland is bulbourethral gland.

7. Due to a sphincter that closes off the bladder, urine cannot be expelled with semen during ejaculation.

8. Breast cancer among men is actually quite common although less prominent in the media.

9. Anabolic-androgenic steroid use has been associated with schizophrenia and other mood disorders.

10. TSE stands for Testicular Self-Examination.

Multiple-Choice

11. Which below is NOT a part of the penis?
    a. corpus spongiosum
    b. urethra
    c. Cowper's glands
    d. connective tissue
    e. corpora cavernosa

12. How does the penis become erect?
    a. The penile bone extends to an erection.
    b. It fills with sperm, making it feel stiff.
    c. The numerous muscles contract, causing the erection.
    d. Sponge-like tissue engorges with blood.
    e. All of the above

13. The corona, frenulum, and meatus are parts of the _______________.
    a. urethra
    b. prostate gland
    c. scrotum
    d. ejaculatory duct
    e. glans penis

14. Which statement below regarding the penis is TRUE?
    a. The penis has a root that extends far into the body.
    b. The penis can only become erect when touched.
    c. The penis contains three bones linked by connective tissue.
    d. The tip of the penis contains very few nerve endings.
    e. All of the above

15. What is an important function of the scrotum?
    a. to manufacture sperm
    b. to keep the testicles cool
    c. to contain the root of the penis
    d. to produce testosterone
    e. All of the above

16. What is the term for sperm production?
    a. cryptorchidism
    b. testicularism
    c. spermatogenesis
    d. inguinal development
    e. semenology

17. Sperm is produced in ________________.
    a. the vas deferens
    b. the prostate gland
    c. seminal vesicles
    d. seminiferous tubules
    e. All of the above

18. What do Sertoli cells do in the male reproductive system?
    a. absorb faulty sperm
    b. provide nutritional substances for the developing sperm
    c. regulate the size of the prostate gland
    d. produce testosterone
    e. accumulate fluid during puberty

19. Which of the following is NOT part of the sperm?
    a. acrosome
    b. flagellum
    c. midpiece
    d. ampulla
    e. head

20. How many sperm does the average human male produce daily?
    a. 100
    b. 200,000
    c. 1 million
    d. 300 million
    e. 500 billion

21. The epididymis sits adjacent to what organ?
    a. penis
    b. testicle
    c. cremaster gland
    d. prostate gland
    e. seminal vesicles

22. Which of the following statements is FALSE regarding the seminal vesicles?
    a. They provide secretions that make up 60-70% of semen.
    b. There are two of them in the male reproductive system.
    c. They provide one source of testosterone.
    d. They are adjacent to the ampulla.
    e. They provide nutrition for traveling sperm.

23. What is the source of the pre-ejaculatory fluid that many men experience during arousal?
    a. epididymis
    b. testicles
    c. Cowper's glands
    d. Prostate gland
    e. There is no pre-ejaculatory fluid.

24. What is the term for male breast enlargement that is common in puberty and old age?
    a. cryptorchidism
    b. mammorphy
    c. priapism
    d. andropause
    e. gynecomastia

25. How does a boy's body change during puberty?
    a. The testicles and penis increase in size.
    b. Metabolism increases.
    c. Bones and muscles develop more rapidly.
    d. The prostate gland and seminal vesicles increase in size.
    e. All of the above

26. Which of the following statements regarding andropause is TRUE?
    a. Men experience andropause in their 40's-50's
    b. Men experience andropause in their 70's-80's
    c. Andropause does not include mood disturbances.
    d. Ejaculatory pressure is unaffected.
    e. It is very similar to female menopause.

27. The inguinal canal is ________________.
    a. another name for the bulbourethral canal
    b. the canal through which the testes descend into the scrotum
    c. a tube that transports spermatozoa from the vas deferens to the urethra
    d. located inside of the penis
    e. the tubule that engorges with blood during an erection

28. Cryptorchidism is ________________.
    a. usually associated with severe pain and swelling of the testicles
    b. a bacterial infection of the scrotum
    c. a parasitic infection of the scrotum
    d. a genetic condition involving an additional X chromosome in biological males
    e. when the testes fail to descend into the scrotum

29. The twisting of a testis on its spermatic cord is called ________________.
    a. testicular torsion
    b. hydrocele
    c. priapism
    d. Peyronie's Disease
    e. an inguinal hernia

30. Which of the following statements about priapism is TRUE?
    a. It is characterized by the testes failing to descend into the scrotum.
    b. The most common cause of priapism is a lack of exercise.
    c. It is characterized by a painful and persistent erection.
    d. Priapism is rarely serious and cannot affect future sexual functioning.
    e. The cause of priapism has been linked to psychological issues.

31. Abnormal calcifications in the penis, which may cause a painful curvature, is called __________________.
    a. priapism
    b. Peyronie's Disease
    c. gynecomastia
    d. penile hypertrophy
    e. cryptorchidism

32. With an inguinal hernia, the intestines push through the opening in the abdominal wall into which structure?
    a. vas deferens
    b. prostate gland
    c. rectum
    d. bladder
    e. inguinal canal

33. What is the name for the condition that causes a scrotal mass when an excessive accumulation of fluid builds up within the tissue surrounding the testicle.
    a. hydrocele
    b. priapism
    c. testicular torsion
    d. scrotal edema
    e. inguinal hernia

34. Which of the following statements below regarding testicular cancer is TRUE?
    a. It cannot be cured once detected.
    b. Early detection is virtually impossible.
    c. It is characterized by a sudden decrease in testicular size.
    d. There are few symptoms until the cancer is advanced.
    e. All of the above

35. What is an uncommon type of cancer in the human male's sexual and reproductive system?
    a. penile cancer
    b. prostate cancer
    c. rectal cancer
    d. inguinal cancer
    e. testicular cancer

36. Benign prostatic hypertrophy __________________.
    a. may block urination
    b. involves a twisted spermatic cord
    c. is usually characterized by a hard mass in the scrotum
    d. my include lesions on the penis and scrotum
    e. involves abnormal calcifications

37. Which group of men has the highest rate of prostate cancer in the world?
    a. African-American
    b. White Eastern European
    c. South East Asian
    d. Inuit
    e. White American

38. Which of the following is NOT a risk factor for prostate cancer?
    a. race
    b. genetics
    c. aging
    d. unprotected sex
    e. high fat diet

39. What is the commonly used abbreviation for the blood test that measures the levels of molecules that are overproduced by prostate cancer cells?
    a. SSA
    b. PPA
    c. PSA
    d. PBA
    e. SAP

40. What is used to treat prostate cancer?
    a. cryosurgery
    b. radiation treatment
    c. radical prostatectomy
    d. All of the above
    e. None of the above

Matching

Column 1		Column 2
A. vas deferens	_____	41. a comma-shaped organ that sits on top of each testicle and holds sperm for 10-14 days while sperm matures
B. priapism	_____	42. one of the three cylinders of erectile tissue in the penis; this one contains the urethra
C. cryptorchidism	_____	43. an 18-inch tube that carries the sperm from the testicles and propels it towards the urethra during ejaculation after it's mixed with other fluid
D. seminal vesicles	_____	44. a condition characterized by the twisting of a testicle on its spermatic cord which can cause pain and swelling
E. corpus spongiosum	_____	45. a condition that occurs when there is an excessive accumulation of fluid within the tissue surrounding the testicle
F. Cowper's glands	_____	46. structures next to the bladder that secrete a fluid that makes up 60-70% of semen
G. Peyronie's Disease	_____	47. a painful and persistent erection that is not associated with sexual excitement and is most commonly caused by drug use
H. testicular torsion	_____	48. two pea-size glands on either side of the urethra that produce a fluid that cleans and lubricates the urethra for sperm; source of pre-ejaculatory fluid
I. epididymis	_____	49. abnormal curvature of the penis which lasts about two years and can make intercourse impossible
J. hydrocele	_____	50. a condition in which the testicles fail to descend into the scrotum during gestation

TEST YOURSELF ANSWER KEY

The Male Sexual and Reproductive System

1. bone, muscle (p. 102)
2. foreskin (p. 105)
3. cremaster (p. 107)
4. seminiferous tubules (p. 107)
5. 72, 300 million (p. 107)
6. epididymis (p. 108)
7. ejaculation (p. 108)
8. prostate (p. 109)
9. Cowper's, urine (p. 109)
10. 5, 15 (p. 110)
11. arteries, blood (p. 110)
12. infertile (p. 110)
13. gynecomastia (p. 110)
14. spermatogenesis (p. 112)
15. testosterone (p. 112)
16. corona, frenulum, and meatus (p. 103)
17. in the Leydig cells located in the testicles (p. 108)
18. They provide nutrition for the traveling sperm. (p. 109)
19. 50-150 million (p. 110)
20. in their 70's or 80's (p. 112)

Male Reproductive and Sexual Health

21. inguinal canal (p. 112)
22. 25 (p. 113)
23. priapism, blood (p. 113)
24. curves (p. 114)
25. Peyronie's Disease (p. 114)
26. reproductive, testicles (p. 115)
27. inguinal hernia (p. 115
28. testicle (p. 115)
29. Testicular (p. 115)
30. testicular, monthly (p. 116)
31. penis (p. 116)
32. benign prostatic hypertrophy (BPH) (p. 117)
33. 50 (p. 117)
34. prostate-specific antigen (PSA) (p. 118)
35. treatments, controversial (p. 118)
36. cryptorchidism (p. 112)
37. drug use (erection drugs, cocaine, marijuana, anticoagulants) (p. 113)
38. 12% (p. 115)
39. a painless testicular mass or a harder consistency of the testes (p. 115)
40. aging, race, a diet high in fat, and a genetic risk (p. 118)

POST TEST ANSWER KEY

True/False

1. T (p. 105)
2. T (p. 107)
3. F (p. 107)
4. F (p. 107)
5. T (p. 108)
6. F (p. 109)
7. T (p. 109)
8. T (p. 111)
9. T (pp. 114-115)
10. F (p. 116)

Multiple-Choice

11. c (p. 103)
12. d (p. 103)
13. e (p. 103)
14. a (pp. 104-106)
15. b (p. 106)
16. c (p. 107)
17. d (p. 107)
18. b (p. 107)
19. d (p. 107)
20. d (p. 107)
21. b (p. 108)
22. c (p. 109)
23. c (p. 109)
24. e (p. 110)
25. e (p. 112)
26. b (p. 112)
27. b (p. 112)
28. e (p. 112)
29. a (p. 113)
30. c (p. 113)
31. b (p. 114)
32. e (p. 115)
33. a (p. 115)
34. d (p. 115)
35. a (p. 116)
36. a (p. 117)
37. a (p. 117)
38. d (p. 118)
39. c (p. 118)
40. d (p. 118)

Matching

41. I (p. 108)
42. E (p. 103)
43. A (p. 108)
44. H (p. 113)
45. J (p. 115)
46. D (p. 109)
47. B (p. 113)
48. F (p. 109)
49. G (p. 114)
50. C (p. 112)

Chapter 5 – Female Anatomy and Physiology

CHAPTER SUMMARY

Girls and women tend to be less familiar with their genitals than boys and men, and genital self-examinations can be a good way to learn more about their bodies. The external sex organs or the "vulva" is made up of the mons veneris, labia majora, labia minora, clitoris, vestibule, and the perineum. The clitoris is an organ made up of erectile tissue with a glans, shaft, and internal crura, and is the only human organ whose sole function is to bring sexual pleasure. The vestibule is the name for the entire region between the labia minora, containing the opening of the urethra and the vagina and the ducts of the Bartholin's glands.

The internal sex organs include the vagina, uterus, cervix, Fallopian tubes, and ovaries. The vagina is an expandable, muscular organ that extends from the cervix of the uterus to the vestibule. The uterus is a muscular organ in the pelvis where implantation of the fertilized ovum takes place and is composed of three layers: perimetrium, myometrium, and endometrium. The cervix is the lower portion of the uterus that contains the opening, leading into the uterus where sperm and menstrual blood can pass. The Fallopian tubes are four-inch long tubules that curl around the ovary to accept ova when they are released. The ovaries are two almond-shaped structures attached to either side of the uterus that produce ova and manufacture hormones. Once a month, during ovulation, an ovary releases an ovum that is swept into the fallopian tube and pushed through by muscular contractions.

Another sexual/reproductive organ, the breasts, or mammary glands, contain fatty tissue and milk-producing glands and are capped by a nipple and surrounded by a round, pigmented area called the areola. The breasts contain between 15-20 lobes, made up of compartments that contain alveoli, the milk-secreting glands.

Puberty in girls lasts from 3 to 5 years and begins around the age of eight but varies based on race and weight. Estrogen leads to the development of various characteristics such as breast and pubic hair growth. Around the age of 11 or 12 girls begin to ovulate. The onset of ovulation typically coincides roughly with the start of menstruation. The average menstrual cycle lasts about 28 days and is divided into four phases: follicular, ovulation, luteal, menstrual. There are a number of menstruation variations including amenorrhea, menorrhagia, dysfunctional uterine bleeding, and dysmenorrhea. Premenstrual syndrome (PMS) refers to physical or emotional symptoms that appear in some women during the latter half of the menstrual cycle. Menopause refers to the cessation of the menstrual cycle, typically between the ages of 49-51.

There are a number of gynecological health concerns that affect women at various ages. Endometriosis occurs when endometrial cells begin to migrate to places other than the uterus may cause painful menstrual periods. Toxic shock syndrome, uterine fibroids, vulvodynia, and other infections can affect women's sexual and reproductive functioning. Breast cancer is the most common cancer in women. Cervical cancer is the second most common cancer of the reproductive tract although the rates have been falling as Pap screening has become more prevalent. Ovarian cancer is the most deadly of all gynecologic cancers because it persists with few symptoms until it reaches an advanced stage.

LEARNING OBJECTIVES

After studying Chapter 5, you should be able to:

1. List and define the female external sexual and reproductive organs.
2. List and define the female internal sexual and reproductive organs.
3. Describe the structure of the clitoris.
4. Explain the physical and societal aspects of the hymen.
5. Summarize the research on the Grafenberg spot and the connection to female ejaculation.
6. Compare and contrast the role the uterus, cervix, and vagina play during pregnancy.
7. Compare and contrast the role the ovaries, Fallopian tubes, and uterus play during ovulation.
8. Outline the origin and pathway of menstrual fluid.
9. Outline the pathway of ova with and without fertilization.
10. Describe the sexual and reproductive capabilities of female breasts.
11. Explain the changes that girls experience during puberty.
12. List and define the four phases of the menstrual cycle.
13. Describe the relationship between menstruation and ovulation and how they are influenced by the negative feedback loop.
14. Compare and contrast variations in menstruation: amenorrhea, menorrhagia, dysfunctional uterine bleeding, and dysmenorrhea.
15. Describe the research in premenstrual syndrome (PMS).
16. Compare and contrast the terms menopause and climacteric.
17. List at least four gynecological health concerns experienced by women.
18. Describe the diagnosis, symptoms, and treatment options for endometriosis.
19. Explain the symptoms and history of toxic shock syndrome.
20. Identify the most common types of cancer in women.
21. Describe the detection and diagnosis procedure for breast, cervical, and ovarian cancer.
22. List the symptoms of breast, cervical, and ovarian cancer.
23. List the treatment options for breast, cervical, and ovarian cancer.

CHAPTER OUTLINE

I. THE FEMALE SEXUAL AND REPRODUCTIVE SYSTEM

- Female Genital Self-Examination

A. **External Sex Organs**

- The correct term for the female external sex organs is "vulva" rather than "vagina".

1. **Mons Veneris** (mons pubis)
    a. The mons veneris is the fatty cushion resting over the front surface of the pubic bone.
    b. It becomes covered with pubic hair after puberty and can be sensitive to stimulation.
2. **Labia Majora** (outer lips)
    a. The labia majora are two longitudinal folds of fatty tissue that extend from the mons and meet at the perineum.
    b. The skin of the outer labia majora is pigmented and covered with hair, while the inner surface is hairless and contains oil glands.
    c. During sexual arousal, the labia majora fill with blood and engorge.
3. **Labia Minora** (inner lips)
    a. The labia minora are two hairless, smaller skin folds situated between the labia majora and the vestibule that differ considerably in appearance in different women.
    b. They contain oil glands and join at the clitoris to form the prepuce or "hood" of the clitoris.
    c. They contain some erectile tissue and serve to protect the vagina and urethra.
4. **Clitoris**
    a. The clitoris is a cylindrical-shaped organ made up of erectile tissue made up of the glans, the shaft, and the internal crura.
    b. It is richly supplied with blood vessels and nerve endings with a higher concentration of nerve fibers than anywhere else on the body, including the tongue or fingertips.
    c. It is the only human organ whose sole function is to bring sexual pleasure.
    d. In some cultures the clitoris and possibly other sexual organs are removed during circumcision and referred to as **clitorectomy** or **infibulation**.
5. **Vestibule**
    a. The vestibule is the name for the entire region between the labia minora, containing the opening of the urethra and the vagina and the ducts of the Bartholin's glands.
    b. **Urethral Meatus**
        - The opening, or meatus, to the urethra lies between the vagina and the clitoris.
        - **Urinary tract infections** are very common and more common among women but are quite treatable and somewhat preventable by things such as staying hydrated and avoiding douches and hygiene sprays.
    c. **Introitus** and **Hymen**
        - The introitus is the entrance of the vagina.
        - The introitus is usually covered at birth by a fold of tissue called the hymen, which varies in thickness and is usually open in the center for the release of menstrual blood.
        - In rare cases, women have an **imperforate hymen** that is solid and doesn't allow for the release of menstrual blood but can be easily treated.

- Many historical and cultural practices have surrounded the hymen due to the view that an intact hymen is a sign of virginity, despite the fact that a hymen can be "broken" by many activities such a riding a bike.

d. **Bartholin's Glands**

- The Bartholin's Glands are glands whose ducts empty into the vestibule in the middle of the labia minor.
- Historically, it was thought that they secreted lubrication but that has proven to be untrue.
- These glands can become infected and form a cyst or abscess.

6. **Perineum**

a. The perineum is the tissue between the vagina and the anus.
b. This tissue may be cut during childbirth during a procedure called an **episiotomy**.

B. **Internal Sex Organs**

1. **Vagina**

a. The vagina is an expandable and muscular organ that extends from the cervix of the uterus to the vestibule.
b. It serves as a passageway for menstrual fluid, babies, and semen.
c. The vagina lubricates through small openings on the vaginal walls during engorgement by mucus produced from glands on the cervix.
d. The first third of the vagina has numerous nerve endings, while the inner two-thirds have none or very few for most women.
e. Grafenberg Spot and Female Ejaculation

- The Grafenberg Spot (G-Spot) is an area about the size of a dime in the lower third of the front part of the vagina.
- Some debate exists whether the G-spot is a separate physiological entity or a sensitive area of the vagina.
- Stimulation of the G-spot leads to pleasant sensation and arousal for some women and can result in orgasms accompanied by the forceful expulsion of fluid, leading to the term female ejaculation.

2. **Uterus**

a. The uterus is a thick-walled, muscular organ in the pelvis between the bladder and the rectum shaped like an inverted pear.
b. It is the source of menstrual fluid and the site for the fetus to develop.
c. It provides a path for sperm to fertilize the ovum and is where implantation of the fertilized ovum takes place.
d. The expandable uterine wall is about one inch thick and is made up of three layers.

- The **perimetrium** is the outer layer made up of thin tissue.
- The **myometrium** is the muscular, middle layer that contracts to expel menstrual fluid and to push the fetus out of the uterus.
- The **endometrium** is the inner layer that responds to fluctuating hormonal levels by building up and shedding with the menstrual cycle.

e. **Cervix**

- The cervix is the lower portion of the uterus that contains the opening, or os, leading into the uterus where sperm and menstrual blood can pass.
- During childbirth, the cervix softens and the os dilates to allow the baby to pass.

- The cervix can be felt at the top end of the vagina and can be seen with a mirror during a pelvic exam.

3. **Fallopian Tubes**

a. The Fallopian tubes or **oviducts** are four-inch long tubules that extend laterally from the sides of the uterus.
b. They curve around to form a trumpet-shaped end called the **infundibulum**, which has fingerlike projections (**fimbriae**) that curl around the ovary to accept ova when they are released.
c. Once a month, during ovulation, an ovary releases an ovum that is swept into the fallopian tube by the fimbriae and pushed through the fallopian tube by muscular contractions.
d. The inner surface of the fallopian tubes is covered by hairlike projections that create a current to move the ovum.
e. The fallopian tubes are the location for fertilization.

4. **Ovaries**

a. The ovaries are two almond-shaped structures attached to either side of the uterus.
b. They produce ova and manufacture hormones.
c. Follicle stimulation hormone (FSH) and luteinizing hormone (LH) are released by the pituitary gland during each menstrual cycle, causing about 20 follicles to begin maturing, but typically only one follicle bursts, releasing an ovum.

C. **Other Sexual Organs**

1. **Breasts**

a. Breasts, or mammary glands, contain fatty tissue and milk-producing glands and are capped by a nipple and surrounded by a round, pigmented area called the areola.
b. The breasts contain between 15-20 lobes, made up of compartments that contain alveoli, the milk-secreting glands.
c. When lactation begins, infant suckling stimulates the pituitary gland to release prolactin, which signals milk synthesis and oxytocin, which allows the milk to be ejected.
d. The breasts are typically seen as an erogenous zone and some women can experience orgasm from breast and nipple stimulation alone.

2. **Other Erogenous Zones**

a. The largest sexual organ of all is the skin.
b. Other areas that are used to increase sexual pleasure include the lips, ears, back of the knees, armpits, base of the neck, anus, and the one organ you can only stimulate indirectly—the brain.

II. THE FEMALE MATURATION CYCLE

A. **Female Puberty**

1. Puberty in girls lasts from 3 to 5 years and begins around the age of eight but varies based on race and weight. African-American girls typically begin puberty earlier than white girls.
2. With puberty, the pituitary gland begins to secrete the hormones FSH and LH which stimulate the ovaries to produce estrogen.
3. Estrogen leads to the development of various characteristics such as breast and pubic hair growth.
4. Around the age of 11 or 12 girls begin to ovulate, and in some women a slight pain or sensation of may be felt called **mittelschmerz**.

5. The onset of ovulation typically coincides roughly with **menarche** (the start of menstruation) although about 80% of menstrual cycles are anovulatory during the first year of menstruation.

B. **Menstruation**

1. Menstruation refers to the monthly release of blood and tissue from the uterine lining.
3. The average menstrual cycle lasts about 28 days, when the uterine lining builds up to prepare for a possible pregnancy.
4. The menstrual cycle is divided into four phases.
   - The **follicular phase** begins after menstruation and is characterized by maturing follicles in the ovaries and the growth of the endometrium in the uterus.
   - During the **ovulation phase**, an ovum is released, usually about the 14th day of the cycle.
   - During the **luteal phase**, a small gland called the **corpus luteum** forms on the ovary and secretes additional progesterone and estrogen, which increases the thickness of the endometrium to prepare it for implantation of a fertilized egg.
   - If fertilization does not occur, the corpus luteum degenerates and hormone levels decrease. The **menstrual stage** then begins, shedding the uterine lining, or **menses**.
5. During menstruation, about 2-4 tablespoons of fluid (mostly blood and mucus) are expelled; unless a woman is using oral contraceptives in which case the amount will be less.
6. The monthly cyclical process involves a **negative feedback loop** where one set of hormones controls the production of another set, which in turn controls the first.

C. **Variations in Menstruation**

1. Amenorrhea is the absence of menstruation.
2. **Primary amenorrhea**, when a woman never begins menstruation, is due to underdeveloped reproductive organs, glandular disorders, poor health, excessive exercise, or emotional factors.
3. **Secondary amenorrhea**, when menstruation stops before menopause, is due to pregnancy, excessive exercise, eating disorders, emotional issues, certain diseases, surgical removal of the ovaries or uterus, or hormonal imbalances typically due to steroids.
3. **Menorrhagia** is excessive menstrual flow and can be treated with oral contraceptives.
4. **Dysfunctional uterine bleeding (DUB)**, when a woman bleeds for long periods of time or intermittently bleeds throughout her cycle, is usually caused by a hormonal imbalance, significant weight loss, eating disorders, stress, chronic illness, or excessive exercise and needs to be treated by a health care provider.
5. **Dysmenorrhea**, or painful menstruation, may be caused by inflammation, constipation, or stress and can be treated by medication or relaxation.

D. **Premenstrual Syndrome**

1. **Premenstrual syndrome (PMS)** refers to physical or emotional symptoms that appear in some women during the latter half of the menstrual cycle.
3. Estimates of the incidence of PMS vary widely depending on how it is defined, but only a small number of women find it debilitating.
4. The diagnosis of PMS has been controversial due to political and historical reasons.
5. A diagnosis of **premenstrual dysphoric disorder (PMDD)** includes severe mood, behavioral, somatic, and cognitive symptoms prior to menstruation.

6. PMDD has been linked to the dysregulation of a neurotransmitter in the brain called serotonin, which has lead to serotonin re-uptake inhibitor drugs as treatment, in addition to lifestyle changes such as exercise and caffeine reduction.

E. **Menstrual Manipulation and Suppression**

1. Menstrual manipulation or menstrual suppression, when women control or stop their periods, has become more common through continuous birth control pills, progesterone intrauterine devices, and injections.
3. The suppression of menstruation has been used for years to treat endometriosis, a condition characterized by severe menstrual cramping and irregular periods.
4. New oral contraceptives are being tested that are used for 84 consecutive days rather than the usual 21-day pills.
5. Birth control pills were originally designed to mimic the typical menstrual cycle, leaving a period of time for bleeding, but the bleeding is not typical menses because there is very little of the endometrium to be shed.

F. **Sexual Behavior and Menstruation**

1. Many cultures have taboos related to engaging in sexual behaviors during menstruation such as in Orthodox Judaism, where women are required to abstain from intercourse for one week after menstruation.
3. Many couples engage in various sexual behaviors during menstruation without concerns or problems, although women may use tampons or diaphragms to catch the menses.

G. **Menopause**

1. **Menopause** refers to the cessation of the menstrual cycle, which typically occurs between the ages of 49-51.
3. **Climacteric** refers to the combination of physiological and psychological changes that develop at the end of a woman's reproductive life, usually including menopause.
4. As women age, their ovaries become less responsive to hormonal stimulation, resulting in decreased hormone production.
5. Decreased estrogen often leads to a reduction in size of the clitoris, labia, uterus, and ovaries in addition to less pubic and head hair, skin wrinkling, and **osteoporosis**.
6. Many women experience menopause with few problems, however, some women may experience hot flashes, headaches, insomnia, decreased sexual desire, and painful intercourse, which can be treated using notional, vitamin, and herbal therapies.
7. Hormone Replacement Therapy
   - **Hormone Replacement Therapy (HRT)** refers to the administration of estrogen and other hormones in the form of creams, pills, or a small adhesive patch.
   - HRT was widely used and touted as a treatment for menopausal symptoms and other diseases, however, in 2002, researchers concluded that there were serious, long-term risks of taking hormones, leading many women to discontinue the treatment.

III. FEMALE REPRODUCTIVE AND SEXUAL HEALTH

- A genital self-exam can help increase a woman's comfort and familiarity with her genitals.
- All women should start routine gynecological examinations once they begin menstruating or before engaging in sexual interactions.
- During a pelvic exam, a health care provider will use a speculum to hold open the walls of the vagina in order to visually inspect the cervix and take a Papanicolaou (Pap) smear from the cervix.

- In addition to the Pap smear, the health care provider will insert two fingers in the vaginal and press down on the abdomen to feel the ovaries and uterus and may also perform a rectal exam by inserting one finger in the anus.

A. **Gynecological Health Concerns**

1. **Endometriosis**

a. Endometriosis occurs when endometrial cells begin to migrate to places other than the uterus, such as the reproductive or abdominal organs.
b. The disease ranges from mild to severe and women may experience symptoms such as painful periods, pelvic or lower abdominal pain, which begin a day or two before menstruation and decrease afterwards.
c. The cause of endometriosis is still unknown, and many women discover endometriosis when they have trouble becoming pregnant.
d. It can be diagnosed through biopsy and treated using hormones, surgery, or laser therapy.

2. **Toxic Shock Syndrome**

a. Toxic Shock Syndrome (TSS) is an acute, fast-developing disease that can result in multi-organ failure, with symptoms such as fever, sore throat, diarrhea, vomiting, muscle aches, and a scarlet-colored rash, and can be fatal if treatment is not received.
b. In the 1980s, TSS gained national attention due to a number of deaths connected to the use of a particular brand of tampons (Rely) which was designed to be kept in the vagina for a long period, leading to the build up of bacteria and eventual infection.
c. The incidence of TSS has decreased in the last decade due to changes in the absorbency and composition of tampons.

3. **Uterine Fibroids**

a. Uterine fibroids are hard tissue masses in the uterus that affect 20-40% of women 35 and older, but are more common among African-American women.
b. Symptoms of uterine fibroids include pelvic pain and pressure, heavy cramping, prolonged or heavy bleeding, constipation, abdominal tenderness or bloating, infertility, recurrent pregnancy loss, frequent urination, and painful sexual intercourse.
c. Treatment includes hormone or drug therapy, laser therapy, surgery, or cryotherapy.

4. **Vulvodynia**

a. Vulvodynia refers to chronic vulvar pain and soreness, with women also reporting itching, burning, rawness, and stinging.
b. The pain can be intermittent or constant and can range from mildly disturbing to completely disabling, leading women to refrain from sexual activities.
c. The cause is not completely known but may involve nerve injuries or allergies.
d. Treatment includes biofeedback, diet modification, drug therapy, oral and topical medications, nerve blocks, vulvar injection, surgery, and/or pelvic floor muscle strengthening.

5. **Infections**

a. Aside from sexually transmitted infections, other infections can afflict women's sexual and reproductive organs.
b. The Bartholin's glands and the urinary tract can become infected, causing discomfort, but can be treated with antibiotics.
c. Douching puts women at risk for vaginal infections since it changes the pH levels in the vagina, destroying healthy bacteria necessary to maintain proper balance.

B. **Cancer of the Female Reproductive Organs**

1. Breast Cancer
    a. Breast cancer is the most common cancer in women, with one in eight American women developing breast cancer in her lifetime.
    b. Early detection is very important so breast self-examinations are recommended.
    c. Some controversy exists regarding whether or not women should receive routine mammography, particularly under the age of 50.
    d. Symptoms of breast cancer include, most commonly, the discovery of a lump in addition to possible breast pain, nipple discharge, changes in nipple shape, and skin dimpling.
    e. Most lumps are not cancerous but should be reviewed by a physician.
    f. Numerous treatment options for breast cancer exist depending upon the severity or type of cancer.
    - In the past a **radical mastectomy** (removal of the breast) was commonly performed, although today a partial or modified mastectomy is performed, leaving many of the underlying muscles and lymph nodes in place.
    - If the tumor is contained and has not spread, a lumpectomy may be performed, removing the tumor along with some surrounding tissue.
    - Radiation and/or chemotherapy are often used in conjunction with the above surgical treatments.

    g. A number of risk factors have been identified for breast cancer.
    - Obesity, levels of physical activity, and alcohol consumption may be risk factors.
    - Family history has been identified as a possible risk factor, although 90% of women do not have a family history of breast cancer.
    - Research is trying to identify women who might be at risk.
    - Early pregnancy before the age of 30 seems to have protective effect although no one is sure why.
    - Debate exists regarding the role oral contraceptives play in the development of breast cancer.
2. Uterine Cancer
    a. Cervical Cancer
    - Cancer of the cervix is the second most common cancer of the reproductive tract, although the rates have been falling as Pap screening has become more prevalent.
    - Pap smears; taken during routine pelvic exams, test for cervical cancer by examining cervical cells that have been scraped off the cervix for abnormalities.
    - Cervical cancer has high cure rates because it starts as an easily identifiable lesion called a **cervical intraepithelial neoplasia (CIN)** usually caused by genital warts (human papilloma virus, HPV).
    - Treatments for CIN include surgery, radiation, or both, which has resulted in declining mortality rates for cervical cancer.
    - Treatment for advanced cervical cancer involves a hysterectomy followed by radiation and chemotherapy.

    b. Endometrial Cancer
    - Cancer of the lining of the uterus is the most frequent gynecological cancer.

- Symptoms include abnormal uterine bleeding and/or spotting, with diagnosis involving a **D & C (dilation and curettage)**.
- Treatment includes surgery, radiation, hormones, and/or chemotherapy.

3. Ovarian Cancer
   a. Ovarian cancer is the most deadly of all gynecologic cancers because it persists with few symptoms until it reaches an advanced stage.
   b. The cause is unknown, although an increased incidence is found among women who undergo early menopause, eat a high fat diet, who are from a higher socioeconomic status, or haven't given birth to children.
   c. Detection has been difficult, with pelvic exams and ultrasounds proving less effective; however, a recent blood test may be available soon.
   d. Early symptoms, if any, may include vague abdominal discomfort, loss of appetite, indigestion, and anorexic symptoms.
   e. Treatment involves removal of the ovaries (with or without removal of the uterus) and radiation and chemotherapy.

TEST YOURSELF

Below you will find fill-in-the-blank and short answer essay questions for topics covered in chapter 5. Check your answers at the end of this chapter.

The Female Sexual and Reproductive System

1. The ______________________ is made up of the mons veneris, the labia majora and labia minor, the vestibule, the perineum, and the clitoris.

2. The skin of the outer ______________________ ______________________ is pigmented and covered with hair, while the inner surface is hairless and contains sebaceous (oil) glands.

3. The ______________________ is richly supplied with blood vessels as well as nerve endings, and the glans is a particularly sensitive receptor and transmitter of sexual stimuli.

4. The ______________________ is the name for the entire region between the labia minora and can be clearly seen when the labia are held apart.

5. The urethra, which brings urine from the bladder to be excreted, is much ______________________ in women than in men, where it goes through the penis.

6. In rare cases a woman has an ______________________ hymen, which is usually detected because her menstrual flow is blocked and can be treated with a simple surgical procedure.

7. The ______________________ glands are bean-shaped glands whose ducts empty into the vestibule in the middle of the labia minora.

8. The ______________________ is the tissue between the vagina and the anus.

9. The ______________________ is approximately four inches in length when relaxed but contains numerous folds that help it expand somewhat like an accordion.

10. Stimulating the ______________________ causes pleasant vaginal sensation in some women and can result in powerful orgasms accompanied by the forceful expulsion of fluid.

11. The inner layer of the uterus, the ______________________, responds to fluctuating hormonal levels, and its outer portion is shed with each menstrual cycle.

12. The ______________________ is the lower portion of the uterus that contains the opening, or os, leading into the body of the uterus.

13. Once a month an ovary releases an ovum that is swept into the fallopian tube by the waving action of the ______________________.

14. At ______________________, the secondary follicle bursts, and the ovum begins its journey down the fallopian tube.

15. When lactation begins, infant suckling stimulates the posterior pituitary gland to release __________________, which signals milk synthesis, and ______________________, which allows the milk to be ejected.

16. What is another name for the opening of the vagina?

17. How have various cultures viewed the hymen throughout history?

18. What is the name of the incision that may be made on the perineum during childbirth to allow more room for the baby's head to emerge?

19. What are the three layers of the uterus starting with the outer layer, middle, and inner?

20. How do ova move through the fallopian tubes to the uterus?

The Female Maturation Cycle

21. Newer research suggests that the onset of puberty in girls may be related to ______________________.

22. ______________________ is responsible for the development and maturation of female primary and secondary sexual characteristics.

23. In a few women, a slight pain or sensation accompanies ovulation, referred to as ______________________.

24. In our culture the age of ______________________ has been steadily falling, and most people believe that there is a difference between being physiologically capable of bearing children and being psychologically ready for sexual intercourse and childbearing.

25. ______________________ is the name for the monthly bleeding that all women of reproductive age experience.

26. During the ______________________ phase, an ovum is released, usually about the 14th day of the cycle.

27. Approximately two days before the end of the normal cycle, the secretion of ______________________ and ______________________ decreases sharply as the corpus luteum becomes inactive and the menstrual stage begins.

28. Menses usually stops about ______________________ to ______________________ days after the onset of menstruation.

29. Some women suffer from ______________________, or excessive menstrual flow.

30. ______________________, or painful menstruation, may be caused by a variety of inflammations, by constipation, or by psychological stress.

31. ______________________ ______________________ refers to physical or emotional symptoms that appear in some women during the latter half of the menstrual cycle.

32. Some physicians have already started prescribing continuous ______________________ ______________________ ______________________, progesterone intrauterine devices, and injections to suppress menstrual periods.

33. While some couples do avoid engaging in ______________________ ______________________ during menstruation, other couples enjoy an active and satisfying sex life throughout their ______________________ ______________________.

34. As women age, their ovaries become less responsive to hormonal stimulation from the ______________________ ______________________, ______________________, resulting in decreased ______________________ production.

35. In some women, hormonal fluctuations can cause ______________________ ______________________, headaches, and ______________________.

36. Describe the hormonal process that begins for girls at puberty.

37. List the four phases of the menstrual cycle.

38. What is the role of the corpus luteum in the menstrual cycle?

39. What is the difference between primary amenorrhea and secondary amenorrhea?

40. What is the difference between menopause and climacteric?

Female Reproductive and Sexual Health

41. In a pelvic exam, the health professional will use a ______________________ to hold open the vagina to examine the cervix.

42. ______________________ occurs when endometrial cells begin to migrate to places other than the uterus.

43. There has been a substantial reduction in the incidence of ______________________ ______________________ ______________________ in the last ten years, which is primarily attributed to the changes in absorbency and composition of tampons available to the consumer.

44. ______________________ ______________________, or hard tissue masses in the uterus, affect from 20-40% of women 35 and older.

45. Treatment for uterine fibroids is hormone or drug therapy, laser therapy, surgery, or ______________________.

46. ______________________ refers to chronic vulvar pain and soreness.

47. Some infections of the female reproductive tract and not necessarily transmitted ______________________.

48. ______________________ changes the pH levels in the vagina and can destroy healthy bacteria necessary to maintain proper balance.

49. ______________________ cancer is the most common cancer in women and is the second most common cause of death from cancer in American women.

50. An important preventive measure for breast cancer is ______________________, which can detect tumors too small to be felt during self-examination.

51. ______________________ ______________________ seems to have a protective effect against getting breast cancer, though no one understands exactly why.

52. Pam smears, taken during routine pelvic exams, test for ______________________ cancer.

53. Cervical cancer has high cure rates because it starts as an easily identifiable lesion, called a(n) ______________________ ______________________ ______________________.

54. ______________________ cancer has few warning signs or symptoms until it reaches an advanced stage.

55. Pelvic exams are not effective in the early diagnosis of ovarian cancer and both blood tests and ultrasound have fairly high ______________________ ______________________.

56. What are some of the symptoms of endometriosis?

57. What was responsible for the number of serious cases of toxic shock syndrome in the 1980s?

58. What is the treatment for vulvodynia?

59. What is the role of family history when it comes to the development of breast cancer?

60. How is ovarian cancer treated?

POST TEST

Below you will find true/false, multiple-choice, and matching quiz items covering the entire chapter. Check your answers at the end of this chapter.

True/False

1. The labia minora contain some erectile tissue and are rich in sebaceous glands.

2. The clitoris has twice the number of nerve endings as the penis.

3. The hymen is the name for the thin fold of tissue that may cover the vaginal opening at birth.

4. An episiotomy is the name for the surgical removal of the uterus.

5. The ejaculation of fluid after sexual stimulation is a phenomenon that has never been documented in women.

6. Girls who are underweight begin menstruating earlier than those who are overweight.

7. Some women experience a slight pain or sensation in the abdomen or pelvis at ovulation.

8. The term, "dysmenorrhea" refers to excessive menstrual flow.

9. Endometriosis refers to women's experience chronic vulvar pain and soreness.

10. Although they are less likely to develop the disease, white women are more likely to die of breast cancer when compared with African-American women.

Multiple-Choice

11. Which of the following is NOT considered part of the vulva?
    a. mons veneris
    b. clitoris
    c. perineum
    d. uterus
    e. labia majora

12. Which of the following statements below regarding the mons veneris is FALSE?
    a. It is Latin for "Mountain of Venus."
    b. It is part of the vulva.
    c. It lies on the anterior wall of the vagina.
    d. It is also called the mons pubis.
    e. It is a mound of fatty tissue over the female pubic bone.

13. The two longitudinal folds of skin extending downward and backward from the mons pubis are called the ____________________.
    a. perineum
    b. clitoris
    c. perimetrium
    d. labia majora
    e. vagina

14. The clitoris is ____________________.
    a. an organ composed of erectile tissue whose sole function is sexual pleasure.
    b. the pigmented ring around the nipple of the breast.
    c. the doughnut-shaped bottom part of the uterus that contains an opening through which menstrual fluid passes.
    d. a part of the cervix that contains numerous glands and follicles.
    e. the smooth muscle layer of the uterus.

15. In women, the urethra meatus is located in the ____________________.
    a. vagina
    b. vestibule
    c. fallopian tubes
    d. cervix
    e. Bartholin's glands

16. The vaginal opening is also called the ____________________.
    a. vestibule
    b. labia
    c. introitus
    d. Grafenberg junction
    e. os

17. The perineum refers to the ____________________.
    a. tissue covering the vaginal opening at birth.
    b. area where the Bartholin's glands empty their contents
    c. hood of the clitoris.
    d. tissue between the vagina and the anus.
    e. region that includes the clitoris, labia, and urethral opening.

18. What muscular organ is approximately four inches in length when relaxed but contains numerous folds that helps it to expand up to four to five times its size?
    a. hymen
    b. cervix
    c. vagina
    d. fallopian tube
    e. vulva

19. Where is the G-spot located?
    a. introitus
    b. vagina
    c. labia majora
    d. cervix
    e. clitoris

20. The muscular, inverted pear-shaped organ in the pelvis that serves as the site of implantation of the fertilized ovum is called the ____________________.
    a. cervix
    b. ovary
    c. G-spot
    d. uterus
    e. fallopian tube

21. The endometrium is ____________________.
    a. the muscular layer of the uterus that contracts to expel menstrual fluid.
    b. the middle layer that pushes the fetus during delivery.
    c. the outer layer of the uterus.
    d. the lower portion of the uterus that contains the opening.
    e. the inner layer of the uterus that responds to fluctuation hormonal levels.

22. What structures are responsible for producing ova and secreting hormones?
    a. fallopian tubes
    b. Inguinal canals
    c. ovaries
    d. oviducts
    e. Bartholin's glands

23. The term "mittelschmerz" refers to ____________________.
    a. the onset of menstruation in puberty
    b. pain or sensation felt by some women at ovulation
    c. the release of estrogen by the ovaries
    d. the growth of endometrial tissue outside the uterus
    e. painful menstruation

24. What is the name for the monthly bleeding which women of reproductive age experience?
    a. ovulation
    b. gynecomastia
    c. mittelschmerz
    d. amenorrhea
    e. menstruation

25. During which menstrual cycle phase do the ovaries, stimulating the regrowth of the endometrium's outer layer, release estrogen?
    a. follicular phase
    b. ovulation
    c. mittel phase
    d. luteal phase
    e. menstrual phase

26. Which phase of the menstrual cycle is the third phase, where a small, pouchlike gland forms on the ovary?
    a. follicular phase
    b. ovulation
    c. mittel phase
    d. luteal phase
    e. menstrual phase

27. During which phase of the menstrual cycle do the endometrial cells shrink and slough off?
    a. follicular phase
    b. ovulation
    c. mittel phase
    d. luteal phase
    e. menstrual phase

28. What is the term used for the absence of menstruation?
    a. menarche
    b. dysmenorrhea
    c. amenorrhea
    d. endometriosis
    e. infibulation

29. Anorexia is a well-known cause of ______________________.
    a. endometriosis
    b. amenorrhea
    c. toxic shock syndrome
    d. uterine fibroids
    e. menorrhagia

30. The term for physical and/or emotional symptoms that appear in some women during the latter half of the menstrual cycle is called ______________________.
    a. toxic shock syndrome
    b. menstrual dysphoria
    c. primary amenorrhea
    d. cryo-menses syndrome
    e. premenstrual syndrome

31. What condition would likely be treated with menstrual suppression?
    a. amenorrhea
    b. urinary tract infections
    c. endometriosis
    d. vulvodynia
    e. uterine fibroids

32. On average, a woman will have approximately how many periods in her lifetime?
    a. 50
    b. 100
    c. 450
    d. 1000
    e. 3000

33. What is the term for the period in which a woman's estrogen production begins to decrease, culminating in the cessation of menstruation?
    a. primary amenorrhea
    b. climacteric
    c. cervical neoplasia
    d. lactation
    e. menopause

34. In the 1980s, a number of women died or lost limbs after contracting a disease that was caused by bacterial build up in the vagina as a result of a particular brand of tampons. What is the name of this disease?
    a. uterine fibroids
    b. gynecomastia
    c. vulvodynia
    d. toxic shock syndrome
    e. endometriosis

35. Which of the following statements regarding uterine fibroids is FALSE?
    a. They can become as large as a basketball.
    b. Excessive menstrual bleeding is the most commonly reported symptom.
    c. They affect 20-40% of women over the age of 35.
    d. They are usually cancerous.
    e. They are hard tissue masses in the uterus.

36. Which is the most common cancer in women?
    a. breast cancer
    b. ovarian cancer
    c. uterine cancer
    d. lung cancer
    e. cervical cancer

37. What percentage of women who develop breast cancer have NO family history of the disease?
    a. 10%
    b. 25%
    c. 50%
    d. 70%
    e. 90%

38. The Pap smear tests for what type of cancer?
    a. cervical
    b. ovarian
    c. breast
    d. vaginal
    e. uterine

39. Cervical cancer starts on an easily identifiable lesion called a ____________________.
    a. papanicolaou abnormality
    b. cervical intraepithelial neoplasia
    c. cryptochordism
    d. cervical fissure
    e. fibroid lesion

40. Which of the following statements below regarding ovarian cancer is FALSE?
    a. The cause of ovarian cancer is unknown.
    b. Pelvic exams are not effective in the early diagnosis of ovarian cancer.
    c. Treatment involves removal of the ovaries.
    d. It is the most deadly of all gynecologic cancers.
    e. Early detection is common since symptoms are typically pronounced.

Matching

Column 1		Column 2
A. amenorrhea	____	41. site of fertilization, joining of the sperm and ovum
B. vulva	____	42. painful menstruation
C. endometriosis	____	43. An acute, fast-developing disease that can result in multi-organ failure, which is caused by the build up of bacteria linked to certain tampons
D. cervix	____	44. the collective term for the female external sex organs
E. uterus	____	45. lower portion of the uterus from which cells are scraped during a Pap smear
F. dysmenorrhea	____	46. an expandable, muscular organ that extends from the cervix of the uterus to the vestibule
G. fallopian tubes	____	47. uterine lining cells begin to migrate to places other than the uterus causing pelvic or abdominal pain
H. menorrhagia	____	48. absence of menstruation
I. vagina	____	49. site of implantation of a fertilized ovum
J. toxic shock syndrome	____	50. excessive menstrual flow

TEST YOURSELF ANSWER KEY

The Female Sexual and Reproductive System

1. vulva (p. 123)
2. labia majora (p. 124)
3. clitoris (p. 124)
4. vestibule (p. 125)
5. shorter (p. 125)
6. imperforate (p. 126)
7. Bartholin's (p. 127)
8. perineum (p. 127)
9. vagina (pp. 128-129)
10. G-spot (p. 129)
11. endometrium (p. 130)

12. cervix (p. 130)
13. fimbriae (p. 130)
14. ovulation (p. 131)
15. prolactin, oxytocin (p. 131)
16. introitus (p. 125)
17. The hymen has been seen as a symbol of "purity," a sign of whether or not a woman has had sexual intercourse, despite the fact that many activities can "break" the hymen. (p. 126)
18. episiotomy (p. 127)
19. perimetrium, myometrium, endometrium (p. 130)
20. The inner surface of the fallopian tubes are covered by hairlike projections whose constant beating action creates a current, moving the ovum along toward the uterus. (p. 130)

The Female Maturation Cycle

21. weight (p. 132)
22. estrogen (p. 132)
23. mittelschmerz (p. 133)
24. menarche (p. 133)
25. menstruation (p. 133)
26. ovulation (p. 134)
27. estrogen, progesterone (pp. 134-135)
28. 3, 7 (p. 135)
29. menorrhagia (p. 137)
30. dysmenorrhea (p. 137)
31. premenstrual syndrome (p. 138)
32. birth control pills (p. 139)
33. sexual intercourse, menstrual cycle (p. 140)
34. anterior pituitary, hormone (p. 140)
35. hot flashes, insomnia (p. 141)
36. When puberty begins, a girl's internal clock signals the pituitary gland to begin secreting the hormones FSH and LH, which stimulate the ovaries to produce estrogen while the girl sleeps. Between the ages of 11 and 14, FSH and LH levels begin to increase during the day as well. (p. 132)
37. follicular, ovulation, luteal, and menstrual (p. 134)
38. Following ovulation, the corpus luteum, a small gland forms on the ovary. It secretes additional progesterone and estrogen for 10 to 12 days, which cause further growth of the cells in the endometrium and increases the blood supply to the lining of the uterus. (p. 134)
39. In primary amenorrhea a woman never begins menstruating, while in secondary amenorrhea, previous typical menses stops before a woman has gone through menopause. (p. 136)
40. Female menopause refers to a woman's final menstrual period, whereas climacteric refers to the period in which a woman's estrogen production begins to decrease, culminating in the cessation of menstruation. (p. 140)

Female Reproductive and Sexual Health

41. speculum (p. 142)
42. endometriosis (p. 143)
43. toxic shock syndrome (p. 143)
44. uterine fibroids (p. 143)
45. cryotherapy (p. 144)
46. vulvodynia (p. 144)
47. sexually (p. 144)
48. douching (p. 144)
49. breast (p. 144)

50. mammography (p. 145)
51. early pregnancy (p. 147)
52. cervical (p. 147)
53. cervical intraepithelial neoplasia (p. 148)
54. Ovarian (p. 148)
55. false negatives (p. 149)
56. Women may experience no symptoms, but the most common symptom is painful periods. Pelvic or lower abdominal pain is also reported. (p. 143)
57. Many of the infected women used a brand of tampons called Rely which was designed to be kept in the vagina over a long period of time. This lead to a build up toxins from bacteria. (p. 143)
58. biofeedback, diet modification, drug therapy, oral and topical medications, nerve blocks, vulvar injections, surgery and/or pelvic floor muscle strengthening (p. 144)
59. Family history may be a risk factor in breast cancer, however, about 90% of women who develop breast cancer do not have any family history of the disease. (p. 147)
60. removal of the ovaries which may be accompanied by radiation and/or chemotherapy (p. 149)

POST TEST ANSWER KEY

True/False		Multiple-Choice		Matching	
1.	T (p. 124)	11.	d (p. 123)	41.	G (p. 130)
2.	T (p. 124)	12.	c (p. 123)	42.	F (p. 137)
3.	T (p. 125)	13.	d (p. 124)	43.	J (p. 143)
4.	F (p. 127)	14.	a (pp. 124-125)	44.	B (p. 123)
5.	F (p. 129)	15.	b (p. 125)	45.	D (p. 130)
6.	F (p. 132)	16.	c (p. 125)	46.	I (p. 128)
7.	T (p. 133)	17.	d (p. 127)	47.	C (p. 143)
8.	F (p. 137)	18.	c (p. 128)	48.	A (p. 136)
9.	F (p. 143)	19.	b (p. 129)	49.	E (p. 129)
10.	F (p. 145)	20.	d (p. 129)	50.	H (p. 137)
		21.	e (p. 130)		
		22.	c (p. 131)		
		23.	b (p. 133)		
		24.	e (p. 133)		
		25.	a (p. 134)		
		26.	d (p. 134)		
		27.	e (p. 135)		
		28.	c (p. 136)		
		29.	b (p. 136)		
		30.	e (p. 138)		
		31.	c (p. 139)		
		32.	c (p. 140)		
		33.	b (p. 140)		
		34.	d (p. 143)		
		35.	d (pp. 143-144)		
		36.	a (p. 144)		
		37.	e (p. 147)		
		38.	a (p. 147)		
		39.	b (p. 148)		
		40.	e (pp. 148-149)		

Chapter 6 – Communication: Enriching Your Sexuality

CHAPTER SUMMARY

Communication fosters mutual understanding, increases emotional intimacy, and helps deepen feelings of love and is one of the most important factors in a satisfying relationship. A lack of communication skills contributes to many serious problems, including violence and abuse. As children acquire language, they also learn new ways of communicating their desires, learning more effective ways of communicating through the use of language.

Researchers and communication experts disagree whether or not there are gender differences in communication. Deborah Tannen is one communication expert who has written extensively on gender differences in communication and coined the term *genderlects*, which refers to the fundamental differences between the way men and women communicate. Tannen suggests that men tend to see the world as a place of hierarchical order where they must struggle to maintain their position. Men interpret comments more often as challenges to their position and attempt to defend their independence. Women, Tannen suggests, see the world more as a network of interactions, and their goal is to form connections and avoid isolation. Women use conversations to establish and maintain intimacy. Women are more likely to use speech patterns that invite discussion and minimize disagreements that tend to decrease the validity and assertiveness of their speech, such as tag questions, disclaimers, question statements, and hedge words.

Researchers disagree about whether there are gender differences, but researchers also disagree on the theories of gender differences, with research suggesting differences may be biological, societal, or psychological. Maltz and Borker suggest a cross-cultural theory of gender differences in communication, proposing that American men and women come from different "sociolinguistic subcultures" and learn different communication rules in childhood. Cautions about the research findings involve the issue that the research has studied only young, well-educated, middle-class Americans and that different cultures differ with respect to communication.

In addition to speech, other forms of communication, such as nonverbal and cyberspace, affect relationships. The majority of our communication with others is nonverbal—through flowers, periods of silence, and also by the way we move our body. Body language helps fill in the gaps in verbal communication. When it comes to cyberspace, text-based conversation can be very intimate and couples can potentially become acquainted faster online than in face-to-face contact. Women were found to have an easier time making their voices heard online than in face-to-face conversations with women using more "smileys" and other emoticons. There can be a risk of "eroticized pseudointimacy" on the web, which can lead to destructive results such as an obsession.

Communicating more effectively requires individuals to put aside their vulnerability and fears and learn to communicate their thoughts to their partners. Self-disclosure is critical in maintaining healthy and satisfying relationships. Research suggests that talking with your partner and sharing feelings helps deepen intimacy and feelings of love; however, there is a risk that individuals can disclose too much before the relationship is stable and communication skills are in place.

LEARNING OBJECTIVES

After studying Chapter 6, you should be able to:

1. Discuss the benefits of good communication in relationships.

2. Discuss the risks of poor communication in relationships.

3. List the characteristics of good communication.

4. Compare and contrast men and women's communication styles based on research findings.

5. List the speech patterns that are more likely to be used by women to invite discussion and minimize disagreements that can decrease assertiveness.

6. Summarize the research on theories of gender differences in communication.

7. Provide examples of nonverbal communication and how the examples affect relationships.

8. Discuss the differences between cyberspace communication and face-to-face communication.

9. Explain the goals and techniques of being a more effective communicator.

10. Describe the ways that self-disclosure can increase intimacy in relationships.

11. Explain why communication about sexual behaviors is important to relationships.

12. Define two types of listening techniques that contribute to healthy communication.

13. Explain how message interpretation can negatively affect a relationship.

14. Identify common defensive techniques used to deny criticism.

15. Compare and contrast nonconstructive communication strategies couples use.

16. Summarize the research on fighting and arguments.

17. Describe why talking about sexuality with a partner can be difficult.

18. Explain how poor self-image and low self-esteem affect relationships.

19. Discuss how different views of sexual behaviors can affect a relationship.

CHAPTER OUTLINE

- The key to the "onion" theory is that as you begin to peel off your layers, so too does your partner. If you share something personal about yourself, your partner will probably do the same.

I. THE IMPORTANCE OF COMMUNICATION

A. **Communication fosters mutual understanding, increases emotional intimacy, and helps deepen feelings of love and intimacy.**

B. **Good communication is one of the most important factors in a satisfying relationship.**

C. **Communication experts have also found that a lack of communication skills contributes to many serious marital problems, including violence and abuse.**

D. **It takes some learning to communicate.**

1. As children acquire language, they also learn new ways of communicating their desires, learning more effective ways of communicating through the use of language.

2. When we communicate with other people, we have three competing goals.

a. The first is to "get the job done"—we have a message for someone and we want to communicate that message.

b. Secondly, we also have a "relational goal"—we want to maintain the relationship and not hurt or offend someone with our message.

c. Finally, we have an "identity management goal"—that is, we want our communication to project a certain image of ourselves.

E. **How Women and Men Communicate**

1. Researchers and communication experts disagree whether or not there are gender differences in communication.

2. The term **genderlects** was coined by Deborah Tannen and refers to the fundamental differences between the way men and women communicate.

a. Tannen suggests that men tend to see the world as a place of hierarchical order where they must struggle to maintain their position.

b. Women, Tannen suggests, see the world more as a network of interactions, and their goal is to form connections and avoid isolation.

c. Tannen's theories and research suggests that men are more likely than women to interrupt when others are speaking.

3. Tannen proposes that when stating an opinion, women often end their statement with a **tag question** such as "It's really cold in here, isn't it?"

4. Research suggests that women also use **disclaimers** ("I may be wrong, but…"), **question statements** ("Will you come with us?"), and **hedge words** such as "sort of," "kind of," "aren't you?"

5. When there is a high degree of conversation, women report more satisfaction with their relationship, while men have less of a need for constant conversation.

6. Theories in Gender Differences

a. Researchers disagree about whether there are gender differences, but researchers also disagree on the reasons for these differences.

b. Maltz and Borker (1982) believe that American men and women come from different "sociolinguistic subcultures" and learn different communication rules in childhood.

c. Research on gender differences in communication has studied only young, well-educated, middle-class Americans, thus findings cannot be generalized to other groups.

d. Different cultures differ with respect to communication, with *individualistic cultures* encouraging their members to have individual goals and values and an independent sense of self, while *collectivist cultures* emphasize the needs of their members over individual needs.

F. **Types of Communication: More Than Words**

1. Nonverbal Communication
    a. The majority of our communication with others is nonverbal—through flowers, periods of silence, and also by the way we move our body.
    b. Nonverbal communication behavior differs widely from culture to culture.
    c. Research suggests that women are better at decoding and translating nonverbal communication.
    d. When it comes to sexual behaviors, nonverbal communication can be a good way to communicate desires, but it can also lead to miscommunication.
2. Cyberspace Communication
    a. There is some research that indicates that there are gender differences in communication styles on the Internet and that, overall, cyberspace remains a male-dominated atmosphere.
    b. Women were found to have an easier time making their voices heard online than in face-to-face conversations with women by using more "smileys" and other emoticons (facial symbols used when communicating online) and can be compared with tag questions.
    c. **Text-based conversation** can be very intimate and couples can potentially become acquainted faster online than in face-to-face contact.
    d. There can be a risk of "eroticized pseudointimacy" on the web, which can lead to destructive results such as an obsession.

G. **Communicating More Effectively**

1. Communicating more effectively requires individuals to put aside their vulnerability and fears and learn to communicate their thoughts to their partners.
2. Timing is an important factor when communicating thoughts and desires.
3. Self-disclosure is critical in maintaining healthy and satisfying relationships.
    a. Research suggests that talking with your partner and sharing feelings helps deepen intimacy and feelings of love.
    b. Women have been found to engage in more self-disclosure than men.
    c. There is a risk that individuals can disclose too much before the relationship is stable and communication skills are in place.
4. Asking for What You Need
    a. Telling your partner what you really want and need during sexual activity can be very difficult.
    b. Not being open about your likes or dislikes is self-defeating, since you may end up feeling resentful of your partner or unhappy in your relationship.

II. CRITICISM: CAN YOU TAKE IT? (AND ALSO DISH IT OUT?)

A. **The Fine Art of Listening**

1. One of the most important communication skills is **nondefensive listening**, which involves focusing your attention on what your partner is saying without being defensive.
2. **Active listening** involves using nonverbal communication to let your partner know that you are attentive and present in the conversation exemplified by eye contact.
3. **Being a More Effective Listener**
    a. Being listened to can make us feel worthy, protected, and cared about.
    b. Effective listening involves several nonverbal behaviors, such as eye contact, nodding, and/or saying, "um hum."

c. When your partner is finished talking, it is important to summarize what your partner has told you as accurately as possible to let your partner know that you heard what he or she was saying and to correct any misunderstandings.

4. Message Interpretation

a. In all conversations the recipient of the message must interpret the intended meaning of the message, yet the message is also dependent upon other factors such as the nature of your relationship and your mood at the time.

b. In one study, women who were preoccupied with their weight were more likely to interpret ambiguous sentences with negative or "fat" meanings, while women who were not preoccupied with their weight did not.

5. Gender Differences in Listening

a. Research has suggested that men and women listen for different things when they engage in conversation.

b. Men are more likely to listen for the bottom line or to find out what action needs to be taken to improve the situation, while women are more likely to listen for details.

B. **Expressing Negative Feelings**

1. The key to expressing negative feelings is managing tension.

2. When softer words are used the disagreement has a better chance of being resolved, but when harsh words are used, chances are the disagreement will build and the tension will escalate.

C. **Accepting Criticism From Someone You Love**

1. Although it would be impossible to eliminate all defensiveness, it's important to reduce defensiveness in order to resolve disagreements.

2. If you are defensive in listening to your partner's criticism, chances are good that you will not be able to hear your partner's message.

3. Keeping our defensiveness in check is another important aspect of good communication.

D. **Non-Constructive Communication: Don't Yell at Me!**

1. **Overgeneralizations,** or making statements like "why do you always…" or, "you never…" generally exaggerate an issue so when communicating, individuals should try to be specific about complaints.

2. When communicating, it's important to stay away from **name-calling** or stereotyping words that will only help escalate anger and frustration.

3. **Overkill** is another common mistake that couples make in conversation when they threaten the worst even when it's not true, for example, "If you don't do that I will leave you."

4. Yelling or screaming can also break down all communication efforts, leading partners to get very defensive and angry.

5. Fighting

a. Couples that disagree are usually happier than those who say that they never fight.

b. Research has shown that happier couples think more positive thoughts about each other during their disagreements, while unhappy couples think negatively about each other.

c. Research suggests that women are more likely to demand a reestablishment of closeness after an argument, while men are more likely to withdraw.

III. ENRICHING YOUR SEXUALITY

A. **Talking with Your Partner about Sex**

1. Sex can be one of the hardest things to discuss, and research reports that the majority of couples show their consent to engage in sexual intercourse by saying nothing at all.
2. Approaching the subject of sex for the first time in a relationship implies moving on to a new level of intimacy, which can be scary.

B. **Self-Esteem: I Like You and I Like Myself**

1. Before anyone else can accept us, we need to accept ourselves since anxieties can interfere with our ability to let go, relax, and enjoy the sexual experience.
2. In the United States, we put a high value on physical attractiveness, which directly affects body image and how attractive we feel.
3. Mental health professionals agree that improving mental health includes improving one's self-acceptance, autonomy, and self-efficacy, which are all important in establishing good sexual relationships.

C. **What Do We Look For in a Partner?**

1. We look for many things in a sexual partner: honesty, fun, sensitivity, good looks, and someone that we can talk to and with whom we can have a good sex life.
2. Research suggest that men often report that physical attractiveness is important to them when looking for a partner, while women are less likely to identify looks as one of the most important characteristics.

D. **What Makes a Good Lover?**

1. People look for many different things in a partner, and what makes someone a good lover to you might not make them a good lover to someone else.
2. Overall, good lovers are sensitive to their partner's needs and desires, can communicate their own desires, and are patient, caring, and confident.
3. Communication is one of the most important aspects of a healthy and satisfying relationship.

E. **Enriching Your Sexuality: It's Not Mind Reading**

1. Good communication skills are an integral part of all healthy relationships and couples who know how to communicate with each other are happier, more satisfied, and have a better chance of making their relationship last.
2. Understanding gender differences in communication can help you minimize disagreements and misunderstandings.
3. It's also important to pay attention to nonverbal cues, since we know that much of our communication is interpreted through our nonverbal behavior.

TEST YOURSELF

Below you will find fill-in-the-blank and short answer essay questions for topics covered in chapter 6. Check your answers at the end of this chapter.

The Importance of Communication

1. The key to the ____________________ theory is that as you begin to peel off your layers, so too does your partner.

2. The research has shown that couples who know how to ____________________ with each other are happier, more satisfied, and have a greater likelihood of making their relationship last.

3. Communication fosters mutual understanding, increases emotional intimacy, and helps deepen feelings of ____________________ and ____________________.

4. Communication experts have found that a lack of ______________________ ______________________ contributes to many serious marital problems, including violence and abuse.

5. As children acquire ______________________, they also learn new ways of communicating their ______________________.

6. Deborah Tannen has termed the fundamental differences between the way men and women communicate as ______________________.

7. Women have been found to use more *rapport-talk*, which establishes ______________________ and connections, while men use more *report-talk*, which imparts ______________________.

8. When stating an opinion, women often end their statement with a ______________________ ______________________ (*It's really cold in here, isn't it?*).

9. These ways of communicating are only trends, and plenty of men and women are good at different ______________________ and ______________________.

10. During these same-sex conversations, girls and boys learn the ______________________ and ______________________ about communication, which follow them through life.

11. ______________________ cultures encourage their members to have individual goals and values and an independent sense of self, while ______________________ cultures emphasize the needs of the members over individual needs.

12. When it comes to sex, ______________________ communication about your likes and needs is far better than ______________________.

13. Women have been found to use more "smileys" and other ______________________ online than men.

14. Given the fact that conversation is reduced to a monitor and a keyboard, ______________________ conversation can still be very intimate and couples can potentially become acquainted faster online than in face-to-face contact.

15. ______________________ lets your partner know what is wrong, how you feel about it, and asks for specific change.

16. What can the lack of communication skills lead to in relationships?

17. What are some of the research findings on gender differences and interrupting?

18. What are at least three examples of speech characteristics that are more likely to be used by women that tend to decrease the speaker's assertiveness?

19. What are some of the cautions when it comes to developing online relationships?

20. What are the reasons that self-disclosure is critical in maintaining healthy and satisfying relationships?

Criticism/Enhancing Your Sexuality

21. The research has shown that the majority of couples spend too much time ____________________ each other and not enough time really listening and making affectionate comments.

22. One of the most important communication skills is ____________________ ____________________, which involves focusing your attention on what your partner is saying without being defensive.

23. Active listening involves using ____________________ communication to let your partner know that you are attentive and present in the conversation.

24. Since we all have ____________________ that our partner can push, it's important to know what these are.

25. Being listened to can make us feel ____________________, protected, and care about.

26. When your partner is finished talking, it is important to ____________________ what your partner has told you as accurately as possible.

27. In all conversations the recipient of the message must ____________________ the intended meaning of the message.

28. Common ____________________ techniques are to deny the criticism, make excuses without taking any responsibility, deflect responsibility, and righteous indignation.

29. ____________________, or making statements like "why do you always…", generally exaggerate an issue.

30. Try to stay away from ____________________ or stereotyping words.

31. ______________________ is when you are frustrated with your partner and threaten the worst.

32. Even though a happy couple is disagreeing about an issue, they still feel ______________________ about each other.

33. ______________________ can be one of the hardest things to discuss.

34. If you have a poor ______________________, or do not like certain aspects of your body or personality, how can you demonstrate to a lover why you are attractive?

35. Mental health professionals agree that improving mental health includes improving one's self-acceptance, ______________________, and self-efficacy.

36. What are two types of listening skills that help improve communication in couples?

37. What are some behaviors that are part of effective listening?

38. Why is it important to accept criticism from a partner?

39. Name three of the mistakes that couples make in their communication patterns.

40. What does research report when it comes to couples and arguing?

POST TEST

Below you will find true/false, multiple-choice and matching quiz items covering the entire chapter. Check your answers at the end of this chapter.

True/False

1. Communication researchers all agree that gender differences in communication exist only among adults under the age of 40.

2. Although men and women tend to think that each other interrupts them more often, research doesn't support more conversation interruptions by either sex.

3. Tag questions and hedge words tend to decrease the speaker's perceived assertiveness.

4. When it comes to nonverbal communication, research suggests that women are more likely to use eye contact than men.

5. The authors recommend keeping communication about sexual likes and dislikes on the nonverbal level to prevent insulting a sexual partner.

6. Poor listeners often think that they understand what a partner is trying to say, but they rarely understand.
7. Making statements that tend to exaggerate a particular issue is called overgeneralizations.

8. The majority of couples show their interest to engage in sexual intercourse by saying nothing at all.

9. Good self-esteem is not related to having satisfying sexual relationships.

10. Talking about sex is one of the best ways to move a relationship to a new level of intimacy and connection.

Multiple-Choice

11. Communication experts have found that a lack of communication skills contributes to _______________________.
    a. abuse
    b. serious marital problems
    c. violence
    d. misunderstandings
    e. All of the above

12. According to research, which of the following below is NOT one of the competing goals when communicating?
    a. relational goal
    b. identity management goal
    c. platonic goal
    d. get the job done
    e. All of the above are competing goals

13. What is the name of the researcher who asserts that women use conversations to establish and maintain intimacy, while men use conversation to establish status?
    a. Virginia Johnson
    b. Jeffrey Klinefelter
    c. Thomas Coleman
    d. Deborah Tannen
    e. Katherine Horney

14. The fundamental differences between the way men and women communicate is referred to as ______________________.
    a. emoticons
    b. beta lenses
    c. stereotones
    d. genderlects
    e. gender typing

15. According to a communication expert who specializes in gender differences, which statement below describes how men view conversation?
    a. Men interpret comments more often as challenges to their position and attempt to defend their independence.
    b. Men view language as an opportunity to bond with friends over common interests.
    c. Men tend to see conversation as a chance to discuss their weakness by incorporating disclaimers.
    d. Men are physically better able to pronounce words, thus they often use conversations as a opportunity to impress others.
    e. All of the above

16. According to a communication expert who specializes in gender differences, which statement below describes how women view conversation?
    a. Women are more likely to view language as an opportunity to comfort others.
    b. Women see the world more as a network of interactions, and their goal is to form connections and avoid isolation.
    c. Women see conversation as a way to brag about their families and friends, often incorporating overkill.
    d. Women tend to view conversation as a chance to illustrate their verbal skills and compete.
    e. All of the above

17. According to communication research, which of the following statements below is TRUE?
    a. Men use more slang when speaking.
    b. When there is a high degree of conversation, women report more satisfaction with their relationship.
    c. Women talk more about psychological states in conversations.
    d. Men tend to have less of a need for constant conversation in a relationship.
    e. All of the above

18. The phrases, "I may be wrong, but…" and "I don't really know, but…" are examples of…
    a. hedge words
    b. gendercons
    c. disclaimers
    d. tag questions
    e. inactive tags

19. Which below is an example of hedge words?
    a. "selfish bastard"
    b. "You always do that."
    c. "You're the best!"
    d. "kind of"
    e. "love" "favorite"

20. What traits do the authors cite as playing a key role in good communication?
    a. understanding and patience
    b. assertiveness and optimism
    c. expressiveness and persuasiveness
    d. eloquence and emotional fluency
    e. silliness and confidence

21. The authors suggest that the origins of gender differences in communication styles may be ______________________.
    a. biological
    b. societal
    c. innate
    d. psychological
    e. All of the above

22. Which below is an example of nonverbal communication?
    a. clenching teeth
    b. smiling
    c. sitting with legs crossed
    d. rubbing eyes
    e. All of the above

23. According to research, who are more likely to use "emoticons?"
    a. women
    b. men
    c. children
    d. senior citizens
    e. All of the above

24. What makes self-disclosure an important communication technique that helps to deepen intimacy in relationships?
    a. It lets your partner know what is wrong.
    b. It leads to vulnerability.
    c. It helps a couple grow together.
    d. It provides a way to solve problems.
    e. All of the above

25. What is one of the risks of self-disclosure discussed by the author?
    a. Self-disclosure may lead to overgeneralizations.
    b. An individual may disclose too much before the relationship is stable.
    c. Question statements become more common after a couple starts self-disclosing.
    d. Sexual intimacy often decreases after self-disclosure.
    e. There are no risks. Self-disclosure is always the right thing to do.

26. Research shows that the majority of couples ____________________.
    a. spend too much time talking
    b. need to practice being more defensive
    c. spend too much time criticizing each other
    d. don't get angry enough
    e. spend too much time making affectionate comments

27. A listening strategy in which the listener focuses attention on what a partner is saying without being defensive is called ____________________.
    a. nonverbal communication
    b. nondefensive listening
    c. emoticon focusing
    d. active disclaiming
    e. nonverbal communication

28. According to the textbook, distressed couples often experience a lack of ____________________.
    a. love
    b. patience
    c. self-esteem
    d. self-actualization
    e. self-restraint

29. What is active listening?
    a. the act of multitalking with focused conversation
    b. a listening technique that involves mimicking a partner's facial expressions while holding hands
    c. a communication technique coined by expert Lucas Picard
    d. a communication/listening technique where the listener uses nonverbal communication to let a partner know the listener is attentive
    e. a listening technique where the listener writes down everything a partner says

30. Which of the following can be a negative form of communication that can often backfire?
    a. advisements
    b. silences
    c. reflections
    d. disclosures
    e. emoticons

31. Which of the following shows a partner that you're tuned in during a discussion?
    a. advisements
    b. emoticons
    c. eye contact
    d. overgeneralizations
    e. disclaimers

32. Which of the following statements reflect the research on gender differences with respect to listening?
    a. Men listen for ways to become intimate, and women listen for how they can improve a situation.
    b. Men listen to gain physical attention, and women listen to try to become more similar to a partner.
    c. Men listen defensively, and women listen to find disagreements
    d. Men listen for ways to fix a problem, and women listen for details.
    e. Men and women both listen in an effort to find the bottom line.

33. Overgeneralizations typically cause ____________________.
    a. nondefensive listening
    b. disclosures
    c. defensiveness
    d. genderlects
    e. All of the above

34. What is the term for a mistake that couples make during arguments when one person threatens the worst but doesn't mean what is said?
    a. hedging
    b. disclaiming
    c. nondefensive advising
    d. emotive expression
    e. overkill

35. What can break down all communication efforts?
    a. nonverbal communication
    b. too much eye contact
    c. nondefensive listening
    d. yelling or screaming
    e. All of the above

36. Couples who say "we never fight" are usually ____________________.
    a. happier than those who say they fight
    b. less happy than those who say they fight
    c. more able to communicate effectively
    d. All of the above
    e. None of the above

37. Why is it so difficult for couples to discuss sexuality openly?
    a. Many people have a sense of shame about sexuality.
    b. Sex typically moves a relationship to a new level.
    c. Children are often taught that sex is "dirty."
    d. Initiating sex can open the way for rejection.
    e. All of the above

38. A man worrying about the size of his penis is an example of ____________________.
    a. report-talk
    b. disclaiming
    c. an emoticon
    d. poor self-image
    e. rapport-talk

39. Being able to communicate personal sexual desires is ____________________.
    a. a trait of individuals with poor self-image
    b. a trait of someone who is adept at employing overgeneralizations
    c. an example of overkill
    d. a key trait of a good lover
    e. an example of nondefensive listening

40. Talking about sex with a partner ____________________.
    a. is a common practice of people with poor self-esteem
    b. is a type of genderlect that should be avoided
    c. should be avoided in favor of nonverbal communication
    d. is one of the best ways to move a relationship to a new level of intimacy and connection
    e. All of the above

Matching

Column 1		Column 2
A. nondefensive listening	____	41. a mistake during arguments when one person threatens the worst but doesn't mean what they say
B. genderlects	____	42. using questions to communicate, decreasing the validity of what is being said
C. tag questions	____	43. listening technique where the listener uses nonverbal communication such as nodding or eye contact
D. overkill	____	44. adding questioning statements at the end of a sentence, such as, "that's an interesting idea, isn't it?"
E. hedge words	____	45. a way of speaking that denies the validity of speech, such as, "I don't really know, but…"
F. question statements	____	46. using phrases such as, "sort of" or "kind of," which decreases the speaker's assertiveness
G. overgeneralizations	____	47. facial symbols used when sending text messages online, such as "smileys"
H. disclaimers	____	48. refers to the fundamental differences between the way men and women communicate
I. active listening	____	49. making statements that tend to exaggerate a particular issue
J. emoticons	____	50. listening strategy where the listener focuses attention without being defensive

Test Yourself Answer Key

The Importance of Communication
1. onion (p. 1)
2. communicate (p. 2)
3. love, intimacy (p. 3)
4. communication skills (p. 3)
5. language, desires (p. 5)
6. genderlects (p. 6)
7. relationships, knowledge (p. 6)

8. tag question (p. 8)
9. techniques, communication (p. 9)
10. rules, assumptions (p. 11)
11. Individualistic, collectivist (p. 12)
12. verbal, nonverbal (p. 14)
13. emoticons (p. 16)
14. text-based (p. 17)
15. Self-disclosure (p. 20)
16. misunderstandings, anger, frustration, violence, and abuse (p. 3)
17. When men interrupt women, they expect to become the primary speaker; when women overlap, they "interrupt" without expecting that they conversation will turn to them. The research shows that men are more likely than women to interrupt when others are speaking. (p. 7)
18. tag questions, disclaimers, question statements, hedge words (p. 8)
19. the risk of eroticized pseudointimacy, overindulgence, and obsession, (p. 17)
20. It helps deepen intimacy and feelings of love. It helps your partner know what is wrong or has caused a problem during a fight. It helps you and your partner know how to change things in the relationship. (p. 19)

Criticism/Enhancing Your Sexuality

21. criticizing (p. 21)
22. nondefensive listening (p. 21)
23. nonverbal (p. 22)
24. buttons (p. 22)
25. worthy (p. 23)
26. summarize (p. 23)
27. interpret (p. 23)
28. defensive (p. 25)
29. overgeneralizations (p. 25)
30. name-calling (p. 26)
31. Overkill (p. 26)
32. positive (p. 27)
33. Sex (p. 28)
34. self-image (p. 29)
35. autonomy (p. 30)
36. nondefensive listening, active listening (pp. 21-22)
37. eye contact, nodding, and/or saying "um hum" (p. 23)
38. It's important to really understand and hear what our partner is trying to tell us and keep our defensiveness in check. (p. 25)
39. overgeneralizations, name-calling, and overkill (pp. 25-26)
40. Disagreements are a common part of relationships. How you handle such disagreements is what is important. Research has shown that happier couples think more positive thoughts about each other during their disagreements, while unhappy couples think negatively about each other. (p. 27)

Post Test Answer Key

True/False		Multiple-Choice		Matching	
1.	F (p. 156)	11.	e (p. 154)	41.	D (p. 167)
2.	F (p. 156)	12.	c (p. 155)	42.	F (p. 157)
3.	T (p. 157)	13.	d (p. 156)	43.	I (p. 165)
4.	T (p. 160)	14.	d (p. 156)	44.	C (p. 157)
5.	F (p. 160)	15.	a (p. 156)	45.	H (p. 157)
6.	T (p. 165)	16.	b (p. 156)	46.	E (p. 157)
7.	T (p. 167)	17.	e (pp. 156-157)	47.	J (p. 160)
8.	T (p. 168)	18.	c (p. 157)	48.	B (p. 156)
9.	F (pp. 168-170)	19.	d (p. 157)	49.	G (p. 167)
10.	T (p. 172)	20.	a (p. 157)	50.	A (p. 164)
		21.	e (p. 157)		
		22.	e (p. 159)		
		23.	a (p. 160)		
		24.	e (p. 164)		
		25.	b (p. 164)		
		26.	c (p. 164)		
		27.	b (p. 164)		
		28.	e (p. 164)		
		29.	d (p. 165)		
		30.	a (p. 165)		
		31.	c (p. 166)		
		32.	d (pp. 166-167)		
		33.	c (p. 167)		
		34.	e (p. 167)		
		35.	d (p. 167)		
		36.	b (p. 168)		
		37.	e (p. 168)		
		38.	d (p. 168)		
		39.	d (p. 170)		
		40.	d (p. 172)		

Chapter 7 – Love & Intimacy

CHAPTER SUMMARY

Love and the ability to form loving, caring, and intimate relationships with others are important for our emotional health, but also for our physical health. The concept of love can be traced back in time to cultures such as the Late Egyptian empire and the Hebrews. Romantic love is the all-encompassing, passionate love of romantic songs and poetry also referred to as passionate love, whereas companionate love includes the development of trust, loyalty, a lack of criticalness, and a willingness to sacrifice for the partner.

Research on love includes the work of John Alan Lee, Robert Sternberg, and Zick Rubin. Lee identified six basic ways to love: eros, ludus, storge, mania, pragma, and agape. Sternberg suggests that different strategies of loving are really different ways of combining the basic building blocks of love: passion, intimacy, and commitment. Sternberg combines these elements into seven forms of love. Rubin was one of the first researchers to try and scientifically measure love by creating a "Love Scale" which measured what he believed to be the three components of attachment: needing, caring, and trusting. A number of researchers have identified theories to explain why humans form attachments and love, including behavioral reinforcement theories, cognitive theories, physiological arousal theories, and evolutionary perspectives.

Love, intimacy, and attachment are learned through life from infancy to adulthood. In infancy, the nature and quality of the bond with the caregiver can have profound effects on the ability of the person to form attachments throughout life. Three attachment styles have been identified that influence adult relationships: secure, anxious/ambivalent, and avoidant. Experimentation with different approaches to others is common during adolescence, when we develop our role repertoire and intimacy repertoire. In adulthood, true intimacy is more difficult to achieve than true love because the emotion of love may be effortless, while the establishment of intimacy always requires effort.

Research on attraction and how we choose partners suggests that individuals of various sexual orientations found that people want partners who have similar interests, values, and religious beliefs, who were honest, intelligent, affectionate, financially independent, dependable, and physically attractive. The ways people think about and express emotion are very different in different cultures. It takes effort and commitments to maintain love, to continually build on and improve the quality of the relationship. When developing intimacy skills, we must take into consideration self-love, receptivity, listening, affection, and trust.

The dark and destructive side of love includes jealousy, compulsiveness, and possessiveness. Jealousy is an emotional reaction to a relationship that is being threatened. A correlation has been found between self-esteem and jealousy; the lower the self-esteem, the more jealousy they feel and in turn the higher their insecurity. Abusive love relationships exist when one partner tries to increase his or her own sense of self-worth or control the other's behavior through withdrawing or manipulating love. Possessiveness indicates a problem of self-esteem and personal boundaries, and can eventually lead to stalking.

LEARNING OBJECTIVES

After studying Chapter 7, you should be able to:

1. Describe some historical depictions of love.
2. Define romantic and companionate love.
3. Summarize the research of John Alan Lee.
4. List the six basic lovestyles or "colors" of love identified by Lee.
5. Summarize the research of Robert Sternberg.
6. List the three elements that make up love according to Sternberg.
7. Identify at least three types of love according to Sternberg's research.
8. Identify at least two scales developed to measure love.
9. Discuss the challenges with measuring love.
10. List and define the theories of why humans love.
11. Compare and contrast behavioral reinforcement theories with evolutionary perspectives.
12. Explain the lessons learned about love in childhood and adolescence.
13. List the three attachment styles that affect intimate relationships in adulthood.
14. Summarize the research what factors determine who humans are attracted to and fall in love with.
15. Summarize the research findings on attraction and love in different cultures.
16. Discuss the research on the relationship between love and sexual behaviors.
17. Describe the ways that couples can developing intimacy skills.
18. Identify the results of love that can be dangerous and destructive.
19. Explain how jealousy is related to personality traits.

CHAPTER OUTLINE

- The "Roseto Effect" is a term used to describe connections among people in a town, Roseto, in Pennsylvania where there was a strong sense of community, love, loyalty, and ties to each other. These connections to each other were found to reduce stress and improve the health and well-being of the townspeople.
- Love and the ability to form loving, caring, and intimate relationships with others are important for our emotional health, but also for our physical health.

I. WHAT IS LOVE?

A. **Love in Other Times and Places**

1. The concept of love can be traced back in time to cultures such as the Late Egyptian empire and the Hebrews.
2. The Middle Ages glorified the modern idea of **romantic love**, including loving from afar, or loving those one could not have, known as **unrequited love**.

II. THE FORMS AND ORIGIN OF LOVE

A. **Romantic vs. Companionate Love**

1. **Romantic Love**

a. Romantic love is the all-encompassing, passionate love of romantic songs and poetry also referred to as passionate love, infatuation, obsessive love.
b. Passionate love blooms in the initial euphoria of a new attachment to a sexual partner, but the passion of that intensity fades after a time.

2. **Companionate** or **Conjugal Love**

a. If a relationship is to continue, romantic love must develop into companionate or conjugal love, which involves feelings of deep affection, attachments, intimacy, and ease with the partner.
b. Companionate love includes the development of trust, loyalty, a lack of criticalness, and a willingness to sacrifice for the partner.
c. It does not have the passionate high and low swings of romantic love, but it can even be a deeper, more intimate love.
d. It can be difficult to evolve from romantic to companionate love due to models of love in the media.

B. **The Colors of Love: John Alan Lee**

1. Psychologist John Alan Lee suggests that more forms of love exist than just romantic and companionate love, and through research he identified six basic ways to love, which he gave Greek and Latin names.

a. Eros: romantic love
b. Ludus: the art of seduction
c. Storge: quiet, calm love that builds over time
d. Mania: consumed by thoughts of the beloved; highs and lows
e. Pragma: realistic love, made the best "deal"
f. Agape: altruistic, selfless, never demanding, patient, and true love

2. Lee points out that two lovers with compatible styles are probably going to be happier and more contented with each other than two without compatible styles.
3. Higher levels of manic and ludic lovestyles are associate with poorer psychological health.
4. Higher levels of storge and eros lovestyles are associated with higher levels of psychological health.

C. **Love Triangles: Robert Sternberg**

1. Sternberg suggests that different strategies of loving are really different ways of combining the basic building blocks of love: passion, intimacy, and commitment.
    a. *Passion* is sparked by physical attraction and sexual desire, and drives a person to pursue a romantic relationship.
    b. *Intimacy* involves feelings of closeness, connectedness, and bondedness in a loving relationship.
    c. *Commitment* is the decision to love someone for the long term.
2. Sternberg combines these elements into seven forms of love.
    a. *Nonlove:* absence of all three elements
    b. *Liking:* intimacy only; friendships
    c. *Infatuation:* passion only
    d. *Empty Love:* commitment only
    e. *Romantic Love:* passion and intimacy
    f. *Companionate Love:* intimacy and commitment
    g. *Fatuous Love:* passion and commitment
    h. *Consummate Love:* passion, intimacy, and commitment; Sternberg identifies this as the ideal form of love
3. Love evolves and changes as individuals mature so a person will experience different forms of love at different times for the same person.

III. CAN WE MEASURE LOVE?

A. **Zick Rubin**

1. Rubin was one of the first researchers to try and scientifically measure love.
2. He thought of love as a form of attachment to another person and created a "Love Scale" that measured what he believed to be the three components of attachment.
    a. Degrees of Needing
    b. Caring
    c. Trusting
3. Rubin's scale proved to be extraordinarily powerful as a tool to measure love.

B. **Keith Davis** and his colleagues created the *Relationship Rating Scale* (RRS), which measures six aspects of relationships such as intimacy, passion, and conflict.

C. **Hatfield and Sprecher** created the *Passionate Love Scale* (PLS), which tries to measure the degree of intense passion.

D. **Theories: Where Does Love Come From?**

1. Behavioral Reinforcement Theories
    a. Behavioral Reinforcement Theories suggest that we love because another person reinforces positive feelings in ourselves.
    b. Research suggested that a rewarding or positive feeling in the presence of another person makes us like them, even if the reward has nothing to do with the other person.
    c. This theory suggest that we like people we associate with feeling good and love people if the association is very good.
2. Cognitive Theories
    a. Cognitive theories of liking and loving are based on an interesting paradox: the less people are paid for a task, the more they tend to like it.
    b. If we are with a person often and find ourselves doing things for them, we ask: *"Why am I with her so often? Why am I doing her laundry? I must like her—I must even love her!*
3. Physiological Arousal Theories

a. These theories suggest that love is a physiological reaction similar to fear, anger, or excitement.
b. Research conducted in 1962 by Schachter and Singer concluded that an emotion happens when there is general physiological arousal for whatever reason and a label is attached to it—and that label might be any emotion.
c. According to this theory, people would be vulnerable to experiencing love (or another emotion) when they are physiologically aroused for whatever reason.
d. Research in 1974 by Dutton and Aron found that men who were approached by female research assistants in a scary environment were more likely to experience attraction.
e. The original 1962 study has been difficult to replicate, leading to little support for the claim that arousal is a necessary condition for an emotional state and the role of arousal has been overstated.

4. Evolutionary Perspectives
a. Sociologists believe love developed as the human form of three basic instincts:
- the need to be protected from outside threats
- the instinct of the parent to protect the child
- the sexual drive

b. They suggest that love is an evolutionary strategy that helps us form the bonds we need to reproduce and pass our genes on to the next generation.
c. This theory suggests that love creates the union that maximizes each partner's chance of passing on their genes to the next generation.

E. **Love from Childhood to Maturity**

1. Childhood
a. In infancy, the nature and quality of the bond with the caregiver can have profound effects on the ability of the person to form attachments throughout life.
b. We tend to relate to others in our love relationships much like we did when we were young.
c. Overall, parental divorce may affect trust and intimacy in a close relationship, but it does not put children at an overall disadvantage in the development of love relationships.

2. Adolescence
a. Experimentation with different approaches to others is common during adolescence, when we develop our **role repertoire**.
c. We also develop an **intimacy repertoire**, a set of behaviors that we use to forge close relationships throughout our lives.
d. Many factors have been found to be associated with the ability to find romantic love in adolescence, such as marital status of the parents, the quality of the parental relationship, and comfort with one's body.
e. Attachment Styles: Attachment research suggests that the type of intimate relationships you form may be due primarily to the type of attachment you formed as a child.
- *Secure* infants believe a caregiver will respond if they cry out or need them, thus the secure adult easily gets close to others and is not threatened when a lover goes away.
- *Anxious/Ambivalent* babies cry more and panic when the caregiver leaves them, and as adults they worry that their partner doesn't really love them or will leave them.

- *Avoidant* babies have caregivers who are uncomfortable with affection and tend to force separation on a child, leaving them as an adult to be uncomfortable with intimacy and trust.

3. Adult Love and Intimacy
    a. True intimacy is more difficult to achieve than true love because the emotion of love may be effortless, while the establishment of intimacy always requires effort.
    b. Attraction
        - One of the most reliable predictors of who a person will form a relationship with is proximity since people are most likely to find partners among the people they know or see around them.
        - The majority of people who fall in love share similarities such as educational levels, attractiveness, and political opinions.
        - Physical attractiveness has been found to be one of the most important influences in forming love relationships.
        - A large percentage of people cite personality as the most important factor in how they choose their partners. Traits such as openness, sociability, sense of humor, and receptivity are also important.
        - Newer research has suggested that both women and men (rather than just women) want partners to have economic resources.
        - Men and women report that at the top of their list is mutual attraction and love.
        - A study of individuals of various sexual orientations found that people want partners who have similar interests, values, and religious beliefs, who were honest, intelligent, affectionate, financially independent, dependable, and physically attractive.
    c. Attraction in Different Cultures
        - Research from David Buss (1989) found that men across 37 cultures valued "good looks" in a partner more than women did, and women valued "good financial prospects" more than men.
        - He assumed all respondents were heterosexual.
        - He also found that men preferred mates who were younger, while women preferred mates who were older.
    d. Intimate Relationships
        - People respond more positively to those who display emotional openness than to people who are only willing to self-disclose thoughts and experiences.
        - Because intimacy makes us vulnerable and because we invest so much in the other person, intimacy can also lead to betrayal and disappointment, anger, and jealousy.
    e. Male and Female Styles of Intimacy
        - Research suggests that women think more about whether a romantic relationship will develop into a long-term committed relationship.
        - Men and women report equally desiring and valuing intimacy, but men grow up with behavioral inhibitions to expressing intimacy.
        - Other researchers suggest that men are just as intimate as women but express intimacy differently.
        - Research with gay men suggests that gay men tend to adopt fewer stereotyped beliefs about gender roles than heterosexual men,

suggesting that values about sex roles is more relevant than biological sex.

f. Intimacy in Different Cultures

- The ways people think about and express emotion are very different in different cultures. For example, passionate love as constructed in Western cultures is unknown in Tahiti.
- In a study of France, Japan, and the U.S.A., intimacy style was directly related to whether the culture was individualistic, collectivistic, or mixed and also to how much the culture had adopted stereotypical views of gender roles.

g. Long-term Love and Commitment

- It takes effort and commitments to maintain love, to continually build on and improve the quality of the relationship.
- Couples who continue to communicate with each other, remain committed to each other and the relationship, and remain interested in and intimate with each other build a lasting bond of trust.

h. Loss of Love

- People experience the loss of love in many different ways.
- Most people are very vulnerable after the loss of a love relationship, becoming vulnerable to self-blame, loss of self-esteem, and distrust of others.

IV. LOVE, SEX, AND HOW WE BUILD INTIMATE RELATIONSHIPS

A. **Love and Sex**

1. Casual sex has become much more common and accepted than it was 35 or 40 years ago.
2. People meeting each other for the first time tend to reveal their levels of attraction by their body language.
3. Sexual desire has been found to be related to "relational maintenance," that is, the higher the sexual desire for the partner, the less likely the couple has thoughts about ending the relationship or being unfaithful to their current partner.
4. Research shows that men in the U.S. have considerably more permissive attitudes towards casual sex than females.
5. When making the decision to initiate a sexual relationship with another person:
    a. clarify your values
    b. be honest with yourself
    c. be honest with your partner

B. **Developing Intimacy Skills**

1. Self-Love
    a. We must first take responsibility to know ourselves (self-intimacy) and then to accept ourselves as we are (self-love).
    b. Self-love is different from conceit or **narcissism**.
    c. Self-love it is not a process of promoting ourselves but of being at ease with our positive qualities and forgiving ourselves for our faults.
2. Receptivity
    a. Receptivity can be communicated through smiling, eye contact, and a warm, relaxed posture, which allows the other person to feel comfortable and makes us approachable.

b. Researchers believe that sexually satisfied couples who are in love have circadian rhythm (daily biological "clocks") cycles that are "in sync" with each other.

3. Listening

a. True communication beings with listening, which enhances intimacy.

b. Research suggests that people were liked better by others when they responded to self-disclosure with emotional support rather than with their own self-disclosure.

4. Affection shows that you feel a sense of warmth and security with your partner.

5. Trust

a. Trust behaviors lead to greater trust in the relationship and more confidence that the relationship will last.

b. Research suggests that a couple that trusts each other expects their partner to care and respond to their needs, now and in the future.

c. A close relationship has a "curative" function—the longer it exists, the more trust can build.

6. Respect is the process of acknowledging and understanding that person's needs, even if you don't share them.

C. **The Dark Side of Love**

1. Jealousy: The Green-Eyed Monster

a. Jealousy is an emotional reaction to a relationship that is being threatened.

b. In a nationwide survey of marriage, counselors found that jealousy is a problem in one-third of all couples in marital therapy.

c. A correlation has been found between self-esteem and jealousy; the lower the self-esteem, the more jealousy they feel and in turn the higher their insecurity.

d. Men and women experience similar levels of jealousy in intimate relationships, yet there is controversy over what triggers jealousy, with some research supporting the idea that men have more jealousy when they believe that their female partner has had a sexual encounter with another man, while women are often more focused on the emotional or relationship aspects of infidelity.

e. Other studies have found that there are definite physiological responses in both men and women when they imagined scenarios of their partner committing either emotional or sexual infidelity.

f. The longer we are in a relationship with someone, our vulnerability to jealousy decreases.

g. Though many people think that jealousy shows that they really care for a person, in fact it is not a complement, but a demonstration of lack of trust and low self-esteem.

2. Compulsiveness: Addicted to Love

a. Being in love can produce a sense of ecstasy, euphoria, and a feeling of well-being, much like a powerful drug.

b. Researchers Peele and Brodsky (1976) suggest that love addiction is more common than most believe and that is based on a continuation of an adolescent view of love that is never replaced as the person matures.

3. Possessiveness: Every Move You Make, I'll Be Watching You

a. Abusive love relationships exist when one partner tries to increase his or her own sense of self-worth or control the other's behavior through withdrawing or manipulating love.

b. Possessiveness indicates a problem of self-esteem and personal boundaries and can eventually lead to **stalking**.

c. Thinking about another person with that level of obsession is a sign of a serious psychological problem, one that should be brought to the attention of a mental health professional.

TEST YOURSELF

Below you will find fill-in-the-blank and short answer essay questions for topics covered in chapter 7. Check your answers at the end of this chapter.

What is Love?

1. The ________________ Effect, as it was known, disappeared and the rates of heart disease began to increase when early family traditions began to diminish in the 1960s.

2. When people love each other, talk to each other and share their inner selves, their ________________ ________________ may actually become stronger than people who isolate themselves or are emotionally withdrawn.

3. One of the great mysteries of humankind is the capacity to ________________, to make attachments with others that involve deep feeling, selflessness, and ________________.

4. The ________________ of love is part of its attraction.

5. The Middle Ages glorified the modern idea of ________________ ________________, including loving from afar, or loving those one could not have, ________________ ________________.

6. ________________ love is the all-encompassing, passionate love of romantic songs and poetry.

7. ________________ love blooms in the initial euphoria of a new attachment to a sexual partner.

8. If the relationship is to continue, romantic love must develop into ________________ or ________________ love.

9. ________________ love may even be a deeper, more intimate love than romantic love.

10. Psychologist John Alan ________________ identified six basic ways to love.

11. Higher levels of ________________ and ________________ lovestyles are associated with poorer psychological health.

12. Robert Sternberg has proposed that love is made up of three elements: ________________, ________________, and ________________.

13. ________________ is sparked by physical attraction and sexual desire, and drives a person to pursue a romantic relationship.

14. ________________ involves feelings of closeness, connectedness, and bondedness in a loving relationship.

15. ________________ is the decision to love someone for the long term.

16. What is romantic love?

17. What is companionate love?

18. How did John Alan Lee develop his theories on love; what were his research methods?

19. Which of the three elements of love (according to Sternberg's research) does romantic love include?

20. Which of the three elements of love (according to Sternberg's research) does consummate love include?

Can We Measure Love?

21. __________________ thought of love as a form of attachment to another person, and created a "Love Scale," which measured what he believed to be the three components of attachment.

22. Hatfield and Sprecher created the __________________ Love Scale (PLS), which tries to measure the degree of intense passion or "longing for union."

23. When you ask people questions about love, they can only answer with their conscious __________________ towards love.

24. One group of theories suggests that we love because another person __________________ positive feelings in ourselves.

25. __________________ theories of liking and loving are based on an interesting paradox: the less people are paid for a task, the more they tend to like it.

26. Schachter and Singer concluded that an emotion happens when there is general __________________ arousal for whatever reason and a label is attached to it.

27. Sociobiologists try to understand the ________________ advantages of human behavior.

28. In ________________, the nature and quality of the bond with the caregiver can have profound effects on the ability of the person to form attachments throughout life.

29. Those who do not experience intimacy growing up may have a harder time establishing ________________ ________________ as adults.

30. ________________ infants tolerate caregivers being out of their sight, because they believe the caregiver will respond if they cry out of need for them.

31. The importance of experimentation to our lives explains why ________________ relationships can be so intense, and fraught with jealousy.

32. One of the most reliable predictors of who a person will date is ________________.

33. Research has found that women tend to give more importance to the ________________ of intimate relationships than men do.

34. ________________ people were more aware of their love feelings, more expressive, and tolerated their partners' faults more than those who score high only on the masculinity scale.

35. The loss of ________________ is a time of mourning. Going through a period of sadness and ________________, as well as anger at the partner, is natural.

36. What are the three components of attachment identified by Zick Rubin?

37. Which theory of why humans' love may involve sweaty palms increased heart rate, fear, anger, or excitement?

38. What are the three basic instincts that sociobiologists believe are connected to love?

39. What are the three types of attachment behaviors identified by Ainsworth that follow an individual throughout life?

40. What are the repertoires that are learned in adolescence, when experimentation helps adolescents move into adulthood?

Love, Sex, and How We Build Intimate Relationships

41. Sex can be an expression of __________________ and __________________ without considering it an expression of passionate love.

42. Entering a relationship with another person takes close __________________.

43. __________________ sex has become much more common and accepted than it was 35 or 40 years ago.

44. People meeting each other for the first time tend to reveal their levels of attraction by their __________________ __________________.

45. Sexual desire has been found to be related to __________________ __________________, that is, the higher the sexual desire for the partner the less likely the couple has thoughts abut ending the relationship or being unfaithful to their current partner.

46. __________________ considered a number of variables before making the decision to have sex with a partner.

47. Developing intimacy begins with __________________ ourselves, and __________________ ourselves.

48. Self-love is different from conceit or __________________.

49. Many of us think we are __________________ to others when actually we are sending subtle signals that we do not want to be bothered.

50. Researchers believe that sexually satisfied couples who are in love have __________________ rhythm cycles that are "in sync" with each other.

51. True communication begins with __________________, and nothing shows you care about another person quite as much as your full attention.

52. __________________ behaviors lead to greater trust in the relationship and more __________________ that the relationship will last.

53. __________________ is the process of acknowledging and understanding that person's needs, even if you don't share them.

54. __________________ is an emotional reaction to a relationship that is being threatened.

55. Being in love can produce a sense of __________________, euphoria, and a feeling of well-being, much like a powerful __________________.

56. What is the current relationship between sex and love in contemporary U.S. culture?

57. What are the skills that can be developed in order to increase intimacy?

58. What are ways to communicate receptivity?

59. How can jealousy be a self-fulfilling prophecy in relationships?

60. What makes possessiveness a negative feature of a relationship?

POST TEST

Below you will find true/false, multiple-choice and matching quiz items covering the entire chapter. Check your answers at the end of this chapter.

True/False

1. Discussions of love can be traced back to the Late Egyptian empire.
2. Another name for "romantic" love is "companionate" love.
3. According to Davies' research on lovestyles, it's more socially desirable for men to have agape.
4. Empty love refers to love without commitment.
5. Cognitive theorists believe that love creates the union that maximizes each partner's chance of passing on their genes to the next generation.
6. Physical attractiveness has been found to be one of the most important influences in forming love relationships.

7. Men and women report equally desiring and valuing intimacy, but men grow up with behavioral inhibitions to expressing intimacy.

8. Culture plays little role in determining how we view love.

9. Receptivity is an important intimacy skill that can be developed.

10. Being in love can produce a sense of ecstasy, euphoria, and a feeling of well-being, much like a powerful drug.

Multiple-Choice

11. Which of the following is NOT related to the Roseto Effect?
    a. community
    b. overall health
    c. educational objectives
    d. rates of heart disease
    e. love

12. Recent research on arguments among intimate couples suggests that the arguments ________________.
    a. may lead to heart disease
    b. decrease feelings of love
    c. often lead to divorce
    d. may weaken the immune system
    e. All of the above

13. Dean Ornish wrote ________________.
    a. *A Treatise on Sexual Communication: Long-Term Commitments*
    b. *Love and Survival: Scientific Basis for the Healing Power of Intimacy*
    c. *Sexuality as Ideology*
    d. *Angels and Demons*
    e. *The Book of Breaking Up*

14. What is the name for loving someone when the love will never be returned?
    a. conjugal love
    b. poignant love
    c. companionate love
    d. unrequited love
    e. glum love

15. Companionate love is ________________.
    a. an intimate form of love with feelings of deep affection, attachment, and intimacy
    b. loving another when love will never be returned
    c. idealized love based on romance and perfection
    d. obsessive love characterized by love sickness and a sense of ecstasy
    e. characterized by strong passion and commitment but very little intimacy

16. Conjugal love includes ________________.
    a. the development of trust
    b. a willingness to sacrifice for the partner
    c. loyalty
    d. a lack of criticalness
    e. All of the above

17. What did John Alan Lee propose in his research on love?
    a. He proposed that love is made up of three different elements combined in different ways.
    b. He explored the relationship between casual sex and commitment.
    c. He identified six basic ways to love.
    d. He developed a "Love Scale" that measures components of attachment.
    e. He identified three different types of jealousy.

18. What lovestyles have research associated with higher levels of psychological health?
    a. romantic and agape
    b. storge and eros
    c. agape and ludic
    d. pragmatic and manic
    e. manic and conjugal

19. According to Davies' research on lovestyles, what type of lovestyle is more socially desirable for women to have?
    a. ludus
    b. mania
    c. agape
    d. eros
    e. pragma

20. What is the name of the love researcher who proposed that love is made up of three elements: passion, intimacy, and commitment?
    a. Robert Sternberg
    b. Joseph Brodsky
    c. Zick Rubin
    d. John Alan Lee
    e. David Tannen

21. Which element identified by research drives a person to pursue a romantic relationship?
    a. intimacy
    b. passion
    c. infatuation
    d. commitment
    e. All of the above

22. Consummate love refers to love with what element(s)?
    a. passion
    b. intimacy
    c. commitment
    d. All of the above
    e. None of the above

23. What researcher created a "Love Scale" that measured three components of attachment?
    a. John Alan Lee
    b. Zick Rubin
    c. Dean Ornish
    d. Joseph Brodsky
    e. Robert Sternberg

24. What is the name of Hatfield and Sprecher's scale that tries to measure the degree of intense passion or "longing for union?"
    a. The Love Triangle Scale
    b. Companionate Love Scale
    c. Passionate Love Scale
    d. Degrees of Intimacy and Love
    e. Colors of Love

25. What theories suggest that we love because another person reinforces positive feelings in ourselves?
    a. emotive constructionist theories
    b. evolutionary theories
    c. cognitive theories
    d. behavioral reinforcement theories
    e. physiological arousal theories

26. Griffitt and Veitch found that people tend to dislike people they meet in a hot, crowded room, no matter what the new person's personality is like. What theories of love origins does this research finding reflect?
    a. emotive constructionist theories
    b. physiological arousal theories
    c. behavioral reinforcement theories
    d. cognitive theories
    e. evolutional theories

27. This group of theories is based on the paradox that the less people are paid for a task, the more they tend to like it.
    a. physiological arousal theories
    b. evolutionary theories
    c. behavioral reinforcement theories
    d. emotive constructionist theories
    e. cognitive theories

28. According to attachment researchers, what type of attachment style in infancy leads adults to be uncomfortable with intimacy and have difficulty trusting others?
    a. secure
    b. anxious/ambivalent
    c. insecure
    d. avoidant
    e. irritable/frustrated

29. According to research, when do we develop our role repertoire and intimacy repertoire?
    a. infancy
    b. childhood
    c. adolescence
    d. early adulthood
    e. later adulthood

30. What is the name researchers have given to developing patterns of behaviors that work in our interactions with others?
    a. intimacy repertoire
    b. repertoire of closeness
    c. range of loving styles
    d. role repertoire
    e. range of attachment and bonding

31. When it comes to attraction and how we find partners, which of the following aspects is least likely to positively affect partner choices?
    a. physical attractiveness
    b. similar educational levels
    c. proximity
    d. similar personality traits
    e. differences in attitudes

32. Which of the following statements below regarding David Buss' research on attraction in different cultures is FALSE?
    a. Financial issues were not a factor when choosing a partner.
    b. Men preferred mates who were younger than they were.
    c. Men valued "good looks" in a partner more than women did.
    d. Women preferred mates who were older than they were.
    e. Women valued "good financial prospect" in a partner more than men did.

33. What is NOT cited as a factor in helping couples maintain their relationships and commitment over time?
    a. trust
    b. sexual intercourse
    c. communication
    d. intimacy
    e. All of these are important factors

34. According to the textbook, when making the decision to initiate a sexual relationship, one should consider the following ________________.
    a. be honest with yourself
    b. clarify your values
    c. be honest with your partner
    d. All of the abovc
    e. None of the above

35. Sexual desire has been found to be related to ________________.
    a. a storgic lovestyle
    b. a sense of community
    c. narcissism
    d. relational maintenance
    e. All of the above

36. Narcissism is a form of ________________.
    a. self-love
    b. agape
    c. communication
    d. self-intimacy
    e. conceit

37. Which of the following statements below about trust is FALSE?
    a. Trust usually develops quickly in a relationship.
    b. Men and women feel less able to trust when their partner is ambivalent.
    c. Trusting behaviors lead to more confidence that the relationship will last.
    d. The longer trust exists, the more it can build.
    e. Trust develops slowly.

38. Which of the following statements regarding jealousy is FALSE?
    a. Jealousy occurs mostly in dating couples.
    b. Women experience more jealousy related to the emotional aspects of infidelity.
    c. Men experience more jealousy when it involves sex.
    d. It is a problem in one-third of all couples in therapy.
    e. People who experience a lack of jealousy tend to be more secure.

39. Which of the following does the textbook discuss as the dark side of love?
    a. pragmatic and ludic lovestyles
    b. stalking and agape
    c. jealousy, compulsiveness, and possessiveness
    d. low self-esteem and flirting
    e. flirting, cheating, and self-blame

40. What the conclusion of love addiction researchers?
    a. Love based on needing someone is the most fulfilling type of love.
    b. Love addiction is more common than most believe.
    c. Love addiction is based on trust and happiness.
    d. Love addiction is a mature form of romantic love.
    e. All of the above

Matching

Column 1		Column 2
A. Robert Sternberg	____	41. a set of behaviors that we use to forge close relationships throughout our lives
B. unrequited love	____	42. idealized love, based on romance and perfection
C. conjugal love	____	43. love with all three elements: commitment, intimacy, and passion
D. intimacy repertoire	____	44. love researcher who believes that love is made up of three elements
E. romantic love	____	45. developing patterns of behaviors that work in our interactions with others
F. John Alan Lee	____	46. love researcher who identified six "colors of love"
G. role repertoire	____	47. love researcher that developed a "Love Scale" to measure three elements of attachment
H. Zick Rubin	____	48. type of love that only involves commitment
I. empty love	____	49. an intimate form of love in which feelings of deep affection, attachment, and intimacy are present
J. consummate love	____	50. a type of love in which the love will never be returned

Test Yourself Answer Key

What is Love?

1. Roseto (p. 176)
2. immune systems (p. 176)
3. love, commitment (p. 177)
4. mystery (p. 177)
5. romantic love, unrequited love (p. 178)
6. Romantic (p. 178)
7. Passionate (p. 179)
8. companionate, conjugal (p. 179)
9. Companionate (p. 179)
10. Lee (p. 179)
11. manic, ludic (p. 180)

12. passion, intimacy, commitment (p. 181)
13. Passion (p. 181)
14. Intimacy (p. 181)
15. Commitment (p. 181)
16. an idealized, passionate love, based on romance and perfection (p. 178)
17. an intimate form of love in which feelings of deep affection, attachment, and intimacy occur (p. 179)
18. He collected statements about love from hundreds of works of fiction and nonfiction, including both ancient and modern authors. He had a panel of professionals in literature, philosophy, and the social sciences sort the thousands of statements on love into categories. (p. 179)
19. passion and intimacy (p. 182)
20. all three elements: passion, intimacy, and commitment. According to Sternberg, it is the ultimate, complete love that humans strive to achieve. (p. 182)

Can We Measure Love?

21. Rubin (pp. 182-183)
22. Passionate (p. 183)
23. attitudes (p. 183)
24. reinforces (p. 183)
25. Cognitive (p. 183)
26. physiological (p. 184)
27. evolutionary (p. 184)
28. infancy (p. 185)
29. intimate relationships (p. 185)
30. Secure (p. 185)
31. adolescent (p. 186)
32. proximity (p. 187)
33. future (p. 190)
34. Androgynous (p. 191)
35. love, depression (p. 195)
36. degrees of needing, caring, and trusting (pp. 182-183)
37. Physiological Arousal Theory (p. 184)
38. the need to be protected from outside threats, the instinct of the parent to protect the child, and the sexual drive (p. 184)
39. secure, anxious/ambivalent, avoidant (p. 185)
40. role repertoire and intimacy repertoire (p. 186)

Love, Sex, and How We Build Intimate Relationships

41. affection, intimacy (p. 195)
42. self-examination (p. 195)
43. Casual (p. 196)
44. body language (p. 196)
45. relational maintenance (p. 196)
46. Women (p. 196)
47. understanding, liking (p. 197)
48. narcissism (p. 197)
49. receptive (p. 197)
50. circadian (p. 197)
51. listening (p. 197)
52. Trusting, confidence (p. 198)
53. Respect (p. 198)
54. Jealousy (p. 198)

55. ecstasy, drug (p. 199)
56. The importance of love as an essential condition for sexual relations has diminished. (p. 196)
57. self-love, receptivity, listening, affection, and trust (pp. 197-198)
58. smiling, eye contact, and a warm, relaxed posture (p. 197)
59. The jealous individual can drive a partner away, which convinces them that they were right to be jealous in the first place. (pp. 198-199)
60. Controlling behavior smothers the relationship, indicating a problem of self-esteem and personal boundaries and can even lead to stalking. (p. 200)

Post Test Answer Key

True/False

1. T (p. 177)
2. F (pp. 178-179)
3. F (p. 180)
4. F (p. 182)
5. F (pp. 183-184)
6. T (p. 187)
7. T (p. 190)
8. F (p. 193)
9. T (p. 197)
10. T (p. 199)

Multiple-Choice

11. c (p. 176)
12. d (p. 176)
13. b (p. 177)
14. d (p. 178)
15. a (p. 179)
16. e (p. 179)
17. c (p. 179)
18. b (p. 180)
19. c (p. 180)
20. a (p. 181)
21. b (p. 181)
22. d (p. 182)
23. b (pp. 182-183)
24. c (p. 183)
25. d (p. 183)
26. c (p. 183)
27. e (p. 183)
28. d (p. 185)
29. c (p. 186)
30. d (p. 186)
31. e (pp. 187-190)
32. a (p. 190)
33. b (pp. 193-194)
34. d (p. 195)
35. d (p. 196)
36. e (p. 197)
37. a (p. 198)
38. a (pp. 198-199)
39. c (pp. 198-200)
40. b (pp. 199-200)

Matching

41. D (p. 186)
42. E (p. 178)
43. J (p. 182)
44. A (p. 181)
45. G (p. 186)
46. F (p. 179)
47. H (p. 182)
48. I (p. 182)
49. C (p. 179)
50. B (p. 178)

Chapter 8 – Childhood and Adolescent Sexuality

CHAPTER SUMMARY

Research on childhood sexuality can be sensitive due to societal reservations about children's sexual development. The U.S. government has sponsored four large-scale, longitudinal studies to examine adolescent behaviors. However, more research is needed to examine other areas such as the frequency of sexual behaviors other than intercourse, same-sex preferences and behavior, and the meaning of sexuality for young people.

Children's physical and psychosexual development is profound as they proceed from infancy to adolescence. Ultrasound has shown that male fetuses have erections in the uterus, and female babies are capable of vaginal lubrication from birth. During early childhood, children begin to ask basic questions about sexuality as toilet training leads to an intense interest in genitals and bodily wastes. Masturbation or self-stimulation is normal and common beginning in infancy, illustrating a natural curiosity about exploring the world and bodies. During early childhood, child sex play often begins with exposing the genitals, touching, and even rubbing up against each other. Children use information from different and even contradictory sources to create a "sexual script."

During adolescence, puberty is a major stage of physiological sexual development, marking the start of reproductive ability. Kissing and petting are the first types of sexual contact that most young people have with partners. In recent years, media has reported that more adolescents are engaging in oral sex at earlier ages due to perceptions that it's safer, less intimate, and not "real" sex. In recent years, the number of adolescents who have engaged in sexual intercourse has decreased for the first time since the 1970s, with research showing that 63% of teens have had sexual intercourse by their 18th birthday. Same-sex sexual contact is common in adolescence, and young people may or may not identify with the labels of heterosexual, lesbian, gay, or bisexual. Adolescents in the United States have the highest rates of pregnancy, childbearing, and abortion among developed countries, and this is primarily due to the fact that they do not use contraception as reliably. Additionally, in the U.S., approximately 4 million adolescents are infected with sexually transmitted infections every year.

Sexuality education in schools remains controversial, with opposition believing that it should be private and take place in homes. With the increasing visibility of sex in the media, sexuality education has increased in colleges and schools for younger students. No federal laws or policies exist requiring U.S. schools to teach sexuality education, although some states do mandate various forms of sexuality education. Comprehensive sexuality programs are developmentally appropriate programs that span a child's schooling and address a wide variety of topics including contraception and sexually transmitted infections. Abstinence-only programs emphasize abstinence from all sexual behaviors without providing much information about contraception or disease prevention. Research results have found that comprehensive sexuality programs have been the most successful at helping adolescents delay their involvement in sexual intercourse and help protect adolescents form sexually transmitted infections and unintended pregnancies.

LEARNING OBJECTIVES

After studying Chapter 8, you should be able to:

1. Discuss some of the large-scale research projects that have been done to examine adolescent behaviors.

2. Summarize the physical development of children in relation to sexuality at various points in the lifespan: infancy, early childhood, middle childhood, and adolescence.

3. Summarize the psychosexual development of children at various points in the lifespan: infancy, early childhood, middle childhood, and adolescence.

4. Summarize the sexual behavior of children at various points in the lifespan: infancy, early childhood, middle childhood, and adolescence.

5. Explain the role that masturbation or self-stimulation plays at various points in the lifespan: infancy, early childhood, middle childhood, and adolescence.

6. Summarize the sexual knowledge and attitudes of children in early and middle childhood.

7. Summarize the research on sexuality and relationships in middle childhood.

8. Discuss adolescent attitudes and behaviors related to oral sex.

9. Discuss adolescent attitudes and behaviors related to sexual intercourse.

10. Explore some of the research on same-sex sexual behaviors and adolescents who identify as gay, lesbian, and bisexual.

11. Compare and contrast the various influences on adolescent sexuality.

12. Discuss the research on contraception use, pregnancy, and abortion among adolescents.

13. Discuss the research on sexually transmitted infections among adolescents.

14. Identify the most popular goals of sexuality education.

15. Identify the effects of comprehensive sexuality education.

16. Compare and contrast various types of sexuality education programs.

17. Discuss the criticisms and limitations of abstinence-only programs.

18. Identify the ways sexuality researchers determine whether or not a sexuality program is successful.

19. Compare and contrast sexuality education in various countries.

CHAPTER OUTLINE

I. RESEARCH ON CHILDHOOD SEXUALITY

A. **Research on childhood sexuality can be sensitive.**

B. **The U.S. government has sponsored four large-scale, longitudinal studies to examine adolescent behaviors.**

1. National Survey of Family Growth (NSFG)
2. National Longitudinal Study of Adolescent Males (NSAM)
3. National Longitudinal Study of Adolescent Health (ADD)
4. Youth Risk Behavior Surveillance System (YRBS)

C. **More future research is needed in the following areas.**

1. the frequency of sexual behaviors other than intercourse
2. differences in gender, ethnicity, race, religion, and social class
3. same-sex preferences and behavior
4. cross-cultural research
5. the meaning of sexuality for young people

II. INFANCY (Birth to Age 2)

A. **Physical Development**

1. Ultrasound has shown that male fetuses have erections in the uterus and female babies are capable of vaginal lubrication from birth.
2. Kinsey established that half of boys between the ages of 3 and 4 could achieve the urogenital muscle spasms of orgasm, though no fluid is ejaculated.
3. Kinsey didn't collect data on girls and orgasm, but there is anecdotal evidence that girls do experience orgasm.

B. **Psychosexual Development**

1. The single most important aspect of infant development is the child's relationship to its parents or caregivers.
2. A baby's crying helps stimulate the secretion of the hormone oxytocin in the mother, which releases her milk for nursing.
3. The need for warmth and contact in infancy was demonstrated in Harlow's (1959) famous experiment.
4. Between the age of 1 and 2, children begin to develop their **gender identity**.
5. Children develop **gender constancy** a bit later when they come to understand that they will not become a member of the other sex sometime in the future.
6. As children develop a strong identification with one gender, they learn about **gender role behaviors** through modeling.

C. **Sexual Behavior**

1. Body pleasures such as nursing or other stimulation can create a generalized neurological response that stimulates the genital response (erections or vaginal lubrication).
2. Masturbation
    a. Many infants start touching their genitals as soon as their hands are coordinated enough, around 3 or 4 months of age.
    b. Masturbation is normal and common at this age, illustrating a natural curiosity about exploring the world and their bodies.

III. EARLY CHILDHOOD (Ages 2 to 5)

A. **Physical Development**

1. Early childhood is a crucial period for overall physical development from walking to talking.

2. Children may learn more in the first few years of childhood about the nature of their bodies than they learn in the entire remainder of their lives.

B. **Psychosexual Development**

1. Toilet training leads to an intense interest in genitals and bodily wastes.
2. During early childhood, children begin to ask basic questions about sexuality.
3. Children are also exploring what it means to be "boys" or girls."

C. **Sexual Behavior**

2. Masturbation

a. Over 70% of mothers in one study reported that their children under six masturbated.
b. Masturbation in early childhood may be deliberate and obvious and may even become a preoccupation.
c. Parental reaction is very important because parents who are tolerant can teach them to respect and take pride in their bodies.
d. Setting socially acceptable limits to touching genitals can be one way parents can teach responsibility.

3. Sexual Contact

a. Childhood sex play often begins with exposing the genitals, touching, and even rubbing up against each other.
b. In one study, 48% of parents reported that their children under 6 years old had engaged in sex play with another child.

D. **Sexual Knowledge and Attitudes**

1. The secrecy surrounding sexuality is typically taught to children during early childhood as they learn about covering up the genitals and not touching them in public.
2. Children are rarely taught the anatomically correct names for their genitals.
3. Due to the visibility of penises, boys and girls express more excitement in boys' bodies, and boys learn more about sexual pleasure and penises, while girls learn very little about their clitorises.

IV. CHILDHOOD THROUGH PREADOLESCENCE (Ages 6 to 12)

A. **Physical Development**

1. Internal pubertal changes begin in the body around the age of 6 or 7, but the visible signs of puberty begin around 9 or 10.
2. In girls, **breast buds** appear and pubic hair growth may begin.
3. In boys, pubic hair growth starts a couple of years later than in girls.

B. **Psychosexual Development**

1. Although Freud believed that children enter a period in childhood when sexual issues remain unimportant, most researchers disagree with that perspective.
2. Research actually shows that sexual interest and activity in societies across the world steadily increases during childhood.
3. It does appear that children become better at hiding their sexual interests and behaviors after learning about societal taboos surrounding sexuality.

C. **Sexual Behavior**

1. Prepubescence continues to be a time of sexual discovery as children learn about adult sexual behaviors and begin to assimilate cultural taboos and prejudices around sexual behaviors.
2. **Sexual Fantasies** are experienced by children as young as 4 or 5, and are not uncommon among prepubescent children, although guilt about fantasies can concern children.
3. **Masturbation** is not uncommon in childhood and early adolescence, with research suggesting that boys masturbate more than girls and reach orgasm more frequently.

4. **Sexual Contact**
    a. Sexual play, both heterosexual and homosexual, is common during this period.
    b. Sex games such as "spin the bottle" and "post office" help children learn about sexuality with friends and through "play."
    c. Research on sexual contact among children in early adolescence is difficult to find due to adult's recollections that may not be accurate and dated research such as Kinsey's data of 1948 and 1953.
    d. Both boys and girls exhibit a range of same-sex sexual behaviors as they move through childhood, from casual rubbing to more focused attention on the genitals.

D. **Sexual Knowledge and Attitudes**
1. Children use information from different and even contradictory sources to create a "sexual script."
2. Sexual scripts refer to sexual messages communicated to children such as "sex is dirty" or "save it for someone you love."
3. Plummer calls the scripting of "absence" when parents don't talk to their children about sexuality, leaving gaps of knowledge and vocabulary.

E. **Sexuality and Relationships**
1. Relationships with Parents and Caretakers
    a. Parents often send contradictory messages to their children about sexuality.
    b. Even parents who want their children to grow up with a healthy view of sexuality are often unsure of the best way to respond to children's sex play.
2. Relationships with Peers
    a. Same-sex peers
    - Friendships provide children in early adolescence the opportunity to learn about acceptable peer-group sexual standards as friendships become more stable, supportive, and intimate.
    - Initial sexual experimentation often takes place among preadolescents of the same sex, with same-sex experimentation being quite common in childhood.

    b. Other-sex peers
    - Preadolescence can be a time when many children recognize their sexual nature and see peers as potential boyfriends and girlfriends.
    - Although preadolescence has traditionally been a time of early sexual contact, such as kissing, it is also the age of first intercourse with one study reported that about 12% of 12-13 year olds reported having sexual intercourse.

    c. Sibling sex
    - Sexual contact with siblings or close relatives such as cousins is another fairly common childhood experience.
    - Research suggests that it can be harmful when there is a large age difference or coercion is used.
    - More often, sibling sexual contact involves mutual sexual curiosity and is harmless.

V. ADOLESCENCE (Ages 12 to 18)

A. **Physical Development**
1. Puberty is a major stage of physiological sexual development, marking the start of reproductive ability.
2. The physiological changes of puberty often lead to awkwardness and discomfort.

3. Maturing earlier or later than average can be challenging for both girls and boys and can contribute to adolescents having negative **body images**.
4. Females
   a. For most girls, the first signs of puberty are the beginnings of breast buds, the appearance of pubic hair, and the widening of the hips.
   b. Menarche is an important aspect of female puberty but can mean different things to different girls, such as being associated with cramps and embarrassment.
5. Males
   a. Spermatogenesis and ejaculation occur early in male puberty, and may even precede secondary sexual characteristics such as body hair and voice changes.
   b. Spontaneous erections, which may have nothing to do with sexual feelings, can be embarrassing for adolescent boys.

B. **Psychosexual Development**
1. Early Adolescence (About Ages 10 to 13)
   a. Friendships during this time are crucial to emotional well-being.
   b. Cliques, dating, and body image become important in adolescence.
2. Middle Adolescence (About Ages 14 to 16)
   a. Adolescents typically increase the frequency of dating during this time.
   b. Gay, lesbian, bisexual, and transgender adolescents may have a tough time fitting in and feel depressed and alone.
3. Late Adolescence (About Ages 17 to Adulthood)
   a. Almost all cultures allow marriage and other adult privileges in late adolescence.
   b. Historically in Western cultures, people began looking for marriage partners during late adolescence, a phenomenon that has become increasingly less common.

C. **Sexual Behavior**
1. Adolescent sexual activity has increased in the past 50 years due to an increasing link between love and sex.
2. Sexual fantasies are often a means of testing out potential sexual situations.
3. Masturbation
   a. Masturbation remains an underreported sexual behavior in adolescence due to the embarrassment factor.
   b. Research suggests that males masturbate more than girls.
4. Sexual Contact
   a. Abstinence
      - Abstinence means different things to different people, such as abstaining from vaginal-penile intercourse or all types of sexual behaviors.
      - SIECUS, the national sexuality education organization, encourages adolescents to delay sexual intercourse until they are physically, cognitively, and emotionally ready for mature sexual relationships and their consequences.
      - More groups and celebrities have emerged in recent years supporting abstinence.
      - Research suggests that boys feel more embarrassment and guilt about their virginity than girls.

   b. Kissing and petting are the first sexual contact that most people have with potential sexual partners.

c. Oral Sex

- In recent years, media has reported that more adolescents are engaging in oral sex at earlier ages due to perceptions that it's safer, less intimate, and not "real" sex.
- Research since Kinsey to current studies has found that rates of **fellatio** (sexually stimulating male genitals with the mouth) and **cunnilingus** (sexually stimulating female genitals with the mouth) have increased among adolescents.
- Some adolescents falsely believe that oral sex does not put them at risk for sexually transmitted infections or HIV/AIDS.

d. Sexual Intercourse

- The decision to have sexual intercourse for adolescents can be a difficult one, and they do it for different reasons.
- Adolescents report that their first intercourse is usually unplanned, although the decision to engage in sexual intercourse is rarely spontaneous.
- In recent years, the number of adolescents who have engaged in sexual intercourse has decreased for the first time since the 1970s, with research showing that 63% of teens have had sexual intercourse by their 18th birthday.
- Research has illustrated racial differences in first intercourse, with African-American males reporting earlier intercourse than males of other races.
- Overall research results must be viewed with some caution due to the issue of self-report and different definitions of sexuality and sexual intercourse.

e. Research on heterosexual anal intercourse is relatively rare with some studies finding that between 15% and 18% of college students report at least one experience with anal intercourse.

f. Same-Sex Sexuality

- Same-sex sexual contact is common in adolescence, and young people may or may not identify with the labels of heterosexual, lesbian, gay, or bisexual.
- Many adolescents who experience same-sex sexual attraction feel rejected and try to hide their feelings, but increasingly gay, lesbian, and bisexual adolescents are taking pride in their identities and starting support groups.

g. Other sexual situations include other types of sexual experiences that adolescents may experience such as prostitution and child pornography.

D. **Influences on Adolescent Sexuality**

1. Peer Influences

a. Peers are often reported as having a strong influence on adolescent sexual behavior, particularly in the decision to engage in sexual intercourse.

b. Research also suggests that adolescents with strong family relationships tend to be less influenced by peers.

2. Relationships with Parents

a. Mothers are more likely to communicate about sexuality with children of both sexes.

b. Both overly strict and overly permissive parents have children who engage in earlier and more frequent sexual intercourse than parents who are moderately strict.

3. Religion

a. In general, more religious youth report fewer incidents of premarital sexual activity and report sex with fewer partners.

b. Major Western religions tend to discourage premarital sex, but religious adolescents have strong ties to people who are more likely to disapprove of early sexual activity.

E. **Adolescent Contraception and Pregnancy**

1. Contraceptive Use

a. Adolescents in the United States have the highest rates of pregnancy, childbearing, and abortion among other developed countries, and this is primarily due to the fact that they do not use contraception as reliably.

b. Research suggests that adolescents who are able to talk to their mothers about sexuality are more likely to use contraception.

c. Birth control responsibility typically falls more on the shoulders of girls.

2. Teenage Pregnancy

a. Pregnancy Rates

- Pregnancy rates among adolescents have declined over the last few years due to increased contraception use.
- The United States still has the highest rates of adolescent pregnancy, abortion, and childbirth of any Western country.
- Although many teen mothers and their babies do fine, they face increased challenges such as lower birthrates, dropping out of school, less financial stability, and children with poorer health, cognitive, and behavioral abilities.

b. Birth Rates

- Pregnancy rates have decreased compared to birth rates due to a high percentage of teen pregnancies that end in abortion.
- Rates of teen pregnancy:
    - 1986: 50 births per 1000
    - 1991: 62 births per 1000
    - 1996: 54 births per 1000
- A large number of adolescent fathers are not involved in their children's lives, but that is changing somewhat, and these boys need family and societal encouragement to stay invested.

c. Abortion

- The legalization of abortion has had the greatest impact on the ratio of teen pregnancies to births.
- The abortion rate among teenagers has decreased in the 1990s, but U.S. adolescents still have the highest abortion rates among developed nations.

d. Race, Poverty, Teen Pregnancy, Birth, and Abortion

- White adolescents have historically had lower birthrates than black adolescents or Latino adolescents, but recent birthrates have declined more steeply among black than among white teenagers.
- American society has been less supportive of adolescent pregnancy in terms of education and job support compared to other countries.

e. What should be Done About Teen Pregnancy?

- American society is extremely conflicted about the issue of sexuality in general.
- The U.S. is far behind other Western countries in providing day care services that would help single or young parents care for their children.

F. **Adolescents and Sexually Transmitted Infections**

a. In the U.S., approximately 4 million adolescents are infected with a sexually transmitted infection every year.

b. Adolescents between 15-19 years old account for one third of all gonorrhea and chlamydia cases in the U.S.

VI. SEXUALITY EDUCATION

A. **The Development of Sexuality Education**

1. Formal sexuality education began in the 20th century due to concern over sexually transmitted infections.

2. In the 1990s, sexuality education was implemented by various national youth groups such as the YMCA and YWCA, with a focus on discouraging early childbearing.

3. Sexuality education in schools remains controversial, with opposition believing that it should be private and take place in homes.

4. With the increasing visibility of sex in the media, sexuality education has increased in colleges and schools for younger students, although no federal laws or policies exist requiring U.S. schools to teach sexuality education. Some states mandate various forms of sexuality education.

B. **Goals of Sexuality Education**

1. Early sexuality education programs focused on increasing knowledge levels.

2. Contemporary sexuality education programs have added values clarification and skills, including communication and decision-making skills.

C. **Why Sexuality Education is Important**

1. Research has found that sexuality education programs make students less permissive about premarital sex and that accurate knowledge about sex may lead to a more positive self-image and self-acceptance.

2. Young people learn about sexuality from a variety of sources such as the media and peers. Many of these are poor sources of sexuality information that perpetuate myths.

D. **Sexuality Education Programs**

1. **Comprehensive sexuality education programs** are developmentally appropriate programs that begin in kindergarten and continue through 12th grade, addressing a variety of topics.

2. **Abstinence-based HIV-prevention sexuality education programs** are programs that emphasize the importance of abstinence but also include information about sexual behavior, contraception, and disease prevention.

3. Sexuality education and abstinence

a. **Abstinence-only programs** emphasize abstinence from all sexual behaviors without providing much information about contraception or disease prevention.

b. **Abstinence-only-until-marriage programs** present marriage as the only morally acceptable context for all sexual activity.

c. Abstinence-only programs often exaggerate the negative consequences of sexual behavior, portraying sexuality as dangerous and harmful.

d. Although many Americans believe that abstinence should be included in sexuality education, they also believe that information on contraception and sexually transmitted infections should also be included.

e. Some of the abstinence-only programs use scare tactics to encourage abstinence, thus some sexuality educators refer to them as "fear-based."
f. Research shows that abstinence-based programs that incorporate broader sexuality information within the context of abstinence have some positive results with respect to affecting young people's sexual behavior.

E. **Heteronormativity in Sexuality Education**
1. A recent criticism of sexuality education programs is that there is an underlying assumption that human sexuality is heterosexual sexuality.
2. Issues relevant to gay, lesbian, and bisexual individuals are often left out of sexuality education in schools or are covered very briefly.

F. **Effects of Sexuality Education**
1. The main way that researchers determine whether or not a sexuality program is successful is by measuring behavioral changes after a program has been presented, measuring vaginal intercourse, pregnancy, and contraceptive use rates.
2. This type of measurement can be somewhat limiting and difficult to determine.
3. Research results have found that comprehensive sexuality programs have been the most successful at helping adolescents delay their involvement in sexual intercourse and help protect adolescents form sexually transmitted infections and unintended pregnancies.
4. Sexuality education programs that teach contraception and communication skills have been found to delay the onset of sexual intercourse, reduce the frequency of sexual intercourse, reduce the number of sexual partners, and increase the use of contraception.
5. Young people who have had sexuality education are not more likely to engage in sexual intercourse than those who have never taken a course.
6. Research has found making a "virginity pledge," promising to remain a virgin until marriage, has helped some teens delay sexual intercourse, but adolescents who do make these pledges are less likely to use contraception than adolescents who have not made a pledge, which has lead researchers to suggest that signing the pledge may make an adolescent unable to accept any consequences if he or she decides to engage in sexual intercourse.
7. Sexuality education research measures changes in attitudes and values, but behavioral measures often receive the most attention.
8. Overall, comprehensive sexuality education programs can increase knowledge levels, affect the attitudes and/or change the behaviors of the students who take them, with the most successful programs being those in which schools and parents worked together in developing the program.

TEST YOURSELF

Below you will find fill-in-the-blank and short answer essay questions for topics covered in chapter 8. Check your answers at the end of this chapter.

Infancy and Childhood

1. It is important to realize that the idea of ____________________ is a recent invention.

2. Several large-scale research studies such as the NSAM and the ADD have helped to shed some light on ____________________ trends in sexual behavior.

3. Our sexual ____________________ becomes functional even before we are born.

4. The single most important aspect of infant development is the child's relationship to his or her ____________________.

5. By age two, children achieve ____________________ ____________________, whereby young children come to understand that they will not become a member of the other sex sometime in the future.

6. ____________________ is common in infancy, as many infants touch their genitals as soon as their hands are coordinated enough to do so, after about three or for months.

7. Self-stimulation is actually more common in ____________________ childhood than ____________________ childhood.

8. During early childhood, children learn that the ____________________ are different than the rest of the body.

9. Children are usually taught about their genitals in order to teach them about elimination, not ____________________.

10. Though the body begins internal changes to prepare for puberty as early as age six or seven, the first outward signs of puberty begin at ____________________ or ____________________.

11. On average, girls experience menarche before boys experience their first ____________________.

12. Children from the age of ____________________ to puberty engage in a variety of heterosexual and homosexual play.

13. While girls' heterosexual activity declines as they reach puberty, they report a steadily increasing rate of ____________________ activity as they approach adolescence.

14. Children get sexual knowledge and information from many different sources and use it to construct a sexual ____________________.

15. As children age and try to determine how they will fare in the world outside the family, their ____________________ increase in importance and become a major catalyst in sexual decision-making.

16. Name the four large-scale studies sponsored by the U.S. government to examine adolescent behaviors.

17. When is the earliest point that the male sexual anatomy becomes functional?

18. What did Harlow's famous experiment with rhesus monkeys demonstrate?

19. What are the first outward signs of puberty in girls?

20. What has research shown about parental attitudes towards masturbation?

Adolescence

21. ____________________ is recognized the world over as a time of transition, as the entrance into the responsibilities and privileges of adulthood.

22. Most societies throughout history have developed ____________________ ____________________ ____________________ around puberty: the Jewish Bar or Bat Mitzvah and Christian Confirmation come to mind.

23. ____________________ is one of the three major stages of physiological sexual development, along with prenatal sexual differentiation and menopause.

24. It may be the combination of nascent sexual exploration, changing bodies, and peer pressure that results in the average adolescent having a negative ____________________ ____________________.

25. The ____________________ body is changing, adding oily skin, fat, sweat, and odor.

26. ____________________ is the hallmark of female puberty.

27. Adolescent boys experience frequent spontaneous ____________________, which may have no association with sexuality, but are nonetheless quite embarrassing.

28. Adolescent boys' increased sexual desire is released through ____________________ ____________________ and increased masturbation.

29. ____________________ is, by far, the most psychologically and socially difficult of the life cycle changes.

30. Among adolescents, ____________________ general self-images tend to worsen, as they grow older, while ____________________ improve.

31. Late adolescence was, until recently, the stage that people in Western cultures were expected to begin their search for ____________________ ____________________ through serious dating.

32. ____________________ is one of the most underreported sexual behaviors in adolescence.

33. Some people who decide to become ____________________ choose to refrain from all sexual behaviors while others may engage in a variety of sexual behaviors, but choose not to engage in penile-vaginal intercourse.

34. In 1979, *The New York Times* reported that high school students believed that ____________________ ____________________ was a less dangerous alternative to sexual intercourse.

35. Teenagers report that their first intercourse is usually ____________________.

36. How has adolescent sexual activity changed over the years?

37. What adolescent sexual behavior has increased in acceptance among young people and has received considerable media attention?

38. According to the research, how do women describe the reasoning behind having sexual intercourse for the first time?

39. Why is it difficult to determine actual figures for adolescent same-sex contact?

40. What has been related to contraceptive use in adolescence?

Sexuality Education

41. Several developments in the United States, including skyrocketing rates of STI's, set the stage for ____________________ ____________________.

42. In the 1950's, ____________________ ____________________ courses began appearing in colleges and universities in the United States.

43. Early sexuality education programs focused primarily on increasing ____________________ levels.

44. ____________________ affects almost all aspects of human behavior and relationships with other persons.

45. Proponents of sexuality education believe that ____________________ ____________________ occurs even when there are no formalized sexuality education programs.

46. ____________________-____________________ programs began in the early 1990s, when there was a proliferation of sexuality education programs that used fear to discourage students from engaging in sexual behavior.

47. Although many Americans believe that abstinence should be included in sexuality education, they also believe that information on ____________________ and ____________________ should also be included.

48. Overall, abstinence-only programs have not been found to significantly change adolescents' values and attitudes about, or their intentions to engage in, ____________________ ____________________.

49. Sexuality educators routinely discuss the importance of ____________________ in their sexuality courses, however, it is ironic that many educators are not permitted to discuss many aspects of human sexuality.

50. A recent criticism of sexuality education programs is that there is an underlying assumption that human sexuality is ____________________ sexuality.

51. If ____________________ behavior is left out of the sexuality education curriculum, in essence this is saying that they are "not fully sexual human beings".

52. Textbooks on human sexuality typically offer one chapter on ____________________ ____________________, discussing primarily what causes it. Conversely, very little, if any, attention is paid to what causes ____________________.

53. The main way that researchers determine whether or not a sexuality program is successful is by measuring ____________________ changes after a program has been presented.

54. Comprehensive sexuality education programs have been found to be the most ____________________ at helping adolescents delay their involvement in sexual intercourse.

55. Over the last few years, ____________________ ____________________, where teens sign pledge cards and promise to remain a virgin until marriage, have become popular.

56. Name the youth-serving programs developed in the early 1900s mainly to demonstrate to young people the responsibilities required in parenting and to discourage early childbearing.

57. What are some of the possible goals of sexuality education?

58. According to research, where does parents' anxiety about sexuality come from?

59. What are the current behavioral measures used in determining whether or not a sexuality program is successful?

60. What do the countries that have the lowest rates of teenage pregnancy, abortion, and childbearing have in common?

POST TEST

Below you will find true/false, multiple-choice and matching quiz items covering the entire chapter. Check your answers at the end of this chapter.

True/False

1. Girls may self stimulate more than is reported since they can often do it more subtly.
2. Research shows that 75% of children knew the terms "penis" and "vagina."
3. Kinsey reported that roughly 50% of adults remembered engaging in some kind of sex play during the preadolescent years.
4. "Sex is dirty" is an example of a sexual script.
5. Sixty-three percent of adolescents have had sexual intercourse by their 18th birthday.

6. Children report far fewer discussions about sex and contraception than their parents report.

7. Teenagers accounted for more than 90% of all abortions in the United States in 1985.

8. As of 2001, there is a federal law that requires schools in the United States to teach sexuality education.

9. The main way that researchers determine whether or not a sexuality program is successful is by measuring behavioral changes after a program has been presented.

10. Sexuality education increases the frequency of sexual behavior.

Multiple-Choice

11. According to the textbook, why is it difficult to carry out research on children's sexuality in American society?
    a. Many people oppose questioning children about sexuality.
    b. Many people believe that research on child sexuality will encourage promiscuity.
    c. People believe that if we do not discuss children's sexuality, it will just go away.
    d. All of the these
    e. None of the above

12. What is gonorrhea?
    a. a sexual behavior commonly practiced by adolescents
    b. a bacterial sexually transmitted infection that can lead to serious complications
    c. the sum total of a person's internalized knowledge about sexuality
    d. a viral sexually transmitted infection that often causes no symptoms
    e. a term used in sexuality education that refers to refraining from sexual behaviors

13. The large government-run sexuality studies that looked at the sexuality of young people focused primarily on what ages?
    a. 15-17
    b. birth to 3
    c. 5-10
    d. 11-12
    e. All of the above

14. What is oxytocin?
    a. a drug used to treat sexually transmitted infections
    b. a hormone that stimulates erection
    c. a hormone that stimulates the release of milk for nursing
    d. a drug given to adolescents who mature too quickly
    e. None of the above

15. What is the name of the researcher who is famous for his experiments with rhesus monkeys that demonstrated the need for warmth and contact in infancy?
    a. Levine
    b. Plummer
    c. Harlow
    d. Guttmacher
    e. Kinsey

16. What is the term for the realization in the young child that one's gender does not normally change over the lifespan?
    a. gender actualization
    b. gender constancy
    c. gender harmony
    d. gender identity
    e. gender hegemony

17. The term semenarche refers to __________________.
    a. masturbation in male infants
    b. semen development
    c. the cessation of semen development in adolescents
    d. first ejaculation in boys
    e. when boys fail to develop sperm

18. According to research, what behaviors are NOT reported among children between the ages of six and puberty?
    a. spin the bottle
    b. homosexual play
    c. sex games
    d. heterosexual play
    e. None of the above

19. What is the term for the sum total of a person's internalized knowledge about sexuality?
    a. sexual script
    b. gender constancy
    c. gender internalization
    d. gender identity
    e. sexual acumen

20. Where do children learn about sexuality?
    a. parents
    b. intimate relationships
    c. peers
    d. siblings
    e. All of the above

21. Which of the following statements below regarding adolescence is FALSE?
    a. Adolescence is partly our emotional and cognitive reactions to puberty.
    b. Adolescence ends when a person achieves a sense of individual identity.
    c. Adolescence begins after the onset of puberty.
    d. Adolescence is a time of transition in many cultures.
    e. Adolescence always ends by the age of 18.

22. At what age does puberty begin for girls?
    a. 7-10
    b. 9-14
    c. 8-13
    d. 13-15
    e. 15-17

23. What is the term for a person's feelings and mental picture of his or her own body?
    a. self respect
    b. body image
    c. gender identity
    d. body status
    e. self representation

24. Which group of girls tend to feel more attractive and positive about their bodies?
    a. girls who feel that the onset of their menstruation occurred on time
    b. girls who feel that the onset of their menstruation occurred early
    c. girls who feel that the onset of their menstruation occurred late
    d. girls where menstruation never occurred
    e. There's no difference among girls and the onset of menstruation.

25. Which of the following statements below regarding boys and puberty is TRUE?
    a. Spermatogenesis and ejaculation occur late in male puberty.
    b. The testicles begin to decrease their production of testosterone.
    c. First ejaculation always occurs after body hair growth.
    d. Pubertal growth in boys tends to be more uneven and sporadic than girls'.
    e. All of the above

26. What is the term for involuntary ejaculation during sleep?
    a. nocturnal emissions
    b. involuntary ejaculation
    c. semenarche
    d. night discharge
    e. semenosis

27. What is one of the most underreported sexual behaviors in adolescence?
    a. sexual intercourse
    b. sexual fantasies
    c. fondling
    d. oral sex
    e. masturbation

28. Which of the following statements below regarding adolescents and oral sex is TRUE?
    a. Adolescents have become more likely to include oral sex in their sexual repertoire.
    b. Many adolescents believe that oral sex protects them from sexually transmitted infections.
    c. The majority of adolescents who engage in oral sex use no barrier protection.
    d. All of the above
    e. None of the above

29. How many teenagers are infected with a sexually transmitted infection every year?
    a. 10,000
    b. 100,000
    c. 500,000
    d. 4 million
    e. 50 million

30. Which program was developed mainly to demonstrate to young people the responsibilities required in parenting and to discourage early childbearing?
    a. YWCA/YMCA
    b. Girl Scouts
    c. Boy Scouts
    d. 4-H clubs
    e. All of the above

31. When did sexuality education courses begin appearing in colleges and universities in the United States?
    a. 1890s
    b. 1920s
    c. 1950s
    d. 1970s
    e. 2000

32. According to research, which of the following is NOT harmful?
    a. lack of sexuality education
    b. unresolved curiosity
    c. knowledge about sexuality
    d. ignorance about sexual issues
    e. All of the above

33. According to research, which statement below regarding students and sexuality education in schools is TRUE?
    a. Many students believe that school-based sex education programs are usually too late for them.
    b. Many students believe that school-based sex education programs are too early and too explicit.
    c. Many students believe that school-based sex education programs are right on time.
    d. There is no research on what students think about the timing of sexuality education.
    e. None of the above

34. Which of the following is TRUE of comprehensive sexuality education programs?
    a. They present marriage as the only morally acceptable context for all sexual activity.
    b. They include a wide variety of topics and help students develop their own skills and learn factual information.
    c. They do not include information on contraception or disease prevention.
    d. All of the above
    e. None of the above

35. Which of the following is TRUE of abstinence-only programs?
    a. They present marriage as the only morally acceptable context for all sexual activity.
    b. They include a wide variety of topics and help students develop their own skills and learn factual information.
    c. They do not include information on contraception or disease prevention.
    d. All of the above
    e. None of the above

36. What is the name of sexuality education programs that often include mottos such as "*Control your urgin' –be a virgin*"?
    a. Abstinence-only
    b. Relationship-based
    c. School-based
    d. Comprehensive
    e. None of the above

37. What is the name of sexuality education programs that often use fear to discourage students from engaging in sexual behaviors
    a. Abstinence-only
    b. Relationship-based
    c. School-based
    d. Comprehensive
    e. None of the above

38. Which of the following is TRUE regarding abstinence-only programs?
    a. They often fail to provide necessary factual information.
    b. They support many myths about human sexuality topics.
    c. They do not significantly change adolescents' intentions to engage in nonmarital sexual activity.
    d. All of the above
    e. None of the above

39. What are the results of sexuality education programs that teach contraception and communication skills?
    a. They increase the frequency of sexual intercourse.
    b. They increase the number of sexual partners.
    c. They decrease the use of contraception.
    d. They increase the use of contraception.
    e. They hasten to onset of sexual intercourse.

40. Which statement below regarding virginity pledge cards is TRUE?
    a. There is no difference in contraceptive use between teens who sign pledge cards and teens who do not.
    b. A teen who has signed a pledge card is more likely to use contraception than a teen who has not made such a pledge.
    c. A teen who has signed a pledge card is less likely to use contraception than a teen who has not made such a pledge.
    d. All of the above
    e. None of the above

Matching

Column 1		Column 2
A. body image	____	41. the main way that researchers determine whether or not a sexuality program is successful
B. Abstinence-only programs	____	42. the act of sexually stimulating the male genitals with the mouth
C. sexual script	____	43. sexuality education programs that present a wide variety of topics to help develop skills and facts
D. nocturnal emissions	____	44. involuntary ejaculation during sleep
E. Comprehensive sexuality education programs	____	45. the realization in the young child that one's gender does not normally change over the lifespan
F. behavioral changes	____	46. a person's feelings and mental picture of his or her own body's beauty
G. gender constancy	____	47. sexuality education programs that do not provide information on contraception or disease prevention
H. fellatio	____	48. the act of sexually stimulating the female genitals with the mouth
I. semenarche	____	49. the experience of first ejaculation
J. cunnilingus	____	50. the sum total of a person's internalized knowledge about sexuality

Test Yourself Answer Key

Infancy and Childhood

1. childhood (p. 204)
2. adolescent (p. 206)
3. anatomy (p. 206)
4. parents (p. 207)
5. gender constancy (p. 207)
6. Self-stimulation (p. 207)
7. early, later (p. 208)
8. genitals (p. 209)
9. sexuality (p. 209)
10. nine, ten (p. 209)
11. ejaculation (p. 209)
12. six (p. 211)
13. same-sex (p. 211)
14. script (p. 212)
15. peers (p. 214)
16. The National Survey of Family Growth, the National Longitudinal Study of Adolescent Males, the National Longitudinal Study of Adolescent Health, and the Youth Risk Behavior Surveillance System (pp. 205-206)
17. in utero (p. 206)
18. the need for warmth and contact in infancy (p. 207)
19. Breast buds appear and pubic hair growth may begin (p. 209)
20. In a study of parental attitudes towards masturbation, Gagnon found about 60% of parents accepted the fact that their children masturbated and said that it was all right. Ironically, less than half wanted their children to have a positive attitude toward masturbation, and that attitude was transmitted to their children. (p. 214)

Adolescence

21. Adolescence (p. 215)
22. rites of passage (p. 215)
23. Puberty (p. 216)
24. body image (p. 216)
25. adolescent's (p. 217)
26. Menarche (p. 217)
27. erections (p. 218)
28. nocturnal emissions (p. 218)
29. Puberty (p. 218)
30. girls', boys' (p. 218)
31. marital partners (p. 220)
32. Masturbation (p. 221)
33. abstinent (pp. 221-222)
34. oral sex (p. 222)
35. unplanned (p. 224)
36. Almost every survey shows that sexual activity has increased overall among teens over the past 50 years. (p. 220)
37. oral sex (p. 223)
38. Nearly half of women said they had sex the first time out of affection for their partner, while a quarter cited curiosity as their primary motivation. Twenty-four percent said they just went along with it, and four percent reported being forced to have sex the first time. (p. 223)

39. Such research relies on self-reports; people may define homosexual differently, deny experiences, or lie due to the stigma on homosexuality that continues to prevail in our society. (p. 226)
40. a good adolescent-mother relationship (p. 228)

Sexuality Education

41. sexuality education (p. 231)
42. sexuality education (p. 232)
43. knowledge (p. 232)
44. Sexuality (p. 233)
45. sexual learning (p. 233)
46. Abstinence-only (p. 234)
47. contraception, STIs (p. 235)
48. premarital sex (p. 235)
49. communication (p. 235)
50. heterosexual (p. 236)
51. same-sex (p. 236)
52. sexual orientation, heterosexuality (p. 236)
53. behavioral (p. 238)
54. successful (p. 238)
55. virginity pledges (p. 238)
56. YMCA, YWCA, Girl Scouts, Boy Scouts, 4H Clubs (p. 231)
57. Knowledge acquisition, improving personal psychological adjustment, and improving relationships between partners (p. 232)
58. Fear, lack of comfort, lack of skills, misinformation (p. 233)
59. The standard behavioral measures include vaginal intercourse, pregnancy, and contraceptive use. (p. 238)
60. They have liberal attitudes toward sexuality, easily accessible birth control services for teenagers, and formal and informal sexuality education programs. (p. 237)

Post Test Answer Key

True/False

1. T (p. 208)
2. F (p. 209)
3. T (p. 211)
4. T (p. 212)
5. T (p. 224)
6. T (p. 228)
7. F (p. 230)
8. F (p. 232)
9. T (p. 238)
10. F (p. 238)

Multiple-Choice

11. d (p. 205)
12. b (p. 205)
13. a (p. 206)
14. c (p. 207)
15. c (p. 207)
16. b (p. 207)
17. d (p. 209)
18. e (p. 211)
19. a (p. 212)
20. e (p. 214)
21. e (p. 215)
22. c (p. 216)
23. b (p. 216)
24. a (p. 217)
25. d (p. 218)
26. a (p. 218)
27. e (p. 221)
28. d (pp. 222-223)
29. d (p. 231)
30. e (p. 231)
31. c (p. 232)
32. c (p. 232)
33. a (p. 233)
34. b (p. 234)
35. c (p. 234)
36. a (p. 234)
37. a (p. 235)
38. d (p. 235)
39. d (p. 238)
40. c (p. 238)

Matching

41. F (p. 238)
42. H (p. 223)
43. E (p. 232)
44. D (p. 218)
45. G (p. 207)
46. A (p. 216)
47. B (p. 232)
48. J (p. 223)
49. I (p. 209)
50. C (p. 212)

Chapter 9 – Adult Sexual Relationships

CHAPTER SUMMARY

Every society has rules to control the ways that people develop sexual bonds with other people. Sociologists view dating as a way for prospective mates to compare the assets and liabilities of eligible partners in order to choose the best available mates; studies show that those who date are in better physical and emotional health than those who do not. Formal dating has given way to more casual dating. When it comes to dating and sex, in the relationships that are rated as the happiest, both partners initiate sex equally, and both feel free to say no if they do not feel like having sex. For individuals who divorce and return to a dating lifestyle, they find that dating patterns have changed a lot, particularly over the last 30 years.

In recent years, cohabitation has increased dramatically. Cohabiting heterosexual couples tend to share some characteristics such as being under the age of 30, living in a metropolitan area, being well-educated, and both being employed but having relatively low incomes. Researchers have suggested a number of advantages and disadvantages to cohabitation. Thirteen states recognize common-law marriage, which means that if a couple lives together for a certain number of years, they are considered married.

Ninety-three percent of Americans say that a happy marriage is one of their most important life goals. Research has found that the quality of the friendship with one's spouse is the most important factor in marital satisfaction for both men and women. Marital quality tends to peak in the first few years of a marriage and then declines until midlife when it rises again. Overall, marriage provides more health benefits to men when compared to women. The reasons sexual behavior decreases in long-term relationships has less to do with getting bored with one's partner than it has to do with the pressures of children, jobs, commuting, housework, and finances. Most older couples report that their marriages improved over time and that the later years are some of the happiest. Although many people think that sexual desire drives an extramarital affair, research has found that over 90% of extramarital affairs occur because of emotional needs not being met within the marital relationship. Some married couples open up their relationships and encourage their partners to have extramarital affairs or to bring other partners into their marital beds, believing that sexual variety and experience enhance their own sexual life.

In 1999 the U.S. Congress enacted the Defense of Marriage Act, which put forth that all states have the right to refuse to recognize same-sex marriages performed in other states. Many gay and lesbian couples marry their partners in ceremonies that are not recognized by the states in which they live.

Currently 60% of marriages in the United States end in divorce. Divorce rates are influenced by changes in the legal, political, religious, and familial patterns in the United States. Couples who marry at a young age often suffer more marital disruption than older couples partly due to emotionally immaturity. One year after a divorce, 50% of men and 60% of women reported being happier than they were during the marriage. Approximately 75% of divorced people remarry, with some people experiencing serial divorce.

LEARNING OBJECTIVES

After studying Chapter 9, you should be able to:

1. Describe the research on dating.
2. Discuss the gender role differences in dating.
3. Identify factors that affect sexual behavior in dating relationships.
4. Compare and contrast contemporary dating with dating from 30 years ago.
5. Describe the research on interracial dating.
6. Compare and contrast the advantages and disadvantages to cohabitation.
7. List and define four different types of marriage.
8. Identify the factors related to marital satisfaction.
9. Compare and contrast the gender role differences when it comes to marital satisfaction.
10. Discuss some of the mental health benefits of marriage.
11. Describe the research on sexual activity in marriage.
12. List the factors that make a marriage that follows the death of a spouse more successful.
13. Identify the characteristics of individuals who engage in extramarital relationships.
14. Define comarital sex.
15. Explore the history and research on same-sex marriage.
16. Identify current divorce patterns in the United States.
17. Identify social and relationship factors that affect divorce.
18. Discuss the adjustment issues that divorced people face.
19. Describe some of the cross-cultural research on adult relationships, marriage, and divorce.
20. Define polygamy.

CHAPTER OUTLINE

- Every society has rules to control the ways that people develop sexual bonds with other people.

I. DATING: FUN OR SERIOUS BUSINESS?

- Sociologists view dating as a way for prospective mates to compare the assets and liabilities of eligible partners to choose the best available mates.
- Studies show that those who date are in better physical and emotional health than those who do not.
- Steady dating in adolescence is associated with higher self-esteem and sex-role identity.
- Dating and courtship behaviors vary in different social classes.

A. **Types of Dating and How Do We Meet?**

1. The various terms for dating such as, "going out," "seeing each other" mean different things to different people and can make research on dating difficult.
2. Formal dating has given way to more casual dating.
3. In one study, 72% of women reported having initiated dates with men within the previous six months.
4. As people get older, it can become more difficult to find ways to meet people.

B. **Sexuality in Dating Relationships**

1. Couples vary greatly in how long they wait to engage in sexual behaviors.
2. If both partners are virgins in a long-term relationship, there is a 50% chance that they will have sexual intercourse with their partner.
3. In recent years, more adults are standing up for virginity instead of being embarrassed about it.
4. Men often initiate sexual behavior, even though men and women often report similar interest levels.
5. In relationships that are rated as the happiest, both partners initiate sex equally, and both feel free to say no if they do not feel like having sex.
6. The majority of couples do not discuss their sexual history (partners and use of condoms) prior to their first sexual intercourse.

C. **Dating After Divorce or Widowhood**

1. Dating is very different from 30 years ago in many ways.
   a. The relationship between the sexes has changed.
   b. Women initiate dates and sexual activity more frequently.
   c. Casual dating is more common.
   d. Sexual activity may be more frequent.
   e. One has to be more worried about sexually transmitted infections.
2. More and more groups and organizations are being created to help divorced and widowed people ease back into the dating scene
3. After a partner dies, the remaining partner is often not interested in marrying again, but they may still be interested in dating.
4. It is estimated that 4% of older couples live together.
5. The likelihood that an older man will choose to date is predicted by his age and social involvement, while older women are influenced more by their health and mobility.

D. **Interracial Dating**

1. In 1967, the Supreme Court struck down state anti-miscegenation laws which outlawed interracial relationships.
2. Since the early 1980s, the number of interracial couples has nearly doubled.
3. In college-aged populations, close to 25% of students report being involved in an interracial relationship, and almost 50% said they'd be open to dating someone from a different race.

4. Although the number of people who approved of interracial relationships exceeded the number who disapproved for the first time in 1990, a large minority does not accept such relationships.

II. COHABITATION: INSTEAD OF, OR ON THE WAY TO, MARRIAGE?

A. In recent years, **cohabitation** (to live together in a sexual relationship when not legally married) has increased dramatically.

B. Cohabitation has become so common that many sociologists regard it as a stage of courtship with more than half of first unions in the early 1990s beginning with cohabitation.

C. Cohabiting heterosexual couples tend to share some characteristics such as being under the age of 30, living in a metropolitan area, being well-educated, and both being employed but having relatively low incomes.

D. There are advantages to cohabitation.

1. It allows couples to move into marriage more slowly.
2. It allows couples to learn more about each other and not be legally or economically tied together.
3. It allows a couple who loves each other to be older, more mature, and more financially stable when they finally marry.
4. It is more realistic than dating because it gives couples the opportunity to learn of their partners' bad habits and idiosyncrasies.

E. There are disadvantages to cohabitation.

1. Parents and relatives may not support the union.
2. Society as a whole tends not to recognize people who live together for purposes of health care, taxes, etc.
3. Partners may want different things out of living together.
4. People can feel cut off from friends, and can become too emeshed and interdependent.

F. Research on cohabitation suggests that those who do marry are at increased risk of divorce, and longer cohabitation has been found to be associated with higher likelihood of divorce.

1. Some researchers suggest the above findings are due to the thought that couples who cohabit for a long period of time develop independently and are simply "playing house" and are not ready for the legal commitment of marriage.
2. Other researchers suggest the above findings maybe due to the fact that the type of people who are willing to cohabit are more likely to divorce when marriage gets difficult since they may be more accepting of divorce and may be less religious and traditional in the first place.

G. Thirteen states recognize **common-law marriage**, which means that if a couple lives together for a certain number of years, they are considered married.

H. There are also cases of individuals who have successfully sued partners they lived with for **alimony** or shared property (called **palimony**).

1. **alimony**—an allowance for support made under court order to a divorced person by the former spouse, usually the chief provider during the marriage
2. **palimony**—an allowance for support made under court order and given usually by one person to his or her former lover or live-in companion after they have separated.

III. MARRIAGE: HAPPY EVER AFTER?

- Ninety-three percent of Americans say that a happy marriage is one of their most important life goals.
- Over the last 30 years, the age at first marriage has been increasing, although this number leveled off in the 1990s.
- Marital satisfaction for men has been found to be related to the frequency of pleasurable activities (doing things together) in the relationship, while for women it was related to the frequency of pleasurable activities that focus on emotional closeness.

- John Gottman has found that the quality of the friendship with one's spouse is the most important factor in marital satisfaction for both men and women.
- Marital quality tends to peak in the first few years of a marriage and then declines until midlife when it rises again.
- Research found that couples married more than 25 years were less likely to be open and use positive affirmations than couples who had been married only a few years.
- The majority of married couples report that their marriages are happy and satisfying with one study finding that 60% of couples reported that their marriages were happy.
- Research where married couples were asked why their marriages lasted found that marriages seem to last most when both partners have a positive attitude toward the marriage, view their partner as a best friend, like their partner as a person, and the belief that marriage is a long-term commitment.
- Marriage has also been found to reduce the impact of several potentially traumatic evens including job loss, retirement, and illness.
- In single men, suicide rates are twice as high as those of married men, and single men have been found to experience more psychological problems such as depression and even nightmares.
- Overall, marriage provides more health benefits to men than women as married women tend to be less healthy than married men.
- Over the last few years, research has found a trend in the mental health benefits of marriage applying equally to men and women most likely due to an increase in equality in marriage.

A. **Marital Sex Changes Over Time**

1. Most couples report a decline in sexual activity over time.
2. The reasons sexual behavior decreases in long-term relationships has less to do with getting bored with one's partner than it has to do with the pressures of children, jobs, commuting, housework, and finances.
3. The frequency of sexual activity and satisfaction with a couple's sex life has been found to be positively correlated. However, it is not known if increased sexual frequency causes more satisfaction, or if increased satisfaction with the marriage is responsible for the increase in frequency of sexual behavior.
4. Frequency and type of marital sex has been found to differ by social class with upper class individuals tending to have marital sex more frequently, using more sexual positions, and practicing more oral sex and other varieties of sexual contact.
5. Some marriages are **asexual**, which means the partners do not engage in sexual intercourse.
6. Masturbation is often viewed as taboo in marital relationships, although married men and women did report masturbating.

B. **Having Children or Remaining Childless**

1. Some research suggests that having children may actually adversely affect overall marital quality
2. Marital happiness is higher before the children come, when it declines steadily until it hits a low when the children are in their teens, and then begins to increase once the children leave the house.

C. **Marriages in Later Life**

1. Most older couples report that their marriages improved over time and that the later years are some of the happiest.
2. Older men are twice as likely to remarry because women outnumber men in older age and also because older men often marry younger women.
3. Marriages that follow the death of a spouse tend to be more successful if the couple knew each other for a period of time prior to the marriage, if their children and peers approve of the marriage, and if they are in good health, financially stable, and have adequate living conditions.

D. **Extramarital Affairs: "It Just Happened"**

1. Research estimates that less than 5% of all societies are as strict about forbidding extramarital intercourse as our society has been.
2. Although extra-marital sex refers to sex outside of marriages, the authors are also referring to extra-relationship sex or unmarried couples who have sex with someone other than their partner.
3. Those who cheat in intimate relationships have been found to have stronger sexual interests, more permissive sexual values, less satisfaction in their intimate relationship, and more opportunities for sex outside the relationship.
4. Half the states in the United States have laws against sex outside of marriage, although these laws are rarely enforced.
5. One study found that 75% of Americans believe that extramarital sex was unacceptable and should not be tolerated.
6. The research on the incidence of extramarital affairs varies greatly.
7. Both non-married heterosexual couples and lesbians who live together have been found to be less deceptive and secretive than married couples.
8. Although many people think that sexual desire drives an extramarital affair, research has found that over 90% of extramarital affairs occur because of emotional needs not being met within the marital relationship.
9. Thompson has identified three types of extramarital affairs with affairs that are both emotional and sexual appear to affect the marital relationship the most, while affairs that are primarily sexual affect it the least.
    a. sexual but not emotional
    b. sexual and emotional
    c. emotional but not sexual
10. Gender and age differences play a role in the type of extramarital sexual relationships that occur.
11. Men are more disapproving of extramarital relationships and less optimistic about the continuation of a marriage after an affair than women.

E. **Open Marriages: Sexual Adventuring**

1. Some married couples open up their relationships and encourage their partners to have extramarital affairs or to bring other partners into their marital beds, believing that sexual variety and experience enhance their own sexual life.
2. **Comarital sex** refers to the consenting of married couples to sexually exchange partners who are often referred to as **swingers** or **polyamorists**.
3. Researchers have explained that open marriages are okay as long as both partners know about it.
4. Most couples that engage in comarital sex have strict rules meant to protect the marriage, with sex seen as separate from the loving relations of marriage

IV. GAY, LESBIAN, BISEXUAL AND TRANSGENDERED RELATIONSHIPS

- GLBT relationships have changed more than heterosexual relationships over the last few decades.

A. **Sexuality in Gay, Lesbian, Bisexual, and Transgendered Relationships**

1. Research suggests that due to gender-role conditioning, lesbians may have more difficulty initiating sex than heterosexual relationships or gay male relationships.
2. Research suggests that gay men engage in sexual behavior more often than lesbian women do.

B. **Same-Sex Marriage**

1. In 1999 the U.S. Congress enacted the Defense of Marriage Act which put forth that all states have the right to refuse to recognize same-sex marriages performed in other states.

2. In 2002, Vermont passed a civil union statute which grants same-sex couples the same benefits and responsibilities given to heterosexual couples, although these civil unions have no legal weight outside of Vermont.
3. Outside the United States in 2003, Canada became the third nation, after the Netherlands and Belgium, to legalize same-sex marriage.
4. Heterosexual marriage is strongly linked to procreation, childbirth, and childrearing, and the U.S. has long regulated marriage in an attempt to protect procreative health.
5. Many gay and lesbian couples marry their partners in ceremonies that are not recognized by the states in which they live.
6. Many groups have been working to get states to set up **domestic partner** acts where same-sex couples who live together in committed relationships can have some of the benefits granted to married couples.

V. DIVORCE: WHOSE FAULT OR NO-FAULT?

- Reflecting the substantial changes in marriage and divorce, by 1985 all states offered couples some type of **no-fault divorce**, which is a divorce law that doesn't allow for blame to be placed on one partner for the dissolution of a marriage.
- Currently 60% of marriages in the United States end in divorce.
- Those who marry early are more likely to divorce eventually, with the majority of divorces taking place among couples who were married between the ages of 20 and 24.
- Interracial marriages have higher divorce rates than marriages within racial groups.
- Usually one partner wants to terminate a relationship more than the other partner.
- When one partner is the initiator in heterosexual relationships, it is usually the women, with one study finding that women initiated 75% of divorces.

A. **Why Do People Get Divorced?**

1. Social Factors Affecting Divorce

a. Divorce rates are influenced by changes in the legal, political, religious, and familial patterns in the United States.
b. Divorce has become more accessible and cheaper.
c. In an attempt to reduce the skyrocketing divorce rates, some states have instituted **covenant marriages**, which are preceded by premarital counseling and have strict rules about divorce.
b. In recent years, divorce has become generally more acceptable in American society.

2. Predisposing Factors for Divorce

a. Couples who marry at a young age often suffer more marital disruption than older couples partly due to emotionally immaturity.
b. Couples who marry because of an unplanned pregnancy are more likely to divorce.
c. Waiting longer before having children promotes marital stability by giving couples time to get to know each other.
d. Catholics and Jews are less likely to divorce than Protestants.
e. People who have been divorced before or whose parents have divorced have more accepting attitudes toward divorce than those who grew up in happy, intact families.

3. Relationship Factors in Divorce

a. Some warning signs when it comes to divorce are communication avoidance; demand and withdrawal patterns of communication, where one partner demands that they address the problem and the other partner pulls away; and little mutually constructive communication.

b. Men and women listed different complaints about their partners.
c. Many couples make the mistake of believing that the little annoyances or character traits that they dislike in the potential spouses will disappear after marriage or that they will be able to change their spouse once married.

B. **Divorce and Sex**

1. Research found that the older a person was at divorce, the less sexual activity occurred afterwards.
2. The more religious a divorced person was, the less likely he or she was to have another sex partner.
3. Divorced persons without children are more likely to have sexual partners than those with children.
4. Men are more likely to have one or more partners, while women are less likely to find new partners.

C. **Adjusting to Divorce**

1. One year after a divorce, 50% of men and 60% of women reported being happier than they were during the marriage.
2. Depression is common in those who divorce but believe that marriage should be permanent.
3. Women often have an increase in depression after a divorce, while men experience poorer physical and mental health.
4. Older people experience more psychological problems because divorce is less common in older populations.
5. Divorced black men and women adjust to divorced more easily and experience less negativity from peers than do whites.
6. Economic adjustment is often harder for women because women's income tends to decline more than men's.
7. Women may also get more emotional support from their friends and coworkers than men do.
8. Approximately 75% of divorced people remarry, with some people experiencing **serial divorce**, which is the practice of divorce and remarriage, followed by divorce and remarriage.
9. Men remarry at higher rates than women.

VI. ADULT SEXUAL RELATIONSHIPS IN OTHER PLACES

A. **Courtship and Arranged Marriages**

1. There are still a few industrialized cultures were **arranged marriages** take place, although those are often in the upper classes. These are marriages arranged by parents or relatives that are often not based on love.
2. In some cultures, courtship was or is a highly ritualized process in which every step is defined by one's family or community.
    a. Venezuela
    b. South Africa
    c. China
    - For 2000 years, marriages in China were arranged by parents and elders, and emotional involvement between prospective marriage partners was frowned upon.
    - After the Communist Revolution of 1949, the Communist leaders established the Marriage Law of the People's Republic of China, which sought to end arranged marriages and establish people's right to choose their spouse freely.
3. Over the last few years, the practice of selling young girls for marriage has risen.

B. **Cohabitation in Other Cultures**

1. Cohabitation is rarer in more traditional societies where, even if a couple has sex before of instead of marriage, social customs would never tolerate an unmarried heterosexual couple living together openly.
2. Asian societies still frown upon cohabitation, although it is sometimes allowed, and it is severely discouraged in Islamic societies.
3. Most Western countries now have substantial numbers of couples who live together, such as France—where by 1990, one out of five couples was living together outside of marriage.

C. **Marriage: A Festival of Styles**

1. Marriage ceremonies take place in every society on earth, but marriage customs vary widely from culture to culture.
2. Most cultures celebrate marriage as a time of rejoicing and have rituals or ceremonies that accompany the wedding process.
3. Sixty percent of marriages worldwide are arranged so the concept of "loving" one's partner may be irrelevant in many societies.
3. Some societies allow the practice of **polygamy**, which is the practice of having more than one spouse at a time.
    a. It typically takes the form of **polygyny,** which is the practice of having multiple wives.
    b. Although polygyny is rarely practice in the United States, there are some small Mormon fundamentalist groups that do practice polygyny. (Although they are not officially recognized by the Mormon Church.)
    c. The thought that polygynous marriages developed in order to increase fertility is disproved by the research, suggesting that wives in polygynous marriages have lower fertility rates.
    d. It may be more likely that polygyny developed as a strategy for men to gain prestige and power by having many wives, while women could gain the protection of a wealthy man.
4. **Polyandry**, the practice of having more than one husband at a time is much less common that polygyny, and when it happens, it is usually for reasons of keeping together inheritance.
5. In a **consanguineous marriages**, a woman marries her own relative to maintain the integrity of a family property.
    a. In the majority of U.S. states, consanguineous marriage is illegal and has been since the late 19th century.
    b. In many Muslim countries in northern Africa, western and southern Asia, north, east, and central India, and the middle Asian republics of the former Soviet Union, marriages take place between relatives between 20% and 55% of the time.
    c. Marriages between certain cousins are legal in many U.S. states.

D. **Extramarital Sex**

1. Extramarital sex is forbidden in many cultures but often tolerated even in cultures where it is technically not allowed.
2. Countries that tolerate extramarital sex often find it more acceptable for men than for women.
3. In some cultures, extramarital relations are replacing polygamy.

E. **Divorce in Other Cultures**

1. In societies such as the United States, Sweden, Russia, and most European countries, divorce is relatively simple and has little stigma.
2. It can be difficult to obtain a divorce in Catholic countries since Catholicism does not allow for divorce.

a. In Latin America, many countries have restrictive divorce legislation, which means that only the wealthy find it easy to divorce because they can fly to Mexico or other countries where divorce is easier.

b. Ireland legalized divorce in 1995, and prior to this time, Ireland was the only country in the western world to constitutionally ban divorce.

3. Many traditional societies had ways to assure that divorces did not disrupt the community, although traditional laws can still be enforced.

a. In Egypt and Israel, according to Islamic law and Jewish law, it is far easier for men to divorce than for women. A man can divorce his wife simply by repudiating her publicly three times, while a woman must go to court to dissolve a marriage.

4. In China, recent changes in divorce law allows a partner to divorce and take everything if a partner was engaging in extramarital sex.

5. Overall, divorce rates seem to be going up in most countries in the world as they modernize and as traditional forms of control over the family lose their power.

TEST YOURSELF

Below you will find fill-in-the-blank and short answer essay questions for topics covered in chapter 9. Check your answers at the end of this chapter.

Dating and Cohabitation

1. Much can be understood about a society just by examining the _____________ and rules it sets up for choosing a mate.

2. Sociologists view _____________ as a way for prospective mates to compare the assets and liabilities of eligible partners to choose the best available mates.

3. Studies show that college-age women who are involved in intimate relationships have better _____________ _____________ than women not involved in intimate relationships.

4. Overall, the research has found that _____________ in relationships is dependent upon how much individual satisfaction there is in the relationship and the cost-benefit ratio of the relationship.

5. In relationships, some couples wait for months, years, or until marriage before they engage in _____________ _____________.

6. Couples who abstain from sexual intercourse tend to hold more _____________ attitudes about sex, and have less prior _____________ experience.

7. In heterosexual relationships, _____________ often initiate sexual behavior, even though men and women often report similar interest levels.

8. In the majority of couples, there is very little _____________ during the transition from thinking about having sexual intercourse to actually having sexual intercourse.

9. After a divorce or the death of a spouse, it can be very difficult to get back into the _____________ scene.

10. In 1967, the Supreme Court struck down state ______________ laws, which outlawed interracial relationships.

11. Since the early 1980s, the number of ______________ couples has nearly doubled.

12. When couples do not openly talk about where they would like their relationship to go, they enter into ______________ ______________ wherein they believe they know what each other wants because they base it on what they themselves want.

13. Today, ______________ has become so common that many sociologists regard it as a stage of courtship.

14. ______________ allows couples to move into marriage more slowly, learn more about each other, and not be legally or economically tied together.

15. Thirteen states recognize ______________ ______________, which means that if a couple lives together for a certain number of years, they are considered married.

16. What traits of a culture can be learned from looking at dating patterns?

17. What is difficult about studying and discussing courtship or dating behavior?

18. How is dating in 2004 different from dating in 1975?

19. What are some ways that cohabiting couples tend to differ from married couples?

20. Cohabiting heterosexual couples tend to share some characteristics. What are some of them?

Marriage and Same-Sex Relationships

21. Ninety-three percent of Americans say that a ______________ ______________ is one of their most important life goals.

22. John Gottman's research has found that the quality of the ______________ with one's spouse is the most important factor in marital satisfaction for both men and women.

23. Researchers have found that marriage seems to be good for a person's ______________.

24. Overall, marriage provides more health benefits to ______________ than ______________.

25. Although many marriages start out high on ______________, these feelings slowly dissipate over time.

26. ______________ is often taboo in marital relationships. The myth is that if a married man ______________, his wife cannot be satisfying him sexually.

27. Some research suggests that having ______________ may actually adversely affect overall marital quality.

28. Most older couples report that their marriages ______________ over time.

29. All societies regulate sexual behavior and use ______________ as a means to control the behavior of its members to some degree.

30. Although many people think that sexual desire drives an extramarital affair, research has found that over 90% of extramarital affairs occur because of ______________ needs not being met within the marital relationship.

31. Affairs that are both ______________ and ______________ appear to affect the marital relationship the most, while affairs that are primarily ______________ affect it the least.

32. ______________ sex is the consenting of married couples to sexually exchange partners.

33. Overall, the majority of ______________ are white, middle class, middle aged, and church going.

34. In 2000, the state of ______________ passed a civil union statute that grants same-sex couples the same benefits and responsibilities given to heterosexual couples.

35. Many groups have been working to get states to set up ______________ ______________ acts, where same-sex couples who live together in committed relationships can have some of the benefits granted to married couples.

36. What do married couples think is important for a good marriage?

37. What is the general trend in marital quality throughout the years of a marriage?

38. Why do some marriages last, while others end in divorce?

39. Describe how frequency and type of marital sex has been found to differ by social class.

40. Many same-sex couples across the Unites States are challenging existing laws that regulate which issues?

Divorce and Relationships in Other Places

41. The liberalization of divorce laws made it easier to obtain a divorce and made it an easier and less _______________ process.

42. When one partner is the initiator in a divorce, it is usually the _______________.

43. A _______________ marriage involves premarital counseling and makes divorce more difficult to get even if the couple decides later they want one.

44. In recent years, divorce has become generally more _______________ in American society.

45. Couples who marry at a young age often suffer more marital disruption than older couples, due in part to _______________ _______________.

46. Many couples make the mistake of believing that the little annoyances or character traits that they dislike in their potential spouses will _______________ after marriage.

47. Divorced persons _______________ children are more likely to have sexual partners than those _______________ children.

48. Women often have an increase in _______________ after a divorce.

49. Typically the length of time between divorce and remarriage is less than _______________ years.

50. In most industrialized countries, mate selection through _______________ is the norm.

51. In ______________, the primary responsibility of each person was supposed to be to his or her extended family.

52. Sixty percent of marriages worldwide are ______________.

53. Some countries allow the practice of polygamy, which usually takes the form of ______________, or having more than one wife.

54. In 1989, ______________ became the first country to allow same-sex marriages.

55. In Egypt the most common reason given for divorce is ______________ by the husband, while among the Hindus of India, the most common reason for divorce is ______________.

56. What is the name for the type of divorce laws that were available in all states by 1985, which doesn't allow for blame to be placed on one partner for the dissolution of marriage?

57. What's the relationship between religion and divorce?

58. What percent of divorced people remarry?

59. In what country did communism affect marriage patterns?

60. What is it called when a woman marries her own relative to maintain the integrity of a family property?

POST TEST

Below you will find true/false, multiple-choice and matching quiz items covering the entire chapter. Check your answers at the end of this chapter.

True/False

1. Dating and courtship behaviors vary in different social classes.

2. In recent years, cohabitation without marriage has decreased dramatically.

3. Common-law marriage means that if a couple lives together for a certain number of years, they are considered married.

4. Ninety-three percent of Americans say that a happy marriage is one of their most important life goals.

5. Research suggests that having children may adversely affect overall marital quality.

6. Over the last few years, attitudes about extramarital sex have become more unaccepting.

7. There are two states in the U.S. where gay and lesbian couples can marry and receive the same benefits and responsibilities given to heterosexual couples once they leave those states.

8. Usually one partner wants to terminate a relationship more than the other partner.

9. Approximately 75% of divorced people remarry.

10. Sixty percent of marriages worldwide are arranged.

Multiple-Choice

11. What does the process of dating serve?
    a. as an important recreational function
    b. as a chance to compare the assets and liabilities of eligible partners
    c. as the opportunity to clarify what we are looking for in a partner
    d. All of the above
    e. None of the above

12. Steady dating in adolescence is associated with _________________.
    a. low self-esteem
    b. increased depression
    c. lower sex-role identity
    d. higher self-esteem
    e. All of the above

13. In what types of heterosexual relationships are men and women equally likely to initiate sex?
    a. relationships where both partners are dissatisfied with the relationship
    b. relationships where the man is less happy than the woman
    c. relationships that are rated as the happiest
    d. relationships where the woman is less happy than the man
    e. None of the above

14. How has dating with respect to gender roles changed in the last 30 years?
    a. Women are more likely to initiate dates and sexual activity.
    b. Men are more likely to initiate dates and sexual activity.
    c. Women are less likely to discuss sexually transmitted infections.
    d. Men are less likely to discuss sexually transmitted infections.
    e. All of the above

15. Who will be more likely to date based on the factors of health and mobility?
    a. adolescent boys
    b. adolescent girls
    c. older women
    d. older men
    e. both older men and women

16. In addition to race, what other factor influences whether people think dating between different people is acceptable?
    a. religion
    b. age
    c. social class
    d. disability
    e. All of the above

17. What do the authors call the situation where without discussion couples believe they know what each other wants because they base it on what they themselves want?
    a. quiet interaction
    b. simple solitude
    c. silent agreement
    d. mute musings
    e. hushed harmony

18. Which of the following is NOT cited as an advantage of cohabitation?
    a. It allows couples to learn more about each other.
    b. It allows couples to be more financially stable when they do marry.
    c. It gives the opportunity to learn of a partner's bad habits.
    d. It allows couples to move into marriage more slowly.
    e. Society tends to recognize people who live together for tax and health care purposes.

19. What is common-law marriage?
    a. a marriage that is preceded by premarital counseling and has strict rules about divorce.
    b. a marriage that is arranged by parents or relatives
    c. a marriage in which a woman marries her own relative to maintain the integrity of family property
    d. a marriage that is not based on love
    e. a marriage existing by mutual agreement by the fact of cohabitation

20. What is an allowance for support made under court order to a divorced person by the former spouse, who is usually the chief provider during the marriage?
    a. polyandry
    b. monetarism
    c. alimony
    d. polygyny
    e. None of the above

21. Of all the variables related to marital satisfaction, what has research suggested as specifically related to women's satisfaction?
    a. personality similarities
    b. housework equality
    c. physical intimacy
    d. frequency of emotional closeness
    e. All of the above

22. What is important in marital satisfaction?
    a. high rewards and low costs
    b. low rewards and high costs
    c. high rewards and high costs
    d. low rewards and high costs
    e. low rewards and no costs

23. What are single men more likely to experience when compared to married men?
    a. feelings of euphoria and relief
    b. contentment
    c. suicide and depression
    d. a healthier reaction to job loss and retirement
    e. There have never been any significant differences between them.

24. Which of the following factors will lead a marriage that follows the death of a spouse to be more successful?
    a. financial stability
    b. good health
    c. the couple knew each other for a period of time prior to the marriage
    d. children and peers approve of the marriage
    e. All of the above

25. What is the term the authors use for *unmarried* couples who have sex with someone other than their partner?
    a. domestic infidelity
    b. extra-domestic
    c. relation infidelity
    d. single deception
    e. extra-relationship

26. What factors do the authors identify as playing a role in the type of extramarital sexual relationships that occur?
    a. gender, race, and region
    b. region and body image status
    c. body image status and self-esteem
    d. age and gender
    e. None of the above

27. What are polyamorists?
    a. members of a couple who exchange sexual partners
    b. a person who cheats repeatedly on a partner without the partner's knowledge
    c. people who travel around the world looking for love
    d. the term given to researchers who study cross-cultural perspectives of love
    e. partners in a common-law marriage

28. In what year did the U.S. Congress enact the Defense of Marriage Act, which put forth that all states have the right to refuse to recognize same-sex marriages performed in other states?
    a. 1910
    b. 1954
    c. 1972
    d. 1980
    e. 1999

29. What is the term for the living situation that was passed in 2000 in Vermont granting same-sex couples many of the same benefits and responsibilities , though only in the state of Vermont?
    a. cohabitation arrangement
    b. consanguineous living
    c. civil union
    d. arranged covenant
    e. palimonious union

30. Who is more likely to initiate a divorce?
    a. women
    b. men
    c. the partner who is more financially stable
    d. the partner who is younger
    e. the partner who is older

31. What is the term for a marriage that is preceded by premarital counseling and has strict rules about divorce?
    a. consanguineous marriage
    b. common-law marriage
    c. domestic marriage
    d. arranged marriage
    e. covenant marriage

32. When it comes to predisposing factors for divorce, which of the following statements below is TRUE?
    a. People who have been divorced before are less likely to get divorced again.
    b. Couples who marry at an older age often suffer more marital disruption than younger couples.
    c. People whose parents have divorced are less likely to get divorced.
    d. Couples who marry because of an unplanned pregnancy are more likely to divorce.
    e. Catholics and Jews are more likely to divorce than Protestants.

33. What is the most frequent complaint of men regarding their wives and their marriages?
    a. their partner is having an affair
    b. their partner is inattentive or neglectful of their needs
    c. their partner spends too much money
    d. their partner doesn't love them
    e. All of the above

34. People who are more religious are ___________________.
    a. more likely to have sex after a divorce
    b. less likely to have sex after a divorce
    c. just as likely to have sex as a less religious person
    d. There is no relationship between religion, divorce, and sex.
    e. None of the above

35. Who is more likely to remarry after a divorce?
    a. women
    b. elderly men and women
    c. men
    d. religious men and women
    e. None of the above

36. What is an arranged marriage?
    a. a marriage that is preceded by premarital counseling and has strict rules about divorce.
    b. a marriage that is planned by parents or relatives that is not based on love
    c. a marriage in which a woman marries her own relative to maintain the integrity of family property
    d. a marriage existing by mutual agreement by the fact of cohabitation
    e. None of the above

37. What lead China to move from a pattern of arranged marriages to an attitude where people should be able to choose their partners freely?
    a. the Communist Revolution
    b. the incidence of HIV/AIDS
    c. Western movies and television
    d. the Internet
    e. All of the above

38. What is the term for the practice of having multiple wives?
    a. polyamory
    b. palifeminy
    c. polygyny
    d. polyandry
    e. palimony

39. What is polyandry?
    a. having multiple husbands at one time
    b. marrying without the legal possibility of divorce
    c. having multiple wives at one time
    d. paying support to a former spouse
    e. exchanging marital partners for sexual activity

40. In what countries can a man can divorce his wife simply by repudiating her publicly three times, while a woman must go to court to dissolve the marriage?
    a. Poland and Latvia
    b. Afghanistan, Pakistan, and India
    c. Mexico, Bolivia, and Peru
    d. Israel and Egypt
    e. India and South Africa

Matching

Column 1		Column 2
A. domestic partner	____	41. an allowance for support made under court order to a divorced person by the former spouse, who is usually the chief provider during the marriage
B. covenant marriage	____	42. the practice of having more than one husband at one time
C. polyandry	____	43. a marriage existing by mutual agreement, or by the fact of their cohabitation, without a civil or religious ceremony.
D. swingers	____	44. a person, other than a spouse, with whom one cohabits who can be either the same or other sex
E. consanguineous marriage	____	45. the practice of having more than one spouse at a time
F. polygamy	____	46. a marriage that is preceded by premarital counseling and has strict rules about divorce
G. common-law marriage	____	47. marriages arranged by parents or relatives that are often not based on love
H. cohabitation	____	48. a married couple who exchanges sexual partners
I. arranged marriage	____	49. a marriage in which a woman marries her own relative to maintain the integrity of family property
J. alimony	____	50. to live together in a sexual relationship when not legally married

Test Yourself Answer Key

Dating and Cohabitation

1. customs (p. 242)
2. dating (p. 242)
3. physical health (p. 243)
4. commitment (p. 244)
5. sexual intercourse (p. 246)
6. conservative, sexual (p. 246)
7. men (p. 246)
8. communication (p. 247)
9. dating (p. 248)
10. anti-miscegenation (p. 249)
11. interracial (p. 249)
12. silent agreements (p. 250)
13. cohabitation (p. 251)
14. Cohabitation (p. 252)
15. common-law marriage (p. 253)
16. The level of patriarchy in society, its ideals of masculinity and femininity, the roles of women and men, the value placed on conformity, the importance of childbearing, the authority of the family, and attitudes toward childhood, pleasure and responsibility (p. 242)
17. There are no agreed-upon words for different levels of commitment. Terms can mean different things to different people. (p. 244)
18. More women initiate dates and sexual activity, casual dating is more common, sexual activity may be more frequent, and one has to be more worried about sexually transmitted infections. (p. 248)
19. Twenty-one percent of females who live with their partner are 2 or more years older than their partner, but only 12% of wives are 2 or more years older than their husband. Most married couples are the same race, but cohabiting partners were twice as likely to be a different race than were married couples. (p. 251)
20. Being under the age of thirty, living in a metropolitan area, being well educated, and both being employed but having relatively low incomes. (pp. 251-252)

Marriage and Same-Sex Relationships

21. happy marriage (p. 253)
22. friendship (p. 254)
23. health (p. 255)
24. men, women (p. 255)
25. passion (p. 255)
26. Masturbation, masturbates (p. 257)
27. children (p. 257)
28. improved (p. 257)
29. marriage (p. 258)
30. emotional (p. 259)
31. emotional, sexual, sexual (p. 259)
32. Comarital (p. 260)
33. swingers (p. 260)
34. Vermont (p. 261)
35. domestic partner (p. 262)
36. Frequency of pleasurable activities (doing things together), being able to talk to each other and self-disclose, physical and emotional intimacy, and personality similarities. (p. 254)

37. Marital quality tends to peak in the first few years of a marriage and then declines until midlife when it rises again. (p. 254)
38. Marriages seem to last most when both partners have a positive attitude toward the marriage, view their partner as a best friend, and like their partner as a person. (p. 254)
39. Upper classes tend to have marital sex more frequently, use more sexual positions, and practice more oral sex and other varieties of sexual contact. (p. 257)
40. Same-sex marriage, separation, child custody, and gay and lesbian adoption. (p. 263)

Divorce and Relationships in Other Places

41. expensive (p. 263)
42. female (p. 264)
43. covenant (p. 265)
44. acceptable (p. 265)
45. emotional immaturity (p. 265)
46. disappear (p. 266)
47. without, with (p. 266)
48. depression (p. 267)
49. four (p. 267)
50. dating (p. 267)
51. China (p. 268)
52. arranged (p. 269)
53. polygyny (p. 270)
54. Denmark (p. 271)
55. infidelity, cruelty (p. 272)
56. no-fault divorce (p. 263)
57. Catholics and Jews are less likely to divorce than Protestants, and divorce rates tend to be high for marriages of mixed religions. Marriages between people having no religious affiliation have particularly high divorce rates. (p. 265)
58. 75% (p. 267)
59. China (p. 268)
60. consanguineous marriage (p. 271)

Post Test Answer Key

True/False

1. T (p. 244)
2. F (p. 251)
3. T (p. 253)
4. T (p. 253)
5. T (p. 257)
6. T (p. 258)
7. F (p. 261)
8. T (p. 264)
9. T (p. 267)
10. T (p. 269)

Multiple-Choice

11. d (pp. 242-243)
12. d (p. 243)
13. c (p. 246)
14. a (p. 248)
15. c (p. 249)
16. e (p. 249)
17. c (p. 250)
18. e (p. 252)
19. e (p. 253)
20. c (p. 253)
21. d (p. 254)
22. a (p. 254)
23. c (p. 255)
24. e (p. 257)
25. e (p. 258)
26. d (p. 259)
27. a (p. 260)
28. e (p. 261)
29. c (p. 261)
30. a (p. 264)
31. e (p. 265)
32. d (pp. 265-266)
33. b (p. 266)
34. b (p. 266)
35. c (p. 267)
36. b (p. 267)
37. a (p. 268)
38. c (p. 270)
39. a (p. 271)
40. d (p. 272)

Matching

41. J (p. 253)
42. C (p. 271)
43. G (p. 253)
44. A (p. 262)
45. F (p. 270)
46. B (p. 265)
47. I (p. 267)
48. D (p. 260)
49. E (p. 271)
50. H (p. 251)

Chapter 10 – Sexual Expression: Arousal and Response

CHAPTER SUMMARY

Human sexuality is a complex part of life, with cultural, psychological, and biological influences shaping how people choose to express their sexuality. The most influential hormones in sexual behavior are estrogen and testosterone. Ethnicity can affect which sexual behaviors we engage in, the frequency of these behaviors, our sexual attitudes, and our ability to communicate about sex.

The term sexual response refers to the series of physiological and psychological changes that occur in the body during sexual behavior. Based on their laboratory work, William Masters and Virginia Johnson proposed a four-phase model of physiological arousal known as the sexual response cycle. The four phases of the sexual response cycle are: excitement, plateau, orgasm, and resolution. The sexual response cycle is similar in females and males, with vasocongestion and myotonia leading to physiological changes in the body. Two other models of sexual response were developed that incorporate psychological aspects. Helen Singer Kaplan's Triphasic Model of sexual response includes sexual desire, excitement, and orgasm. David Reed's model blends features of both Masters and Johnson and Kaplan's models and involves four phases, including seduction, sensation, surrender, and reflection.

A variety of sexual behaviors are prevalent among adults, including sexual fantasies, masturbation, manual sex, oral sex, and sexual intercourse. Foreplay is common but often defined in different ways among different people. Many researchers believe that sexual fantasies are normal and healthy and that they may be a driving force behind human sexuality. Masturbation can decrease anxiety and help people to learn what kind of pressure and manipulation gives them pleasure. Manual sex refers to the physical caressing of the genitals during solo or partner masturbation and has become more popular over the years as a form of safer sex because there is typically no exchange of bodily fluids. The majority of Americans report that they engage in oral sex at least occasionally. Recent research has found that the average for frequency of sexual intercourse is about once a week. There are numerous positions for sexual intercourse including the most common: male-on-top, female-on-top, side-by-side, and rear-entry. Sex therapists often recommend the female-on-top position for couples who are experiencing difficulties with early ejaculation since the female-on-top position can extend the length of erection for men and facilitates female orgasm.

Whether or not a man or woman remains sexually interested and active has to do with a variety of factors, including his or her age, physical health, medications, level of satisfaction with life, and the availability of a partner. Despite the changes, the research reveals that older men and women who have remained sexually active throughout their aging years have a greater potential for a more satisfying sex life later in life.

In the United States, many believe that they term, "safer sex" primarily involves wearing a condom and reducing the number of sexual partners. However, sexuality educators also use "safer sex" to refer to specific sexual behaviors that are safer to engage in because they protect against the risk of acquiring sexually transmitted infections. Alcohol use has clearly been linked to unsafe sexual behaviors due to impaired judgment.

LEARNING OBJECTIVES

After studying Chapter 10, you should be able to:

1. Discuss how hormones affect sexual behavior.
2. Explain what aspects of sexuality are affected by ethnicity.
3. Discuss the research on the frequency of sexual behaviors.
4. List the four phases of Masters and Johnson's sexual response cycle.
5. Identify the physiological changes that occur during each phase for women and men.
6. Compare and contrast the models of Masters and Johnson, Helen Singer Kaplan, and David Reed.
7. Discuss the role of sexual fantasies in adults' lives.
8. Identify the most commonly reported sexual fantasies for men and women.
9. Identify some of the benefits of masturbation.
10. Compare and contrast masturbation among women and men.
11. Explain how and why the patterns of manual sex are changing.
12. List the different types of oral sex.
13. Discuss the values people have about oral sex.
14. List several positions for sexual intercourse.
15. Identify the advantages and disadvantages of various positions for sexual intercourse.
16. Discuss the research and patterns of anal intercourse.
17. Explain the physiological changes that occur for men and women as they age.
18. Identify the factors that help older adults to maintain satisfying sexual experiences.
19. Discuss the challenge with defining "safer sex."
20. Describe the research on safer sex and college populations.

CHAPTER OUTLINE

I. HORMONES AND OUR SEXUAL BEHAVIOR

A. The most influential hormones in sexual behavior are estrogen and testosterone.

B. In humans, despite the fact that hormones have an enormous effect on sexual behavior, there is also a very strong influence from learned experiences, as well as by one's social, cultural, and ethnic environment.

II. ETHNICITY AFFECTS OUT SEXUALITY

A. Ethnicity and sexuality join together to form a barrier, a "sexualized perimeter," which helps us decide who we let in and who we keep out.

B. Ethnicity can also affect which sexual behaviors we engage in, the frequency of these behaviors, our sexual attitudes, and our ability to communicate about sex.

III. FREQUENCY OF SEXUAL BEHAVIORS

A. **Celibacy** and **abstinence** are when a person chooses not to engage in sexual intercourse.

B. As for total number of sex partners, men often report a substantially greater number of partners with whom they have had sexual intercourse than do women, although men tend to over report the number of partners they have, while women may underreport the number of partners.

IV. STUDYING OUR SEXUAL RESPONSES

- The term, **sexual response**, refers to the series of physiological and psychological changes that occur in the body during sexual behavior.

A. **Masters and Johnson's Four-Phase Sexual Response Cycle**

1. Based on their laboratory work, William Masters and Virginia Johnson proposed a four-phase model of physiological arousal known as the **sexual response cycle**.
2. The four phases of the sexual response cycle are: **excitement**, **plateau**, **orgasm**, and **resolution**
3. Sexual Response Cycle in Women
    - Sexual response patterns vary among women.

    a. Excitement Phase
    - The first stage begins with **vasocongestion**, an increase in the blood concentrated in the genitals and/or breasts.
    - **Transudation** refers to the vaginal walls lubricating.
    - The **tenting effect** is when the walls of the vagina expand.
    - Nipple erections may occur and the areolas and breasts enlarge.
    - During sexual arousal in some women, the labia majora thin out and become flattened and may pull slightly away from the **introitus** (the vaginal opening).
    - The labia minora often turn bright pink and begin to increase in size.
    - Vasocongestion may also cause the clitoris to become erect.
    - A **sex flush** resembles a rash that usually begins on the chest and, during the plateau stage, spreads from the breasts to the neck and face to the rest of the body.

    b. Plateau Phase
    - Breast size continues to increase during the plateau phase, and the nipples may remain erect.
    - The clitoris retracts behind the clitoral hood.

- In women who have not had children, the labia majora are difficult to detect, due to the flattened-out appearance, whereas the labia minora often turn bright red.
- In women who have had children, the labia majora become very engorged with blood and turn a darker red.

c. Orgasm Phase

- At the end of the plateau phase, vasocongestion in the pelvis creates an **orgasmic platform** in the lower third of the vagina, labia minora, and the uterus. When this pressure reaches a certain point, a reflex in the surrounding muscles is set off, causing contractions.
- Myotonia, or muscle contractions, cause pleasurable orgasmic sensations.
- In women, contractions typically last longer than in men, possibly due to the fact that vasocongestion occurs in the entire pelvic region in women and is very localized in men.
- During orgasm the body may jerk and/or spasm which may involve the face, hands, and feet.

d. Resolution Phase

- During the last phase of the sexual response cycle, the body returns to the preexcitement conditions.
- The blood leaves the genitals and erections disappear, muscles relax, and heart and breathing rates return to normal.
- Some women can return to the orgasmic stage and experience **multiple orgasms**.
- Masters and Johnson suggested that multiple orgasms are more likely to occur from manual stimulation of the clitoris, rather than from penile-vaginal intercourse.
- The clitoris returns to its original size but often remains extremely sensitive for several minutes.

4. Sexual Response Cycle in Men

- The sexual response cycle in males is similar to that of females, with vasocongestion and myotonia leading to physiological changes in the body.

a. Excitement Phase

- The excitement phase in men is often very short.
- **tumescence**-the swelling of the penis due to vasocongestion (an increase of blood)
- **detumescence**-the decrease of blood flow in the penis
- During the excitement phase, the testicles also increase in size, becoming up to 50% larger. The **dartos** and **cremastic** muscles pull the testicles closer to the body to avoid injury.

b. Plateau Phase

- All of the changes that occur during the excitement phase continue during the plateau phase.
- Some men may experience a sex flush and/or nipple erections.
- Just prior to orgasm, the glans penis becomes engorged and a few drops of pre-ejaculatory fluid may appear on the head of the penis.

c. Orgasm Phase

- Orgasm and ejaculation do not always occur together.

- Some men are capable of anywhere from two to sixteen orgasms prior to ejaculation.
- If orgasm and ejaculation occur at the same time, ejaculation can occur in two stages. During the first stage, which lasts only a few seconds, there are contractions in the vas deferens, seminal vesicles, and prostate gland.
- The contractions lead to **ejaculatory inevitability** whereby just prior to orgasm there is feeling that ejaculation can no longer be controlled.
- Next, the semen is forced out of the urethra by muscle contractions.
- Multiple orgasms in men are possible.

d. Resolution Phase

- Directly following ejaculation, the glans of the penis decreases in size.
- Men go into a **refractory stage** during which they cannot be restimulated to orgasm for a certain time period. The refractory period gets longer as men get older.

B. **Other Models of Sexual Response**

1. Helen Singer Kaplan's **Triphasic Model** of sexual response includes sexual desire, excitement and orgasm.
2. Sexual desire is a psychological phase, while excitement and orgasm involve physiological processes, including vasocongestion of the genitals and muscular contractions during orgasm.
3. Sexual desire was of paramount importance to Kaplan because, without sexual desire, the other two physiological functions would not occur.
4. David Reed's ESP model blends features of Masters and Johnson and Kaplan's models and uses four phases, including seduction, sensation, surrender, and reflection.

V. ADULT SEXUAL BEHAVIORS

A. **Foreplay—The Prelude?**

1. People define foreplay in different ways.
2. For the majority of heterosexuals, foreplay is often defined as everything that happens before sexual intercourse.

B. **Sexual Fantasy—Enhancement or Unfaithfulness?**

- Many researchers believe that sexual fantasies are normal and healthy and that they may be a driving force behind human sexuality.
- Liberal attitudes and more sexual experiences have been found to be associated with longer and more explicit sexual fantasies.
- Those who do not have sexual fantasies have been found to experience a greater likelihood of sexual dissatisfaction and sexual dysfunctions.
- Men have also been found to have more **sexual cognitions** or thoughts about sex than women.
- In the past decade or so, research on men and women's sexual fantasies has shown the fantasies are becoming more similar.
- In fact, women have been reporting more graphic and sexually aggressive fantasies than they have had in the past.
- Fantasies can help enhance masturbation, increase sexual arousal, help a person to reach orgasm, and allow a person to explore various sexual activities that he or she might find taboo or too threatening to engage in.

1. College Students and Sexual Fantasy

- The majority of college students use sexual fantasies, with many feeling that fantasizing is normal, moral, common, socially acceptable, and more beneficial than harmful.
- Some feel a considerable amount of guilt, which can cause a person to be less likely to engage in more intimate types of sexual behavior.
- In one study, students felt a considerable amount of jealousy about their partner's sexual fantasies and believed that engaging in sexual fantasy was equivalent to unfaithfulness.

2. Women's Sexual Fantasies
    - Many women report that they use sexual fantasy to increase their arousal, self-esteem, sexual interest, or to relieve stress and cope with past hurts.
    - The most common sexual fantasies for women include sex with current partner, reliving a past sexual experience, engaging in different sexual positions, having sex in rooms other than the bedroom, and sex on a carpeted floor.
    - Female sexual fantasies tend to be more romantic than male fantasies.
    - Some women incorporate force into sexual fantasies, but it does not mean they want the fantasy to come true.
3. Men's Sexual Fantasies
    - Men's sexual fantasies tend to be more active and aggressive than women's.
    - They generally include visualizing more body parts, specific sexual acts, group sex, a great deal of partner variety, and less romance.
    - The five most common sexual fantasies for men include engaging in different sexual positions, having an aggressive partner, getting oral sex, having sex with a new partner, and having sex on the beach.

C. **Masturbation—A Very Individual Choice**

- For a period in the 19th and early 20th centuries in the United States and Europe, there were myths that masturbation resulted in insanity, death, or even sterility.
- Masturbation can decrease sexual tension and anxiety and provide an outlet for sexual fantasy, while also helping people to learn what kind of pressure and manipulation gives them pleasure.
- **Mutual masturbation** is partners masturbating themselves or each other simultaneously.
- Research indicates that the largest gender difference in sexual behavior is in the incidence of masturbation, with college men reporting more masturbation than women.
- Research has found that people who are having regular sex with a partner masturbate more than people who are not having regular sex.
- In some religious and cultural traditions, masturbation is discouraged or forbidden.

1. Female Masturbation
    - Research has found that women reached the most intense orgasms through masturbation.
    - There is a strong masturbation taboo that makes women feel it is not acceptable to admit they masturbate or can reach orgasm alone.
    - Women are more likely to report guilt about their masturbation activities compared to men.
2. Male Masturbation
    - Like women, some men feel very comfortable with masturbating while others do not.

D. **Manual Sex—A Safer Sex Behavior**

- **Manual sex**, also referred to as "hand jobs," refers to the physical caressing of the genitals during solo or partner masturbation.
- Manual sex has become more popular over the years as a form of safer sex because there is typically no exchange of bodily fluids.

1. Manual Sex on Women
2. Manual Sex on Men

E. **Oral Sex—Not So Taboo**

- Oral sex has been practiced throughout history, and for many people, oral sex is an important part of sexual behavior.
    - **cunnilingus**-oral sex on a woman
    - **fellatio**-oral sex on a man
- The majority of Americans report that they engage in oral sex at least occasionally, with many men and women engaging in oral sex prior to their first experience with sexual intercourse.
- Some people engage in **sixty-nine** (simultaneous oral sex) although this position can be challenging for some couples.
- **Analingus**, or rimming, another form of oral sex, involves oral stimulation of the anus.
- Some people feel like oral sex is more intimate, while others feel like it is less intimate.
- It is possible to transmit viruses to a person during oral sex, especially when a person has a cold sore (herpes).

1. Cunnilingus
    - In the United States, women have been inundated with negative messages about the cleanliness of their vulvas and vaginas.
    - Anxiety about the cleanliness of their vulvas and women's lack of familiarity with their genitals, contribute to many women's discomfort with oral sex.
2. Fellatio
    - Performing fellatio and swallowing the ejaculate is not considered a form of safe sex.
    - An average ejaculation is approximately one to two teaspoons and consists mainly of fructose, enzymes, and different vitamins, containing approximately five calories.

F. **Heterosexual Sexual Intercourse**

- Recent research has found that the average for frequency of sexual intercourse is about once a week.
- Most heterosexual couples engage in sexual intercourse almost every time they have sex, and when most people think of "sex," they think of vaginal-penile intercourse.
- Many women need additional clitoral stimulation to reach orgasm.
- Vaginal lubrication helps intercourse feel more pleasant for both partners.
- Although many men may wish to time their orgasm with their partners, this can be difficult to do.
- Research suggests that the majority of couples do not have eye contact during sexual intercourse.

1. Positions for Sexual Intercourse
    a. **Male-on-Top**

- This is also called the "missionary" or "male superior" position and is one of the most common positions for sexual intercourse.
- Disadvantages to this position include that it can be uncomfortable if a woman is heavy or pregnant, it makes it difficult to provide clitoral stimulation, a woman may be prevented from moving her hips or controlling the strength and/or frequency of thrusting, and it may be difficult for man to control his erection and ejaculation, especially if he has erectile difficulties.

b. **Female-on-Top**

- This is also called "female superior" and has become more popular in the last decade.
- By leaning forward, a woman has greater control over the angle and degree of thrusting and can get more clitoral stimulation.
- Since **intromission** (the insertion of a penis) can be difficult in this position, many couples prefer to begin in the male-on-top position and roll over.
- Sex therapists often recommend this position for couples who are experiencing difficulties with premature ejaculation or a lack of orgasms because the female-on-top position can extend the length of erection for men and facilitates female orgasm.
- Some women feel shy or uncomfortable with this position, and some men do not receive enough penile stimulation in this position to maintain an erection.

c. **Side-by-Side**

- This position takes the primary responsibility off both partners and allows them to relax during sexual intercourse.
- Sometimes couples in this position have difficulties with intromission.

d. **Rear-Entry**

- There are several variations of the rear-entry position.
- These positions provide the best opportunity for direct clitoral stimulation and can also provide direct stimulation of the g-spot.
- It can also be a good position for women who are in the later stages of pregnancy or who are heavy.

2. Anal Intercourse

- Anal stimulation is very pleasurable for many people and is practiced by heterosexual, gay, lesbian, and bisexual men and women.
- Men and women can experience orgasms during anal intercourse, especially with simultaneous penile or clitoral stimulation.
- Research shows that approximately 24% of adults have engaged in anal sex at least once.
- During anal intercourse, the **anal sphincter** muscle must be relaxed, which can be facilitated by manual stimulation.
- Research has shown that the risk of contracting HIV through unprotected anal intercourse is greater than the risk of contracting HIV through unprotected vaginal intercourse.
- Any couple who decides to engage in anal sex should never transfer the penis from the anus to the vagina without changing the condom or washing the penis, hands, or sex toys.

G. **Same-Sex Sexual Techniques**
   1. Gay Men
   2. Lesbians

VI. SEXUAL BEHAVIORS LATER IN LIFE

A. Older adults may find that their preferences for certain types of sexual behaviors change as they age. For example, they engage in sexual intercourse less and oral sex more.
B. Whether or not a man or woman remains sexually interested and active has to do with a variety of factors, including his or her age, physical health, medications, level of satisfaction with life, and the availability of a partner.
C. The most prevalent sexual problems in older women are not medical complaints but rather a lack of tenderness and sexual contact.
D. Many older couples who experience problems in sexual functioning may not understand that they may have some options.
E. After raising a family, one or both partners may also feel that recreational sex is inappropriate.
F. As people age, they inevitably experience changes in their physical health, and some of these changes can affect normal sexual functioning.
   1. Older men often need more stimulation to have an erection and orgasm; experience an increased refractory period; less firm erections; decrease volume of ejaculate; and reduced intensity of ejaculation.
   2. In older women, physical changes can lead to decreases in vaginal lubrication; painful intercourse; reduction in vaginal elasticity; and painful orgasm.
G. Despite the changes, the research reveals that older men and women who have remained sexually active throughout their aging years have a greater potential for a more satisfying sex life later in life.
H. Masturbation can become a more important outlet for older people due to lack of an interested partner.

VII. SAFER SEX BEHAVIORS

A. In the United States, many believe that they term "safer sex" primarily involves wearing a condom and reducing the number of sexual partners.
B. However, sexuality educators also use "safer sex" to refer to specific sexual behaviors that are safer to engage in because they protect against the risk of acquiring sexually transmitted infections.
C. There are no sexual behaviors that protect a person 100% of the time—with the exception of solo masturbation and sexual fantasy.
D. "Safer sex" is a more appropriate term than "safe sex" since we do know there are some sexual behaviors that are safer than others, and no behaviors are completely safe.
E. Research has shown that only about half the time college students are aware of their partner's prior sexual history, past condom use, and HIV status.
F. While there has been a gradual increase in condom use in recent years, effective safer sex negotiation is more an exception than the rule in dating couples.
G. Women are more likely than men to insist on condom use.
H. Alcohol use has clearly been linked to unsafe sexual behaviors due to impaired judgment.
I. In one study, 75% of college students had made decisions while under the influence of alcohol that they later regretted.

TEST YOURSELF

Below you will find fill-in-the-blank and short answer essay questions for topics covered in chapter 10. Check your answers at the end of this chapter.

Sexual Responses

1. Human sexuality is a complex part of life, with ______________, ______________, and ______________ influences shaping how people choose to express their sexuality.

2. The various endocrine glands secrete ______________ into the blood stream, which carries them throughout the body.

3. In men, testosterone is produced in the ______________ and ______________ ______________.

4. In women, testosterone is produced in the ______________ ______________ and ______________.

5. In men, decreases in ______________ can lead to lessening sexual desire and decreases in the quality and quantity of erections.

6. In most animals, the ______________ controls and regulates sexual behavior chiefly through hormones.

7. ______________ is when a person chooses not to engage in sexual intercourse.

8. Based on their laboratory work, William Masters and Virginia Johnson proposed a four-phase model of physiological arousal known as the ______________ ______________ ______________.

9. The two primary physical changes that occur during the sexual response cycle are ______________ and ______________.

10. According to Masters and Johnson, the first phase of sexual response is the ______________ phase.

11. Vasocongestion causes the vaginal walls to begin lubrication, a process called ______________.

12. During the excitement phase, the walls of the vagina, which usually lie flat together, expand. This has also been called the ______________ ______________.

13. In the plateau phase during sexual arousal in women who have not had children, the ______________ ______________ are difficult to detect, due to the flattened-out appearance. The ______________ ______________, on the other hand turn a brilliant red.

14. All orgasms in women are thought to be the result of direct or indirect ______________ stimulation, even though orgasms might feel different at different times.

15. During the resolution phase of sexual response, when the body is returning to its prearousal state, men go into a ______________ ______________, during which they cannot be restimulated to orgasm for a certain time period.

16. What are the most influential hormones in sexual behavior?

17. Describe the effect of estrogen and testosterone on sexual desire in aging women.

18. How does ethnicity and race affect our sexuality?

19. Why is it difficult to compare the numbers of sexual partners between men and women?

20. What are the four phases of the Masters and Johnson sexual response cycle?

Adult Sexual Behaviors

21. Studies on religion and sexuality have found that people with high levels of religiosity were more likely to hold _______________ attitudes about sex.

22. Overall, research has shown that not only is _______________ _______________ an important component to a happy marriage, but it is also linked to satisfaction, love, and commitment in sexually active dating couples as well.

23. Today, many researchers not only believe that _______________ _______________ are normal and healthy, but that they may be a driving force behind human sexuality.

24. Those who do not have sexual _______________ have been found to experience a greater likelihood of sexual dissatisfaction and sexual dysfunction.

25. In the past decade or so, research on men's and women's sexual fantasies has shown that fantasies are becoming more _______________.

26. For a period in the nineteenth and early twentieth centuries in the United States and Europe, there was a fear that _______________ caused terrible things to happen.

27. _______________ fulfills a variety of different needs for different people at different ages, and it can decrease sexual tension and anxiety and provide an outlet for sexual fantasy.

28. _______________ masturbation can be very pleasurable, although it may make reaching orgasm difficult since it can be challenging to concentrate both on feeling aroused and pleasuring your partner.

29. Research shows that people who are having regular sex with a partner masturbate _______________ than people who are not having regular sex.

30. In some religious traditions, masturbation is _______________ or forbidden.

31. Generally, people think of _______________ sex as something that happens before sexual intercourse, but it has become more popular over the years as a form of safer sex.

32. Oral sex, also called _______________ and _______________, has been practiced throughout history.

33. Analingus, or _______________ involves oral stimulation of the anus.

34. Performing _______________ and swallowing the ejaculate is not considered a form of safe sex.

35. Verbal _______________ is needed to ensure that both partners are happy with the pace of intercourse.

36. According to research, what are the five most common sexual fantasies for women?

37. Describe the frequency and intensity of orgasm in female masturbation.

38. What is manual sex?

39. What percent of men and women report engaging in oral sex prior to first sexual intercourse?

40. What position do sex therapists often recommend for couples who are experiencing difficulties with early ejaculation or a lack of orgasms?

Sexual Behaviors Later in Life/Safer Sex

41. The most prevalent sexual problems in older women are not medical complaints but rather a lack of tenderness and _______________ _______________.

42. Hormonal changes associated with menopause are _______________ influential than the effects of society, psychological, and partner-related issues.

43. Many people, young and old, are not aware of the _______________ changes that can affect sexual functioning.

44. Many of the decreases in sexual functioning later in life are exacerbated by sexual _______________.

45. _______________ may continue among the elderly so that they can reassure themselves that they are not the asexual persons that society labels them.

46. What Masters and Johnson referred to as the _______________ syndrome is when a widow or widower becomes abstinent for a period of time as a result of the grief he or she feels over their partner's loss or perhaps because the person has never had other sexual partners and feels it is too late to start.

47. The stereotype that sex _______________ with age is not inevitably true.

48. All sexually active people should be aware of the _______________ associated with various sexual behaviors.

49. Research has shown that college students are aware of their partner's prior sexual history, past condom use and HIV status only about _______________ the time.

50. Overall, condom use on college campuses remains _______________.

51. Even though most people feel anxious about the possibility of acquiring an _______________, sexual activity has increased in the past several years and there have been very few increases in heterosexual safer sex behaviors.

52. One research study found that of those people with multiple partners, less than _______________% use condoms.

53. One behavior that has been clearly linked to unsafe sexual behaviors is _______________ _______________.

54. _______________ is key to safer sex relationships.

55. Not only will _______________ result in safer sex, it will also result in a healthier relationship.

56. What are some factors that have to do with whether or not men or women remain sexually interested and active in their later years?

57. Describe why sexual intercourse would stop in a marital relationship among older people.

58. As men age, what are some of the changes that affect normal sexual functioning?

59. As women age, what are some of the changes that affect normal sexual functioning?

60. What are some things that can enhance sexuality throughout the lifespan?

POST TEST

Below you will find true/false, multiple-choice and matching quiz items covering the entire chapter. Check your answers at the end of this chapter.

True/False

1. Masters and Johnson's model of physiological arousal refers only to sexual intercourse.
2. During the resolution stage of the sexual response cycle, women cannot be stimulated to orgasm.
3. Men can experience multiple orgasms during the refractory stage.
4. David Reed's model of sexual response includes seduction and surrender.
5. Many researchers believe that sexual fantasies are normal and healthy.
6. Although it's rare, masturbation can cause insanity.

7. Due to physical changes, most older men and women are unable to enjoy a satisfying sex life.

8. Masturbation can become an important outlet for older adults.

9. About 95% of the time, college students are aware of their partner's HIV status.

10. Seventy-five percent of college students have made decisions while under the influence of alcohol that they later regretted.

Multiple-Choice

11. What are the most influential hormones in sexual behavior?
    a. oxytocin and progesterone
    b. pitocin and androgen
    c. estrogen and testosterone
    d. gonadotropin and luteinizing hormone
    e. androgen and protesting

12. What is term for the series of physiological and psychological changes that occur in the body during sexual behavior?
    a. sexual response
    b. gendered functioning
    c. sexual scripts
    d. genital cycling
    e. bodily phasing

13. How many stages are in Masters and Johnson's sexual response cycle?
    a. 2
    b. 3
    c. 4
    d. 5
    e. 6

14. What is the tenting effect?
    a. when the vaginal walls begin to lubricate
    b. an increase in blood flow to the genitals during the sexual response
    c. the feeling that ejaculation cannot be stopped just before ejaculation
    d. when the walls of the vagina expand
    e. an increase in muscle contractions during the sexual response cycle

15. During what stage does the sex flush begin?
    a. plateau
    b. orgasm
    c. resolution
    d. arousal
    e. excitement

16. Which of the following statements is true of the orgasmic phase in women?
    a. The clitoris returns to its original size.
    b. Transudation begins.
    c. The tenting effect takes place.
    d. Vasocongestion in the pelvis creates an orgasmic platform.
    e. All of the above

17. During what stage of the sexual response cycle in women does the clitoris return to its original size but remain extremely sensitive?
    a. plateau
    b. orgasm
    c. resolution
    d. arousal
    e. excitement

18. What is the name of the muscle that helps to pull the testicles closer to the body during the excitement phase of the sexual response cycle in men?
    a. cremastic
    b. testos
    c. sartomis
    d. matony
    e. kegel

19. During what stage of Masters and Johnson's sexual response cycle do men experience ejaculatory inevitability?
    a. plateau
    b. resolution
    c. orgasm
    d. refractory
    e. excitement

20. What is the name for the period in which men cannot be restimulated to orgasm?
    a. orgasmic platform
    b. refractory stage
    c. reflection phase
    d. ejaculatory inevitability
    e. surrender phase

21. Who added a sexual desire phase to a model of sexual response, incorporating a psychological element to the response cycle?
    a. Helen Singer Kaplan
    b. Ernest Grafenberg
    c. Alfred Kinsey
    d. Masters and Johnson
    e. David Reed

22. How do sexual fantasies affect a person's sexuality?
    a. They can help a person to reach orgasm.
    b. They increase sexual arousal.
    c. They allow a person to explore sexual activities that might be taboo.
    d. They can help enhance masturbation.
    e. All of the above

23. What's another term for thoughts about sex?
    a. emoticons
    b. sexual scripts
    c. sexicons
    d. sexual cognitions
    e. sex beliefs

24. Which of the following sexual fantasies are reportedly more common among men when compared to women?
    a. having sex on the beach
    b. having an aggressive partner
    c. getting oral sex
    d. sex with a new partner
    e. All of the above

25. What are some of the benefits of masturbation?
    a. It can provide an outlet for sexual fantasy.
    b. It can be exciting for couples to use during sexual activity.
    c. It can decrease sexual tension and anxiety.
    d. It can allow people the opportunity to experiment with their bodies to see what feels good.
    e. All of the above

26. What is the term for partners masturbating themselves or each other at the same time?
    a. mutual masturbation
    b. simultaneous masturbation
    c. synchronized masturbation
    d. concurrent masturbation
    e. None of the above

27. What is manual sex?
    a. oral sex on a woman
    b. the physical caressing of the genitals during solo or partner masturbation.
    c. using a sex manual to explore new positions
    d. oral sex on a man
    e. oral sex on the anus

28. What is the term for engaging in oral sex simultaneously?
    a. manual sex
    b. analingus
    c. fellatio
    d. sixty-nine
    e. cunnilingus

29. What is the term for oral stimulation of the anus?
    a. manual sex
    b. analingus
    c. fellatio
    d. sixty-nine
    e. cunnilingus

30. What is a disadvantage to the male-on-top or "missionary" position of sexual intercourse?
    a. It may be difficult for man to control his erection and ejaculation, especially if he has erectile difficulties
    b. It makes it difficult to provide clitoral stimulation.
    c. It can be uncomfortable for a women if she is heavy or pregnant.
    d. A woman may be prevented from moving her hips or controlling the strength and/or frequency of thrusting.
    e. All of the above

31. What is intromission?
    a. oral stimulation of the anus
    b. another term for the female-on-top position for sexual intercourse
    c. the insertion of a penis
    d. another term for manual sex
    e. being delicate, fragile, and vulnerable with no sexual urge

32. What is a good position for sexual intercourse when a women is in the advanced stages of pregnancy?
    a. male-on-top
    b. side-by-side
    c. female-on-top
    d. sixty-nine
    e. missionary

33. Who participates in anal intercourse?
    a. heterosexual women
    b. gay men
    c. bisexual women
    d. heterosexual men
    e. All of the above

34. What did researchers typically believe were the two biggest issues affecting female sexuality later in life?
    a. lack of a partner and generational issues
    b. medical and health issues
    c. boredom and hormonal changes
    d. lack of tenderness and loss of reproductive capacity
    e. All of the above

35. What is one of the most important factors dictating whether or not a woman continues sexual activity?
    a. medical complaints
    b. hormonal changes
    c. level of comfort with sexuality
    d. whether she has an available and interested partner
    e. insufficient lubrication causing painful intercourse

36. Which of the following statements below regarding sexuality and aging is FALSE?
    a. Many older couples who experience problems in sexual functioning may not understand that they may have some options to continue an active sex life.
    b. Masturbation can become a more important outlet for older people due to a lack of an interested partner.
    c. In older women, physical changes can lead to increases in vaginal lubrication.
    d. As people age, they inevitably experience changes in their physical health
    e. Older men often need more stimulation to have an erection and orgasm.

37. What do many Americans believe safer sex primarily involves?
    a. practicing abstinence
    b. engaging in safer sex behaviors
    c. using condoms and reducing the number of sexual partners
    d. solo masturbation and sexual fantasy
    e. not engaging in anal intercourse

38. What can individuals do to decrease the risk associated with sexual behaviors and practice safer sex?
    a. avoid unprotected intercourse
    b. decrease the number of sexual partners
    c. learn more about the backgrounds of sexual partners
    d. use condoms
    e. All of the above

39. Who is more likely to insist on condom use?
    a. Men are more likely than women to insist on condom use.
    b. The partner who is more "in love" is less likely to insist on condom use.
    c. Women are more likely than men to insist on condom use.
    d. The older person in the relationship is more likely to insist on condom use.
    e. The younger person in the relationship is more likely to insist on condom use.

40. What behavior is one of the most important factors that is repeatedly linked to unsafe sexual behavior?
    a. drinking alcohol
    b. watching television regularly
    c. serial divorce
    d. falling in love
    e. All of the above

Matching

Column 1		Column 2
A. sexual cognitions	____	41. when the walls of the vagina expand during sexual arousal
B. Kaplan	____	42. created four-phase model of physiological arousal of sexual response cycle
C. detumescence	____	43. the insertion of a penis
D. sexual response	____	44. when the walls of the vagina begin to lubricate
E. tenting effect	____	45. a decrease in vasocongestion in the penis
F. Masters and Johnson	____	46. created a triphasic model of sexual response
G. intromission	____	47. an increase of blood in the genitals
H. vasocongestion	____	48. thoughts about sex
I. transudation	____	49. the series of physiological and psychological changes that occur in the body during sexual behavior
J. myotonia	____	50. muscle tension that leads to orgasmic contractions

Test Yourself Answer Key

Sexual Responses

1. cultural, psychological, biological (p. 277)
2. hormones (p. 277)
3. testes, adrenal glands (p. 277)
4. adrenal glands, ovaries (p. 277)
5. testosterone (p. 277)
6. brain (p. 277)
7. Celibacy or Abstinence (p. 278)
8. sexual response cycle (p. 279)
9. vasocongestion, myotonia (p. 279)
10. excitement (p. 279)
11. transudation (p. 279)
12. tenting effect (p. 279)

13. labia majora, labia minora (p. 280)
14. clitoral (p. 282)
15. refractory stage (p. 285)
16. estrogen, testosterone (p. 277)
17. Although estrogen decreases in aging women, testosterone levels often remain constant. This may result in an increase in sexual desire. (p. 277)
18. They can affect which sexual behaviors we engage in, the frequency of these behaviors, our sexual attitudes, and our ability to communicate about sex. (p. 278)
19. Men tend to overreport the number of partners they have, while women may underreport the number of partners (p. 278)
20. excitement, plateau, orgasm, resolution (p. 279)

Adult Sexual Behaviors

21. conservative (p. 286)
22. sexual satisfaction (p. 286)
23. sexual fantasies (p. 287)
24. fantasies (p. 287)
25. similar (p. 287)
26. masturbation (p. 291)
27. Masturbation (p. 291)
28. Mutual (p. 291)
29. more (p. 292)
30. discouraged (p. 292)
31. manual (p. 294)
32. cunnilingus, fellatio (p. 295)
33. rimming (p. 295)
34. fellatio (p. 297)
35. communication (p. 298)
36. sex with current partner, reliving a past sexual experience, engaging in different sexual positions, having sex in rooms other than the bedroom, and sex on a carpeted floor (p. 287)
37. The average woman is able to reach orgasm in 95% or more of her masturbatory attempts, and female masturbation has been found to produce the most intense orgasms in women. (p. 292)
38. Manual sex refers to the physical caressing of the genitals during solo or partner masturbation. (p. 294)
39. In one study, 70% of males reported performing cunnilingus prior to their first sexual intercourse, whereas 57% of females reported performing fellatio prior to their first sexual intercourse. (p. 295)
40. female-on-top (p. 301)

Sexual Behaviors Later in Life/Safer Sex

41. sexual contact (p. 305)
42. less (p. 305)
43. physical (p. 306)
44. inactivity (p. 307)
45. Masturbation (p. 307)
46. widow/widower (p. 308)
47. worsens (p. 308)
48. risks (p. 309)
49. half (p. 309)
50. low (p. 309)
51. STI (p. 309)
52. 50 (p. 310)

53. drinking alcohol (p. 310)
54. Communication (p. 310)
55. openness (p. 310)
56. age, physical health, medications, level of satisfaction with life, availability of a partner (p. 305)
57. It is usually because of the male's refusal or inability to continue. This is often due to the existence of an erectile problem. (p. 305)
58. Men often need more stimulation to have an erection and orgasm; experience an increased refractory period; less firm erections; decreased volume of ejaculate; and reduced intensity of ejaculation. (p. 307)
59. In aging women, physical changes can lead to decreases in vaginal lubrication, painful intercourse, reduction in vaginal elasticity, and painful orgasm. (p. 307)
60. physical fitness, good nutrition, adequate rest and sleep, a reduction in alcohol intake, positive self-esteem (p. 308)

Post Test Answer Key

True/False

1. F (p. 279)
2. F (p. 282)
3. F (p. 285)
4. T (p. 285)
5. T (p. 287)
6. F (p. 291)
7. F (p. 305)
8. T (p. 307)
9. F (p. 309)
10. T (p. 310)

Multiple Choice

11. c (p. 277)
12. a (p. 279)
13. c (p. 279)
14. d (p. 279)
15. e (p. 280)
16. d (pp. 281-282)
17. c (p. 283)
18. a (p. 283)
19. c (p. 284)
20. b (p. 285)
21. a (p. 285)
22. e (p. 287)
23. d (p. 287)
24. e (pp. 287-290)
25. e (p. 291)
26. a (p. 291)
27. b (p. 294)
28. d (p. 295)
29. b (p. 295)
30. e (p. 300)
31. c (p. 300)
32. c (p. 301)
33. e (p. 302)
34. b (p. 305)
35. d (p. 305)
36. c (pp. 305-307)
37. c (p. 308)
38. e (p. 309)
39. c (pp. 310)
40. a (p. 310)

Matching

41. E (p. 279)
42. F (p. 279)
43. G (p. 300)
44. I (p. 279)
45. C (p. 283)
46. B (p. 285)
47. H (p. 279)
48. A (p. 287)
49. D (p. 279)
50. J (p. 279)

Chapter 11 – Sexual Orientation

CHAPTER SUMMARY

Sexual orientation refers to the gender or genders that a person is attracted to emotionally, physically, sexually, and romantically. Behavior, romantic love, and self-identification are different ways to classify sexual orientation, and they don't always correspond with each other. In an attempt to better understand homosexuality and identify homosexuals, Alfred Kinsey developed a 7-point scale, The Kinsey Continuum, ranging from exclusively heterosexual behavior (0) to exclusively homosexual behavior (6).

Many theories have developed to explain why there are different sexual orientations. Biological theories look at genetics, hormones, birth order, and physiology. Developmental theories include the psychoanalytic views that homosexuality is the result of overly possessive mothers and absent fathers and the idea that gender-role nonconformity leads to homosexuality. Sociological theories are constructionist and try to explain how social forces produce homosexuality in a society, suggesting that the idea of "homosexuality" is a product of a particular culture at a particular historical moment. One interactional theory combines elements of biology and sociology to explain homosexuality.

The attitudes towards same-sex behaviors varied throughout history in different places. In the late 19th century, physicians and scientists began to suggest that homosexuality was not a sin but an illness that would spread like a contagious disease. Same-sex sexual behavior is found in every culture, and its prevalence remains about the same no matter how permissive or repressive that culture's attitude is toward it. Cross-cultural attitudes and behaviors related to same-sex sexual behaviors suggest that trying to pigeonhole people into the restrictive, Western model of sexual orientation seems inadequate

The presumption of heterosexuality is transmitted to young people continually by society, and overall, gay, lesbian, and bisexual youth have been found to experience higher levels of stigmatization and discrimination than heterosexual youth. Disclosure to self, family, friends, and the public (coming out) play an important role in identity development and psychological adjustment.

Gay men, lesbians, and bisexuals face a variety of issues throughout the lifespan. Contrary to the image of gay and lesbian couples having a dominant and a submissive partner, such relationships are actually characterized by greater role flexibility and partner equality than are heterosexual relationships. Increasing numbers of GLBT individuals are choosing to be parents, and all the scientific evidence suggests that children who grow up with gay or lesbian parents do as well emotionally, cognitively, socially, and sexually as do children from heterosexual parents. It is estimated that retirement housing specifically designed for lesbian, gay and bisexual seniors will increase in the coming years.

The term homophobia is used to refer to strongly negative attitudes toward homosexuals and homosexuality, while the term heterosexism refers to the presumption of heterosexuality. Homophobia creates an atmosphere where it is seen as permissible to harass, assault, and even kill homosexuals.

LEARNING OBJECTIVES

After studying Chapter 11, you should be able to:

1. Summarize Kinsey's model of sexual orientation, the Kinsey Continuum.
2. Compare and contrast the Kinsey Continuum with the Klein Sexual Orientation Grid.
3. Discuss the research findings around measuring the incidence of same-sex behavior and individuals who identify as gay, lesbian, bisexual, and transgender.
4. List the major theories used to explore the reasons behind sexual orientation.
5. Compare and contrast the major theories used to explore the reasons behind sexual orientation.
6. Discuss some of the research around same-sex sexual behavior in history.
7. Identify the different values and behavioral patterns around same-sex sexual behavior in other cultures.
8. Compare and contrast Western views of same-sex sexual behavior with other cultures.
9. Discuss the experiences of gay, lesbian, and bisexual youth.
10. Identify benefits and challenges of coming out for lesbians, gay men, and bisexuals.
11. Identify the factors that characterize same-sex relationships as discussed in research findings.
12. Summarize the research findings on children who grow up with lesbian, gay, or bisexual parents.
13. Discuss the challenges faced by gay, lesbian, and bisexual seniors.
14. Identify some of the prominent organizations available for lesbians, gay men, and bisexuals.
15. Define homophobia and heterosexism.
16. List the characteristics of homophobic individuals.
17. Discuss some of the research around hate crimes against gay, lesbian, bisexual, and transgender individuals.
18. Identify some unique challenges faced by bisexual individuals.
19. Define situational homosexuality.
20. Explore the relationship between major religions and homosexuality.

CHAPTER OUTLINE

- **Sexual orientation** refers to the gender or genders that a person is attracted to emotionally, physically, sexually, and romantically.
- **Heterosexuals** are predominantly attracted to members of the other sex (**straight**).
- **Homosexuals** are predominantly attracted to members of the same sex.
- **Bisexuals** are attracted to both women and men.
- **GLBT** refers to people whose identity is gay, lesbian, bisexual, or transgendered.

I. WHAT DETERMINES SEXUAL ORIENTATION?

- Behavior, romantic love, and self-identification are different ways to classify sexual orientation, and they don't always correspond with each other.

A. **Models of Sexual Orientation: Who Is Homosexual?**

1. Alfred Kinsey and his colleagues studied sexual orientation.
   a. Kinsey developed a 7-point scale, The Kinsey Continuum, ranging from exclusively heterosexual behavior (0) to exclusively homosexual behavior (6).
   b. This scale was the first to suggest that people engaged in complex sexual behaviors were not reducible simply to "homosexual" and "heterosexual."
   c. Some criticisms of Kinsey's continuum are that Kinsey didn't take emotions and fantasies into enough consideration and that it is static over time.
2. Klein Sexual Orientation Grid (KSOG)
   a. This model took the Kinsey Continuum further by including seven dimensions: attraction, behavior, fantasy, emotional preference, social preference, self-identification, and lifestyle.
   b. Each dimension is measured for the past, present, and the ideal.

B. **Measuring Sexual Orientation: How Many Are We?**

1. There continues to be controversy about how many gay men, lesbians, and bisexuals there are today.
2. Estimates range from 2-4% to greater than 10% in males, 1-3% in females, and 3% for bisexuals.
3. Surveys indicate that the frequency of homosexual behavior in the United States has remained constant over the years in spite of the changes in the social status of homosexuality.
4. The challenge with measuring these groups is that researchers are measuring different things such as behavior, self-identification, etc.

II. WHY ARE THERE DIFFERENT SEXUAL ORIENTATIONS?

- **Essentialism** suggests that homosexuals are innately different from heterosexuals, a result of either biological or developmental processes.
- **Constructionists** suggest that homosexuality is a social role that has developed differently in different cultures and times, and therefore nothing is innately different between homosexuals and heterosexuals.

A. **Biological Theories: Differences are Innate**

1. Genetics
   a. Beginning with Franz Kallman in 1952, who studied identical twins, a number of researchers have tried to show that there is a genetic component to homosexuality.
   b. Despite some research findings that homosexuality in men may be more common in identical twins, cautions exist due to the problems with separating out environmental factors and genetics.

c. Another genetics researcher, Dean Hamer, found that homosexual males tended to have more homosexual relatives on their mother's side, and he identified a gene to support this theory.

2. Hormones

a. Prenatal Factors

- Some research has suggested that homosexual men had lower levels of androgens than heterosexual men during sexual brain differentiation.
- Early hormone levels were found to influence sexual orientation.
- Other researchers have concluded that the evidence for the effect of prenatal hormones on both male and female homosexuality is weak.

b. Adult Hormone Levels-Overall studies so far do not support the idea of adult hormone involvement in sexual orientation.

3. Birth Order

a. Birth order research attempts to examine the effects of sibling placement.

b. Fraternal birth order could contribute to a homosexual orientation due to placental cells in the uterine endometrium influencing later gestations, and children born later could develop an immune response that influences the expression of key genes during brain development.

4. Physiology

a. Research in the 1990s suggested differences in the hypothalamus between gay men and heterosexual men.

b. Other studies in physiology have looked at other structures such as the inner ear and finger length, finding some differences between lesbians and heterosexual women.

B. **Developmental Theories: Differences Are Learned**

1. Freud and the Psychoanalytic School

a. Freud thought everyone is inherently bisexual, yet he also viewed male heterosexuality as the result of normal maturation.

b. Freud saw homosexuality as partly **autoerotic**: by making love to a body like one's own, one is really making love to a mirror of oneself.

c. Later psychoanalytic theorists, Irving Bieber in particular, refuted some of Freud's views and emphasized that male homosexuality was the result of overly possessive mothers and absent fathers and this **triangulation** drove the boy to the arms of his mother, which inhibited his normal masculine development.

c. Bieber's subjects were all in therapy, and the majority didn't have a family pattern that fit his theories.

d. Evelyn Hooker used psychological tests to combat the psychoanalytic view of homosexuality and to show that homosexuals were as well adjusted as heterosexuals.

e. **Reparative therapy**, or conversion therapy, has included techniques such as aversive conditioning, drug treatment, electroconvulsive shock, brain surgery and hysterectomies. The American Psychological Association officially denounces conversion therapy and states that there is no scientific evidence to support that it can change sexual orientation.

2. Gender-Role Nonconformity

a. **Gender-role nonconformity** studies are based on the observation that boys who exhibit cross-gender traits, that is, who behave in ways more characteristic of girls of that age, are more likely to grow up to be gay, while girls who behave in typical male behaviors are more likely to grow up to be lesbians.

b. As children, gay men on average have been found to be more feminine than straight men, while lesbians have been found to be more masculine. Remember though that these findings are correlational, meaning there is no cause-and-effect relationship.
c. These gender-role nonconformity studies do not tell whether these boys are physiologically or developmentally different or whether society's reaction to their unconventional play encouraged them to develop a particular sexual orientation.
d. Many, if not most, gay men were not effeminate as children, and not all effeminate boys grow up to be gay.

3. Peer Group Interaction
a. Storms suggests that those who develop early begin to become sexually aroused before they have significant contact with the other sex.
b. His theory suggests that these early developers are more likely to focus their emerging erotic feelings on boys, same-sex peers.

C. **Behaviorist Theories**
1. Behavioral theories of homosexuality consider it a learned behavior.
2. The rewarding or pleasant reinforcement of homosexual behaviors or the punishing or negative reinforcement of heterosexual behavior may lead to a homosexual identity.

D. **Sociological Theories: Social Forces at Work**
1. Sociological theories are constructionist and try to explain how social forces produce homosexuality in a society.
2. The idea of "homosexuality" is a product of a particular culture at a particular historical moment.
3. The idea that people are either "heterosexual" or "homosexual" is not a biological fact but simply a way of thinking that evolves as social conditions change.

E. **Interactional Theory: Biology and Sociology**
1. Social psychologist Daryl Bem has proposed that biological variables, such as genetics, hormones, and brain neuroanatomy, do not cause certain sexual orientations, but rather they contribute to childhood temperaments that influence a child's preferences for sex-typical or sex-atypical activities and peers.
2. Bem's "Exotic-becomes-Erotic" theory suggests that sexual feelings evolve from experiencing one gender as more exotic, or different from oneself, than the other sex.

III. HOMOSEXUALITY AND HETEROSEXUALITY IN OTHER TIMES AND PLACES

A. **Homosexuality in History**
1. The Ancient World
a. Before the 19th century, men who engaged in same-sex sexual behaviors were accused of **sodomy** or **buggery**, which were seen as sex crimes and not considered part of a person's fundamental nature.
b. Homosexual activity was common, and homosexual prostitution was taxed by the state.
c. Lesbianism was rarely explicitly against the law in most ancient societies.
d. Contrary to popular belief, homosexuality was not treated with concern or much interest by either early Jews or early Christians.

2. The Middle Ages
a. Homosexual relations were not forbidden even though there were many sexual codes based on Church teachings.
b. By 1300, homosexuality was punishable by death in Europe, influencing the Western world's view of homosexuality for the last 700 years.

3. The Modern Era

a. From the 16th century on, individuals who engaged in same-sex sexual behaviors were subject to periods of tolerance and periods of severe repression.
b. Early American colonists viewed same-sex sexual behavior negatively.
c. Openly homosexual communities appeared now and then.
d. During the 19th and early 20th centuries **passing women** disguised themselves as men, entered the workforce and even married women.
e. In the late 19th century, physicians and scientists began to suggest that homosexuality was not a sin but an illness that would spread like a contagious disease.
f. In Nazi Germany, homosexuals were imprisoned and murdered along with Jews, Gypsies, epileptics, and others.
g. The view of homosexuality as a master status encouraged homosexuals to band together and press for recognition of their civil rights as a minority group.
h. Western, predominantly Christian, societies have often existed without the hostility to homosexuality that characterizes modern America, and Christianity itself has had periods of tolerance.

B. **Homosexuality in Other Cultures**

- Same-sex sexual behavior is found in every culture, and its prevalence remains about the same no matter how permissive or repressive that culture's attitude is toward it.

1. Latin American Countries
a. In Central and South America male gender roles are defined through what makes one a man, or **machismo**, and that is defined by being the penetrator, the active partner in sex.
b. A man is not considered a homosexual if he takes the active, insertive role, even if it is with other men.

2. Arab Cultures
a. It is not uncommon to see men holding hands or walking down the street arm-in-arm on Arab streets, but for the most part, male homosexuality is taboo.
b. Gay men in the Arab world often limit their interactions with other men to sex.

3. Asian Countries
a. In 2001, the Chinese Psychiatric Associate removed homosexuality from its list of mental disorders.
b. Buddhism does not condemn homosexuality so Buddhist countries generally accept it.
c. In Thailand, there are well known politicians who live openly as gay men.

4. Sambia
a. The Sambia tribe of Papua New Guinea has been described in depth by Gilbert Herdt, where a boy's life has historically involved **sequential homosexuality**.
b. Sambians believed that mother's milk must be replaced by man's milk (semen) for a boy to reach puberty and so at the age of 7, boys began a process of oral sex with older men in order to ingest semen.
c. After puberty, the boy switches to the receiving role until he reaches 19, when he engages in primarily heterosexual relations.
d. Although this was a common practice for many years, the practice has faded in Sambia as Western influences have changed the cultural traditions.

5. The lesson of cross-cultural studies of homosexuality suggest that trying to pigeonhole people or ways of life into our restrictive, Western "homosexuality-heterosexuality-bisexuality" model seems inadequate.

IV. GAYS, LESBIANS, AND BISEXUALS THROUGHOUT THE LIFE CYCLE

A. Growing Up Gay, Lesbian, or Bisexual

1. The presumption of heterosexuality is usually transmitted to young people continually by parents, friends, television, movies, newspapers, magazines, and even the government.
2. Because group sports and heterosexual dating are important to male adolescents forming peer group bonds, young gay or bisexual boys can feel unattached and alienated.
3. The pressure and alienation may come slightly later in life for lesbians and bisexual girls because same-sex affection is more accepted for girls and because lesbians tend to determine their sexual orientation later than gay men.
4. Overall, gay, lesbian, and bisexual youth have been found to experience higher levels of stigmatization and discrimination than heterosexual youth, which maybe responsible for the higher levels of mental health issues among GBL youth.

B. Coming Out to Self and Others

1. **Coming out** is the need to establish a personal self-identity and communicate it to others.
    a. Vivienne Cass has proposed a six-stage model of gay and lesbian identity formation:
    - Identity Confusion
    - Identity Comparison
    - Identity Tolerance
    - Identity Acceptance
    - Identity Pride
    - Identity Synthesis
2. Disclosure to self, family, friends, and the public plays an important role in identity development and psychological adjustment.
3. Gay men and lesbians have been coming out at earlier ages in the past few years, with the average age at 16.
4. The National Longitudinal Study of Adolescent Health has found that homosexual and bisexual youth are more likely than heterosexual youth to think about and commit suicide.
5. Lesbian and gay youth who have a positive coming out experience also have been found to have a higher self-concept, lower rates of depression, and better psychological adjustment than those who have a negative experience.
6. Youth who are rejected by their parents have been found to have increased levels of isolation, loneliness, depression, suicide, homelessness, prostitution, and sexually transmitted infections.
7. The *Federation of Parents and Friends of Lesbians and Gays (PFLAG)* helps parents accept a child's sexual orientation along with engaging in advocacy and support for gays, lesbians, and bisexuals.
8. Between 14% and 25% of gay men and about 33% of lesbians marry the other sex at some point, either before they recognize that they are gay or lesbian or because they want to fit into heterosexual society.

C. Life Issues: Partnering, Sexuality, Parenthood, Aging

1. Looking for Partners
    a. A variety of social institutions allow lesbians and gay men to meet each other more easily.
    b. Adults can meet others at **gay bars** or clubs that cater primarily to same-sex couples, gay support or discussion groups, and also through the Internet.
    c. Gay magazines like the *Advocate* carry personal ads, and ads for dating services, travel clubs, resorts, etc.

2. Same-Sex Couples
   a. Contrary to the image of gay and lesbian couples having a dominant and submissive partners, such relationships are actually characterized by greater role flexibility and partner equality and lower levels of jealousy than are heterosexual relationships.
   b. Research suggests that female partners tend to have a better grasp of relationship problems, which may contribute to the higher levels of relationships satisfaction among lesbian couples.
   c. Men have been socialized to be independent and to withdraw from conflict, which may put gay men at a disadvantage in relationships.
   d. There are no states that allow same-sex couples to become legally married, although Vermont has a state law that recognizes "civil unions" of gay and lesbian couples that gives the same protections accorded married couples.
   e. In 2003, the supreme court of the U.S. struck down the Texas anti-sodomy law, which could have far reaching consequences for same-sex couples.
3. Gay and Lesbian Sexuality
   a. Gay men and lesbians tend to see their community as broad, with sexuality as only one component.
   b. Masters and Johnson found in 1979 that arousal and orgasm in homosexuals was physiologically no different than in heterosexuals.
   c. Masters and Johnson found that gay and lesbian couples were less goal oriented than heterosexual couples.
4. Gay and Lesbian Parents
   a. The 2000 U.S. Census Bureau figures revealed that there are 601,209 same-sex families, but the actual numbers are thought to be much higher.
   b. No significant differences have been found between the offspring of lesbian and straight mothers, including the sexual orientation of their children.
   c. All the scientific evidence suggests that children who grow up with one or two gay and/or lesbian parents do as well emotionally, cognitively, socially, and sexually as do children from heterosexual parents.
   d. Three states (Florida, Mississippi and Utah) specifically bar homosexuals from adopting children, and several states making it very difficult for homosexuals to adopt children.
   e. Since same-sex marriages are not yet legally recognized in the United States, gay couples may have trouble gaining joint custody of children, and the nonbiological parent may not be granted parental leavc and may not be able to get benefits for the child through their workplace.
5. Gay and Lesbian Seniors
   a. The number of gay, lesbian, and bisexual seniors will climb to over 4 million by the year 2030.
   b. Important issues for GLBT seniors include survivor benefits, lack of health insurance, social security issues, and assisted living needs.
   c. Studies have found that 52% of nursing home staff reported intolerant or condemning attitudes toward homosexual and bisexual residents.
   d. It is estimated that retirement housing specifically designed for lesbian, gay and bisexual seniors will increase in the coming years.
   e. An advocacy organization called SAGE, or Senior Action in a Gay Environment, has been helping aging gay, lesbian, and bisexual individuals and couples.
6. Gay, Lesbian, and Bisexual—Specific Problems

a. Gay, lesbian, and bisexual youth and adults have been found to have higher rates of substance abuse, alcohol-related problems, and more widespread use of marijuana and cocaine that heterosexual youth and adults.
b. Gay men have been found to earn 11% to 27% less than straight men with the same qualifications, while lesbians earn about the same as straight women.

D. **Gay, Lesbian, and Bisexual Organizational Life**

1. Because many organizations misunderstand the needs of homosexuals and bisexuals, gay and lesbian social services, medical, political, entertainment, and even religious organizations have been formed.
2. The National Gay and Lesbian Task Force (NGLTF) and its associated Policy Institute advocate for gay civil rights, lobbying Congress for such things as health care reform, AIDS policy reform, and hate-crime laws.
3. The Lambda Legal Defense and Education Fund pursues test-case litigation of concern to the gay and lesbian community.
4. New York City has the Harvey Milk School, the first and largest accredited public school in the world devoted to the educational needs of GBLT and questioning youth.
5. Many high schools today have Gay-Straight Alliances (GSAs) to help encourage tolerance, and the *Gay, Lesbian, and Straight Education Network (GLSEN)* is a national organization that works to reduce heterosexism and homophobia in all K-12 schools.
6. Most major cities have their own lesbian and gay newspaper.

V. HOMOPHOBIA AND HETEROSEXISM

A. **What is Homophobia?**

1. Homophobia is used to refer to strongly negative attitudes toward homosexuals and homosexuality.
2. People who hold negative views are less likely to have had contact with homosexuals and bisexuals, are likely to be older and less well education, are more likely to be religious and to subscribe to a conservative religious ideology, have more traditional attitudes toward sex roles and less support for equality of the sexes, are less permissive sexually, and are more likely to be authoritarian.
3. Homosexuals may also experience internalized homophobia when they harbor negative feelings about homosexuality, leading to decreased levels of self-esteem and increased levels of shame and psychological distress.
4. **Heterosexism** refers to the presumption of heterosexuality.
   a. Even those with no negative feelings towards homosexuality are often unaware that businesses will not provide health care and other benefits to the partners of homosexuals.
   b. Heterosexism can be a passive lack of awareness rather than active discrimination.

B. **Hate Crimes Against Lesbian, Gay, and Bisexual Persons**

1. Homophobia creates an atmosphere where it is seen as permissible to harass, assault, and even kill homosexuals.
2. Hate crimes are those motivated by hatred of someone's religion, sex, race, sexual orientation, disability, or ethnic group.
3. Homosexuals are victimized four times more often than the average American, with estimates that 80% of GLB youth being verbally abused.
4. A large number of hate crimes go unreported.

C. **Why Are People Homophobic?**

1. One theory as to why people are homophobic is that it may be a function of personality type since rigid, authoritarian personalities are more likely to be homophobic.

2. Another common suggestion is that straight people fear their own suppressed homosexual desires or are insecure in their own masculinity and femininity.
3. Perhaps people are simply ignorant about homosexuality and would change their attitudes with education.
4. People also tend to confuse sexual orientation with gender identity and may react negatively when they see males violating gender roles.

D. **How Can We Combat Homophobia and Heterosexism?**

- Ethnocentrism refers to the belief that all standards of correct behavior are determined by one's own cultural background, leading to racism, ethnic bigotry, and even sexism and heterosexism.

1. Laws
   a. As of 2000, 22 states and the District of Columbia punish perpetrators of hate crimes motivated by sexual orientation, although they way they are punished varies.
   b. Monitoring or recording hate crimes does not necessarily mean putting any resources into improving enforcement or prevention.
2. The Media
   a. Portrayals of GLBT individuals has become more positive on television and in movies.
   b. Now mainstream bookstores and movie theaters are carrying books and movies with lesbian and gay themes.
3. Education
   a. Discussion of educating about homosexuality in schools can encounter strong opposition by certain parent groups.
   b. Teaching about homosexuality can include discussion of famous authors and artists (Walt Whitman or Leonardo Da Vinci) and their lives as GLBT individuals.

VI. DIFFERENCES AMONG HOMOSEXUAL GROUPS

A. **Lesbians: Sexism Plus Homophobia**

1. Research focusing on lesbian life and the lesbian community in particular lags far behind research on gay men.
2. Scholarship by lesbians tends to be strongly political, in part because lesbians deal with both sexism and heterosexism.
3. Lesbian and bisexual women have lower rates of preventative care than heterosexual women.

B. **Bisexuals: Just a Trendy Myth?**

1. People who identify as bisexual often first identified as heterosexuals, and their self-labeling generally occurs later in life than either gay or lesbian self-labeling.
2. There was an absence of research on bisexuality due to the fact that researchers believed that sexuality was composed of only two opposing forms of sexuality: homosexuality and heterosexuality.
3. Even some sexuality scholars have claimed that bisexuality is a myth, or an attempt to deny one's homosexuality.
4. Bisexuals have begun to speak of **biphobia**, which exists in both the heterosexual and gay and lesbian communities.
3. Bisexuality can be seen as simply a lack of prejudice and full acceptance of both sexes.
4. In **sequential bisexuality**, a person has sex exclusively with one gender followed by sex exclusively with the other.
5. **Contemporaneous bisexuality** refers to having sexual partners of both sexes during the same time period.

6. It is difficult to determine what percent of people are bisexual since many people who engage in bisexual behavior do not self-identify as bisexual.

C. **Minority Homosexuals: Culture Shock?**

1. Homosexuality is not accepted by many ethnic groups, and the gay and lesbian community does not easily accommodate expressions of ethnic identity.
2. Minority GLBT youth have been found to experience greater psychological distress than non-minority GLBT youth.
3. Gay African-Americans can find their situation particularly troubling, as they often have to deal with the heterosexism of the African-American community and the racism of the homosexual and straight communities.
4. Although many African-American lesbians report positive relationships and pleasant feelings about their sexual relationships, more than half also report feeling guilty about these relationships.

D. **Same-Sex Behavior in Prison**

1. Same-sex sexual contact between inmates, although prohibited, still occurs in prisons today; however, research has found that the majority of this sexual activity is consensual.
2. **Situational homosexuality** refers to the practice of engaging in same-sex sexual behaviors with the plan to return to heterosexual relationships exclusively once they are released from jail or leave the confines of a ship.
3. Inmates speak of loving their inmate partners, and relations can become extremely intimate, even among those who return to a heterosexual life upon release.

VII. HOMOSEXUALITY IN RELIGION AND THE LAW

A. **Homosexuality and Religions**

1. Religion has generally been considered a bastion of antihomosexual teachings and beliefs, while only traditional Judaism and Christianity have strongly opposed homosexual behavior.
2. Christian religions that are more on the liberal side include the United Church of Christ and the Unitarian Universalist Association.
3. Mainline Christian religions, such as Presbyterians, Methodists, Lutherans, and Episcopalians, have more conflict over the issue of sexual orientation, resulting in both liberal and conservative views.
4. While Orthodox Jews believe that homosexuality is an abomination forbidden by the Torah, conservative Jews are more likely to welcome all sexual orientations, and reform Jews tend to be the most accepting towards gays, lesbians, and bisexual members.
5. Recently religious scholars have begun to promote arguments based on religious law and even scripture that argue for a more liberal attitude toward homosexuality.
6. Buddhism encourages relationships that are mutually loving and supportive.

B. **Homosexuality and the Law**

1. Throughout history, laws have existed in the Western world that prohibited homosexual behavior.
2. All 50 states outlawed homosexual acts until 1961.
3. Even in long-term, committed, same-sex couples, partners are routinely denied the worker's compensation and health care benefits normally extended to a spouse or dependants.
4. Gay and lesbian couples are denied tax breaks, Social Security benefits, and rights of inheritance, all of which are available to married heterosexual couples.
3. Why Do Laws Discriminate Against Homosexuals?
    a. Some members of the legal community continue to hold predominantly negative views of gays, lesbians and bisexuals.

b. The efforts of local, grassroots gay organizations, as well as the national efforts of groups like the Lambda Legal Defense and Education Fund, are working to fight discrimination and victimization in the United States.

VII. BEYOND OUR ASSUMPTIONS: A FINAL COMMENT

TEST YOURSELF

Below you will find fill-in-the-blank and short answer essay questions for topics covered in chapter 11. Check your answers at the end of this chapter.

Sexual Orientation and Other Times and Places

1. The acronym ________________ refers to people whose identity is gay, lesbian, bisexual, or transgendered.

2. The ________________ continuum was the first scale to suggest that people engaged in complex sexual behaviors that were not reducible simply to "homosexual" and "heterosexual".

3. Some researchers suggest that people's emotions and fantasies are the most important determinants of ________________ ________________.

4. The ________________ ________________ ________________ Grid tries to take the Kinsey Scale further by including seven dimensions.

5. ________________ suggests that lesbian and gay people are innately different from heterosexuals, a result of either biological or developmental processes.

6. ________________ suggest that homosexuality is a social role that has developed differently in different cultures and times, and therefore nothing is innately different between homosexuals and heterosexuals.

7. Most theories on sexual orientation ignore ________________ or do not offer enough research to explain why ________________ exists.

8. Biological theories are ________________; that is, they claim that differing sexual orientations are due to differences in physiology.

9. Many studies have compared blood ________________ levels in adult gay men with those in adult male heterosexuals, and most have found no significant differences.

10. ________________ theories focus on a person's upbringing and personal history to find the origins of homosexuality.

11. The American Psychological Association has said that there is no evidence to support that therapy can ________________ sexual orientation.

12. ________________ ________________ studies are based on the observation that boys who exhibit cross-gender traits, that is, who behave in ways more characteristic of girls of that age, are more likely to grow up to be gay, while girls who behave in typical male behaviors ore more likely to grow up to be lesbians.

13. ________________ theories suggest that concepts like homosexuality, bisexuality, and heterosexuality are products of our social imagination and are dependent upon how we as a society decide to define things.

14. Before the nineteenth century, men who engaged in homosexual acts were accused of ________________ or buggery, which were simply seen as sex crimes and not considered part of a person's fundamental nature.

15. In Central and South America, people do not tend to think in terms of homosexuality and heterosexuality but ________________ and ________________.

16. What are some ways that we categorize a person's sexual orientation?

17. What is the Kinsey Scale?

18. What are the seven dimensions of the Klein Sexual Orientation Grid?

19. What are the two basic types of sexual orientation causation theories?

20. Describe Evelyn Hooker's famous research on sexual orientation.

Gays, Lesbians and Bisexuals Throughout the Life Cycle

21. ________________ ________________ refers, first of all, to acknowledging one's sexual identity to oneself.

22. First awareness of sexual orientation typically occurs around the age of ________________, but most youth do not tell anyone until about the age of ________________.

23. Lesbian and gay youth who have a ________________ coming out experience also have been found to have a higher self-concept, lower rates of depression, and better psychological adjustment than those who have a negative experience.

24. Contrary to the image of gay and lesbian couples having a ________________ and a ________________ partner, such relationships are actually characterized by greater role flexibility and partner equality than are heterosexual relationships.

25. While gay and lesbian marriages are legal in ________________, ________________, and ________________, there are no states in the United States that allow same-sex couples to become legally married.

26. In July of 2000, Vermont's law established ________________ ________________, which provided gay and lesbian couples with the same protections, benefits, and responsibilities accorded married couples under state law.

27. ________________ may become the first state in the nation to give full marriage rights to lesbian and gay couples.

28. No significant differences have been found between the offspring of lesbian and straight mothers, including the ________________ ________________ of their children.

29. Three states, ________________, ________________, and ________________, specifically bar lesbians and gays from adopting children.

30. Gay and lesbian couples today are creating new kinds of ________________, and the social system is going to have to learn how to deal with them.

31. Many studies have found that having ________________ ________________ prior to the senior years often helps a gay or lesbian senior to feel more comfortable with their life and their sexuality.

32. In 2000, the ________________ ________________ ________________ ________________ ________________ ________________ released the first comprehensive report to address public policy regarding issues that confront aging gay and lesbian seniors.

33. An advocacy organization called SAGE, or ________________ ________________ ________________ ________________ ________________ ________________, has been helping aging gay, lesbian, and bisexual individuals and couples.

34. The ________________ ________________ ________________ ________________ ________________ ________________ and its associated Policy Institute advocate for lesbian and gay civil rights, lobbying Congress for such things as a Federal Gay and Lesbian Civil Rights Act, health care reform, AIDS policy reform, and hate-crime laws.

35. The ________________ ________________ ________________ ________________ ________________ ________________ pursues test-case litigation of concern to the lesbian and gay community.

36. What is coming out?

37. What is the name of the organization that helps parents learn to accept their children's sexual orientation?

38. What three states specifically bar lesbians and gays from adopting children?

39. What are some problems that gay and lesbian parents encounter that heterosexual parents do not face?

40. What are some issues confronting aging gay and lesbian seniors?

Homophobia and Heterosexism

41. Research has shown that lesbian and gay people who have _________________ _________________ also have decreased levels of self-esteem and increased levels of shame and psychological distress.

42. Even bigger than the problem of homophobia for most gay men and lesbians is _________________.

43. Hate crimes against lesbian and gay people occur _________________ times more often than crimes against the average American.

44. People tend to confuse sexual orientation with _________________ _________________.

45. As of 2000, 22 states and the District of Columbia punish perpetrators of _________________ _________________ motivated by sexual orientation.

46. In 1998 the _________________ _________________ _________________ _________________ _________________ _________________ was passed, which requires college campuses to report all hate crimes.

47. An important step to stopping heterosexism is _________________.

48. Scholarship by lesbians tends to be strongly political, in part because lesbians have to deal with both _________________ and _________________.

49. Recently, _________________ have begun their own "coming out", declaring that their sexual identity is different from both lesbian and gay people and heterosexuals.

50. _________________ bisexuality refers to having sexual partners of both sexes during the same time period.

51. _________________ homosexuality is found in places where men must spend long periods of time together, such as on naval ships.

52. _________________ has generally been considered a bastion an anti-lesbian and gay teachings and beliefs.

53. In the United States, _________________ has been illegal since colonial days, and it was punishable by death until the late eighteenth century.

54. In mid-2003 the Supreme Court overturned the _________________ anti-sodomy law that made consensual sex between same-sex couples illegal.

55. Theories of sexual orientation change as _________________ changes.

56. What are some things that people who hold negative views of lesbian and gay people are likely to have in common?

57. What motivates people to be homophobic?

58. What is the difference between sexual orientation and gender identity?

59. Why has there been an absence of research on bisexuality?

60. What is sequential bisexuality?

POST TEST

Below you will find true/false, multiple-choice and matching quiz items covering the entire chapter. Check your answers at the end of this chapter.

True/False

1. Reparative therapy, or conversion therapy, to change sexual orientation, has included techniques such as aversive conditioning, drug treatment, electroconvulsive shock and brain surgery.

2. Engaging in gender-role nonconformity in childhood will usually cause a boy to grow up to be gay.

3. Buddhism does not condemn homosexuality so Buddhist countries generally accept it.

4. Youth who are rejected by their parents have been found to have increased levels of isolation, loneliness, depression, suicide, homelessness, prostitution, and sexually transmitted infections.

5. Lesbian and gay relationships are characterized by greater role flexibility and partner equality than are heterosexual relationships.

6. Gay men consistently earn more money than straight men with the same qualifications.

7. The Gay, Lesbian, and Straight Education Network is a national organization that works to reduce heterosexism and homophobia in all K-12 schools.

8. Some sexuality scholars have claimed that bisexuality is a myth, or an attempt to deny one's homosexuality.

9. When it comes to same-sex sexual relationships in prison, inmates speak of loving their inmate partners and engaging in consensual sexual behaviors.

10. All Christian religions believe that homosexuality is an abomination.

Multiple-Choice

11. Who developed the 7-point continuum, ranging from exclusively heterosexual behavior (0) to exclusively homosexual behavior (6)?
    a. Franz Kallman
    b. Evelyn Hooker
    c. Irving Bieber
    d. Vivienne Cass
    e. Alfred Kinsey

12. When it comes to determining the numbers of individuals with same-sex behaviors or identification, who consistently is the highest?
    a. lesbians and bisexuals
    b. lesbians
    c. gay men
    d. bisexuals
    e. All three groups are the same.

13. What group of theories suggest that homosexuality is a social role that has developed differently in different cultures and times, and therefore nothing is innately different between homosexuals and heterosexuals?
    a. continuist
    b. constructionist
    c. homosexist
    d. reparativist
    e. essentialist

14. What is a major way researchers have studied the genetics of sexual orientation?
    a. measuring blood androgen levels in gay and heterosexual men
    b. examining identical twins
    c. asking adult gay men and lesbians about their play choices as children
    d. comparing the finger lengths of lesbians and heterosexual women
    e. All of the above

15. Assessing the birth order of gay men falls into what theoretical perspective?
    a. Biological Theories
    b. Constructionist Theories
    c. Sociological Theories
    d. Interactional Theories
    e. Developmental Theories

16. Who used psychological tests to combat the psychoanalytic view of homosexuality to show that homosexuals were as well adjusted as heterosexuals?
    a. Franz Kallman
    b. Evelyn Hooker
    c. Irving Bieber
    d. Vivienne Cass
    e. Alfred Kinsey

17. How might a constructionist view gender-role nonconformity studies?
    a. These boys have a physiological difference that leads them to identify with feminine things.
    b. Genetics may play a role in leading boys to be more feminine and to grow up to be gay.
    c. Society's reaction to boys' cross-gender play lead them to identify with being gay or bisexual.
    d. All of the above
    e. None of the above

18. What is the name of Daryl Bem's interactional theory of sexual orientation development?
    a. Sissy-Boy Syndrome
    b. Bio-Social Continuum
    c. Peer Role Relations
    d. Exotic-becomes-Erotic
    e. Genetics-as-Society

19. What cultures tends to not consider a man a homosexual if he takes the active, insertive role in anal intercourse with other men.?
    a. Arab societies
    b. African countries
    c. Asian communities
    d. Latin American countries
    e. Eastern European communities

20. What do cross-cultural studies of homosexuality suggest about homosexuality?
    a. Engaging in any same-sex sexual behavior is viewed as taboo or criminal in all non-Western cultures.
    b. Homosexuality as a distinct category is recognized all over the world in all countries.
    c. Trying to pigeonhole people or ways of life into restrictive, Western "homosexuality-heterosexuality-bisexuality" model is inadequate.
    d. There are many cultures where same-sex sexual behavior does not occur.
    e. All of the above

21. The presumption of heterosexuality is usually transmitted to young people continually by ___________________?
    a. television
    b. parents
    c. government
    d. magazines
    e. All of the above

22. What is "coming out?"
    a. the practice of fellatio carried out in Latin American Countries
    b. the practice of winking young gay men use to identify each other
    c. the need to establish a personal self-identity and communicate it to others
    d. the formation of body image experienced by young gay men, lesbians, and bisexuals
    e. None of the above

23. Who proposed a popular, six-stage model of gay and lesbian identity formation?
    a. Franz Kallman
    b. Evelyn Hooker
    c. Irving Bieber
    d. Vivienne Cass
    e. Alfred Kinsey

24. What has been linked to a positive coming out experience for lesbian and gay youth?
    a. better psychological adjustment
    b. a higher self-concept
    c. lower rates of depression
    d. All of the above
    e. None of the above

25. What organization is specifically designed to help parents accept a child's sexual orientation and also engages in advocacy and support for gays, lesbians, and bisexuals?
    a. PFLAG
    b. GSA
    c. GLAAD
    d. LAMBDA
    e. GLSEN

26. Which statement below regarding lesbian and gay relationships is true?
    a. Lesbian couples report higher levels of relationship satisfaction.
    b. Most gay and lesbian couples have higher levels of jealousy than heterosexual couples.
    c. Most same-sex couples have relationships where partners take clear dominant and submissive roles
    d. All of the above
    e. None of the above

27. In 2003, the supreme court of the U.S. struck an anti-sodomy law in which state?
    a. Florida
    b. Wisconsin
    c. Texas
    d. Hawaii
    e. Vermont

28. Which statement below regarding gay and lesbian parenting is TRUE?
    a. Compared to straight fathers, gay fathers have more permissive disciplinary guidelines.
    b. Children who grow up with gay or lesbian parents are more likely to identify as homosexual adults.
    c. Children of same-sex parents have been found to aspire to occupations more typical for their gender than do children of heterosexuals.
    d. Gay fathers are less involved in their children's activities when compared to heterosexual fathers.
    e. Children who grow up with gay or lesbian parents do as well emotionally, cognitively, socially and sexually as do children from heterosexual parents.

29. What percent of nursing home staff reported intolerant or condemning attitudes toward homosexual and bisexual residents?
    a. 14%
    b. 31%
    c. 52%
    d. 74%
    e. 96%

30. What does the National Gay and Lesbian Task Force (NGLTF) lobby Congress for?
    a. hate-crimes legislation
    b. health care reform
    c. lesbian and gay immigration
    d. AIDS policy reform
    e. All of the above

31. What is the term for strongly negative attitudes toward homosexuals and homosexuality?
    a. homophobia
    b. homoism
    c. gaytred
    d. heterosexism
    e. gayrevulsion

32. What is NOT a characteristic of people who hold negative views of lesbians, gay men and bisexuals?
    a. more likely to be authoritarian.
    b. less likely to have had contact with homosexuals and bisexuals
    c. more support for equality of the sexes
    d. more likely to be religious
    e. more likely to be older

33. What is the term for the presumption of heterosexuality?
    a. homophobia
    b. homoism
    c. gaytred
    d. heterosexism
    e. heterosumption

34. What is the term used for crimes perpetrated because of someone's sexual orientation, race, or religion?
    a. ethno-violence
    b. hate crimes
    c. cultural bias crimes
    d. identity abuse
    e. individuality violence

35. What the term for having sex exclusively with one gender followed by sex exclusively with the other?
    a. serial bisexuality
    b. sequential bisexuality
    c. contemporaneous bisexuality
    d. consecutive bisexuality
    e. situational bisexuality

36. What is the term for having sexual partners of both sexes during the same time period?
    a. serial bisexuality
    b. sequential bisexuality
    c. contemporaneous bisexuality
    d. consecutive bisexuality
    e. situational bisexuality

37. What is situational homosexuality
    a. the practice of engaging in same-sex sexual behaviors when one is in specific situations such as at bars or in particular neighborhoods
    b. engaging in same-sex sexual behaviors only while intoxicated
    c. engaging in same-sex sexual behaviors only while associating with homosexuals, while preferring heterosexual encounters
    d. engaging in same-sex sexual behaviors while excluded from heterosexual encounters with the plan to return to heterosexual relationships exclusively
    e. None of the above

38. Regarding the issue of religion and homosexuality, what is one of the most accepting churches that promotes itself as having a primary, affirming ministry to gays, lesbians, bisexuals and transgendered persons?
    a. Orthodox Judaism
    b. Metropolitan Community Church
    c. Catholic
    d. Southern Baptists
    e. None of the above

39. Which of the following statements below regarding Judaism and perspectives on homosexuality is TRUE?
    a. Orthodox Jews believe that homosexuality is an abomination forbidden by the Torah
    b. Conservative Jews are more likely to welcome all sexual orientations
    c. Reform Jews tend to be the most accepting towards gays, lesbians, and bisexual members
    d. All of the above
    e. None of the above

40. How many states outlawed homosexual acts until 1961?
    a. 3
    b. 15
    c. 31
    d. 50
    e. They've never been outlawed in any state.

Matching

Column 1		Column 2
A. Developmental Theories	____	41. constructionist theories that try to explain how social forces produce homosexuality in a society
B. Klein Sexual Orientation Grid	____	42. This scale was the first to suggest that people engaged in complex sexual behaviors, which were not reducible simply to "homosexual" and "heterosexual."
C. Biological Theories	____	43. emphasized that male homosexuality was the result of overly possessive mothers and absent fathers
D. Coming out	____	44. the presumption of heterosexuality and the social power used to promote it
E. Evelyn Hooker	____	45. biological variables contribute to childhood temperaments that influence a child's preferences for sex-typical or sex-atypical activities and peers
F. Kinsey Continuum	____	46. This model included seven dimensions of sexual orientation and measured them for the past, present and the ideal
G. Interactional Theories	____	47. used psychological tests to combat the psychoanalytic view of homosexuality to show that homosexuals were as well adjusted as heterosexuals
H. Sociological Theories	____	48. the need to establish a personal self-identity and communicate it to others
I. heterosexism	____	49. suggests differences are innate, a result of factors such as genetics or hormones
J. Irving Bieber	____	50. suggests differences are learned such as psychoanalytic or gender-role theories

Test Yourself Answer Key

Sexual Orientation and Other Times and Places

1. GLBT (p. 314)
2. Kinsey (p. 316)
3. sexual orientation (p. 316)
4. Klein Sexual Orientation (p. 316)
5. Essentialism (p. 318)
6. Constructionists (p. 318)
7. bisexuality, bisexuality (p. 318)
8. essentialist (p. 319)
9. androgen (p. 320)
10. Developmental (p. 321)
11. change (p. 322)
12. Gender-role nonconformity (p. 322)
13. Sociological (p. 323)
14. sodomy (p. 324)
15. masculinity, femininity (p. 327)
16. through their behavior, through their sexual fantasies, based on who they love (p. 314)
17. It is a seven-point scale used to describe sexual orientation, ranging from exclusively heterosexual behavior (0) to exclusively homosexual behavior (6). (p. 316)
18. attraction, behavior, fantasy, emotional preference, social preference, self-identification, and lifestyle (p. 316)
19. essentialist and constructionist (p. 318)
20. Hooker used psychological tests, personal histories, and psychological evaluations to show that homosexuals were as well adjusted as heterosexuals and that no real evidence existed that homosexuality was psychopathology. (pp. 321-322)

Gays, Lesbians and Bisexuals Throughout the Life Cycle

21. Coming out (p. 329)
22. ten, sixteen (p. 329)
23. positive (p. 331)
24. dominant, submissive (p. 332)
25. Netherlands, Belgium, and Canada (p. 332)
26. civil unions (p. 332)
27. Massachusetts (p. 332)
28. sexual orientation (p. 333)
29. Florida, Mississippi, Utah (p. 333)
30. families (p. 333)
31. come out (p. 333)
32. National Gay and Lesbian Task Force (p. 333)
33. Senior Action in a Gay Environment (p. 334)
34. National Gay and Lesbian Task Force or NGLTF (p. 335)
35. Lambda Legal Defense and Education Fund (p. 335)
36. It is the need to establish a personal self-identity and communicate it to others (p. 329)
37. The Federation of Parents and Friends of Lesbians and Gays (PFLAG) (p. 331)
38. Florida, Mississippi, and Utah. (p. 333)
39. trouble gaining joint custody of a child, the nonbiological parent may not be granted parental leave, and may not be able to get benefits for the child through their workplace (p. 333)
40. survivor benefits, lack of health insurance, social security issues, assisted living needs (pp. 333-334)

Homophobia and Heterosexism

41. internalized homophobia (p. 336)
42. heterosexism (p. 336)
43. four (p. 338)
44. gender identity (p. 338)
45. hate crimes (p. 339)
46. *Hate Crimes Right to Know Act* (p. 339)
47. education (p. 339)
48. sexism, heterosexism (p. 340)
49. bisexuals (p. 341)
50. Contemporaneous (p. 341)
51. Situational (p. 344)
52. Religion (p. 344)
53. sodomy (p. 345)
54. Texas (p. 345)
55. society (p. 346)
56. They are likely to be older and less well educated, are more likely to be religious and to subscribe to a conservative religious ideology, have more traditional attitudes toward sex roles and less support for equality of the sexes, are less permissive sexually, and are more likely to be authoritarian. (p. 336)
57. Rigid, authoritarian personalities, fear of their own suppressed same-sex desires, insecurity in their own masculinity or femininity, and ignorance about lesbian and gay people (p. 338)
58. Sexual orientation refers to who your sexual partners are; gender identity has to do with definitions of masculinity and femininity. (p. 338)
59. This absence was due to the fact that researchers believed that sexuality was composed of only two opposing forms of sexuality: heterosexuality and homosexuality. (pp. 340-341)
60. In sequential bisexuality, the person has sex exclusively with one gender followed be sex exclusively with the other. (p. 341)

Post Test Answer Key

True/False

1. T (p. 322)
2. F (p. 322)
3. T (p. 327)
4. T (p. 331)
5. T (p. 332)
6. F (p. 335)
7. T (p. 335)
8. T (p. 341)
9. T (p. 344)
10. F (p. 344)

Multiple Choice

11. e (p. 316)
12. c (p. 317)
13. b (p. 318)
14. b (p. 319)
15. a (p. 320)
16. b (pp. 321-322)
17. c (p. 323)
18. d (p. 324)
19. d (p. 327)
20. c (p. 328)
21. e (p. 328)
22. c (p. 329)
23. d (p. 330)
24. d (p. 331)
25. a (p. 331)
26. a (p. 332)
27. c (p. 332)
28. e (p. 333)
29. c (pp. 333-334)
30. e (p. 335)
31. a (p. 336)
32. c (p. 336)
33. d (p. 336)
34. b (p. 338)
35. b (p. 341)
36. c (p. 341)
37. d (p. 344)
38. b (p. 344)
39. d (p. 345)
40. d (p. 345)

Matching

41. H (p. 323)
42. F (p. 316)
43. J (p. 321)
44. I (p. 336)
45. G (p. 323)
46. B (p. 316)
47. E (p. 321)
48. D (p. 329)
49. C (p. 319)
50. A (p. 321)

Chapter 12 – Pregnancy and Birth

CHAPTER SUMMARY

Conception is an incredible process, and human bodies are biologically programmed to help pregnancy occur in many different ways. Because the ovum can live for up to 17 hours, and the majority of sperm can live up to 34 hours in the female reproductive tract, pregnancy may occur if intercourse takes place either a few days before or after ovulation. The fertilized ovum is referred to as a zygote that evolves into a blastocyst and eventually into an embryo. Physical signs of pregnancy include missing a period, breast tenderness, frequent urination, and morning sickness. Pregnancy tests measure for a hormone in the blood and urine called human chorionic gonadotropin (hCG).

Infertility is defined as the inability to conceive (or impregnate) after one year of regular sexual intercourse without the use of any form of birth control and may result from a multitude of factors related to the female, male, or both. There are many options available for infertile individuals or couples, but many of the options are time consuming, expensive and do not guarantee success. In vitro fertilization, gamete intra-fallopian tube transfer (GIFT), and zygote intra-fallopian tube transfer (ZIFT) are some of the more popular procedures used in conjunction with fertility drugs that report varying levels of success.

Pregnancy can vary from 38 to 40 weeks in length, and it is divided into 3 three-month periods called trimesters. Throughout the three trimesters, numerous physiological and psychological changes take place as the fetus and mother develop. Fathers have begun playing a larger role in pregnancy and childbirth, and they often undergo many psychological and even physical changes along with their female partners. Pregnant women need to monitor their exercise and nutrition levels to maintain a healthy pregnancy. Light exercise has been found to result in a greater sense of well being, shorter labor, and fewer obstetric problems. It is recommend that pregnant women avoid caffeine, nicotine, alcohol, marijuana, and other drugs. Over the last few years it has become common for women to delay having children until they are 35 or older. Sexual intercourse during pregnancy is safe for most mothers and the developing child up until the last several weeks of pregnancy. There are some problems that can occur with pregnancies such as ectopic pregnancies, spontaneous abortions, chromosomal abnormalities, Rh incompatibility, and toxemia.

Childbirth is a busy time as the mother's body undergoes physical preparations and the woman or a couple prepares for the psychological and social changes. A variety of decisions must be made such as where to give birth, birthing positions, and drugs during childbirth. There are three stages of childbirth that begin with cervical effacement and dilation, then expulsion of the fetus, and lastly, expulsion of the placenta. Problems during childbirth include prematurity, breech births, stillbirths, and cesarean sections. After childbirth, a woman or a couple must undergo the postpartum period, which involves both physical and psychological changes. Postpartum depression, sexual activity, and breast-feeding are all postpartum issues that must be traversed during this busy time.

LEARNING OBJECTIVES

After studying Chapter 12, you should be able to:

1. Describe the process of conception.
2. List the early signs of pregnancy.
3. Explain how pregnancy tests work.
4. Discuss some of the historical and contemporary techniques used for sex selection.
5. Define infertility
6. List the different infertility options.
7. Compare and contrast the different infertility options.
8. Describe the physiological changes that take place during pregnancy for the fetus and the pregnant woman.
9. Explain what kind of exercise and nutrition is recommended for pregnant women.
10. Identify the risks associated with alcohol and nicotine ingested during pregnancy.
11. Identify some of the specific situations related to pregnancy over the age of 30.
12. Discuss the role of engaging in sexual behavior during pregnancy.
13. List and define some of the major problems that can occur during pregnancy.
14. Describe the social, psychological, and physiological changes that occur as a woman or a couple prepares for childbirth.
15. Discuss the pros and cons around birthplace choices and birthing positions.
16. Identify what happens during the three stages of childbirth.
17. Identify some of the problems that can occur during childbirth.
18. List some of the postpartum experiences that can affect women and/or couples.
19. Discuss the recommendations for postpartum sexual activity.
20. Explain the importance of breast-feeding.

CHAPTER OUTLINE

I. FERTILITY

A. **Conception: The Incredible Journey**

1. During ovulation, a **mucus plug** in the cervix disappears, making it easier for sperm to enter the uterus and the cervical mucus changes in consistency (becoming thinner and stretchy), making it easier for sperm to move through the cervix.
2. After ejaculation, semen thickens to help it stay in the vagina, but it thins again after it moves into the uterus.
3. Thirty percent of the time a pregnancy results when a fertile woman engages in sexual intercourse, although a significant number of these pregnancies end in **spontaneous abortion**.
4. Because the ovum can live for up to 17 hours, and the majority of sperm can live up to 34 hours in the female reproductive tract, pregnancy may occur if intercourse takes place either a few days before or after ovulation.
5. The sperm secretes a chemical that allows the sperm to enter the ovum for fertilization, and the outer layer of the ovum undergoes a physical change, making it impossible for any other sperm to enter. The entire fertilization process takes about 24 hours and usually occurs in the ampulla section of the Fallopian tube.
6. The fertilized ovum is referred to as a **zygote**.
7. About 12 hours after the genetic material from the sperm and ovum join together, the first cell division begins and it is called a **blastocyst**, which will divide in two every 12-15 hours, doubling in size.
8. About 3-4 days after conception, the blastocyst enters the uterus, and on about the 6th day after fertilization, the uterus secretes a chemical that dissolves the hard covering around the blastocyst, allowing it to implant in the uterine wall.
9. The blastocyst divides into three layers that will develop into all the bodily tissues.
10. From the 2nd through the 8th weeks, the developing human is referred to as an **embryo**, and soon a membrane called the **amnion** begins to grown over the developing embryo and the amniotic cavity begins to fill with amniotic fluid.
11. The **placenta**, which is the portion that is attached to the uterine wall and connects to the fetus by the **umbilical cord**, supplies nutrients to the developing fetus, aids in respiratory and excretory functions, and secretes hormones necessary for the continuation of the pregnancy.
12. **Fraternal twins** result from two ova and are **dizygotic**, and they can be either of the same or different sex.
13. **Identical twins** occur when a single zygote completely divides into two separate zygotes and are also called **monozygotic** twins.

B. **Early Signs of Pregnancy**

1. Physical signs include missing a period, breast tenderness, frequent urination, and morning sickness.
2. Fifty to eighty percent of pregnant women experience some form of nausea due to the increase in estrogen that may irritate the stomach lining.
3. **Pseudocyesis** is a false pregnancy where a woman may experience some of the signs of pregnancy usually with a psychological basis.
4. Expectant fathers can experience a condition called **couvade** where they experience the symptoms of their pregnant partners, including nausea, vomiting, increased appetite, etc.

C. **Pregnancy Testing: Confirming the Signs**

1. Pregnancy tests (including over the counter) measure for a hormone in the blood called human chorionic gonadotropin (hCG), which is produced during pregnancy and can be identified in the blood or urine 8-9 days after ovulation.
2. Home pregnancy tests can be inaccurate if taken too soon after conception, and if done after the 12th week, **false negatives** may occur due to low levels of hCG.
3. **Radioimmunoassay (RIA)** blood tests are the most accurate since they can detect hCG within a few days after conception and can monitor pregnancies that may be in jeopardy.
4. Most physicians calculate a due date from the first day of the last menstrual period rather than the day of ovulation or fertilization.
5. The standard for due date calculation is called the **Naegeles Rule**: subtract 3 months from the 1st day of the last period and add 15 days for a single birth and 10 days for multiple births.

D. **Sex Selection: Myth and Modern Methods**

1. Although more males are conceived, a higher percentage of male fetuses are spontaneously aborted or die before birth than female fetuses.
2. Historically, many myths have existed related to choosing the sex of a child such as tying up the right or left testicle.
3. Some people want to choose the sex of their child to reduce the chance of having a child with a genetic condition that affects one sex more often.
4. Although there are no 100% methods of sex selection right now, there are some expensive methods available through artificial insemination.
5. During the 16th or 17th week of pregnancy, an **amniocentesis** can be performed to determine sex and heath issues. A small sample of amniotic fluid is drawn out of the uterus through a needle inserted in the abdomen.
6. In India, male children are valued much more, and female infanticide has become common, resulting in the lowest ratio of girls to boys.

II. INFERTILITY

A. **What is Infertility?**

1. **Infertility** is defined as the inability to conceive (or impregnate) after one year of regular sexual intercourse without the use of any form of birth control.
2. 40% of the time it's the female's problem 30% the male, 20% both, and 10% it is unknown.
3. Emotional reactions to infertility can include depression, anxiety, anger, self-blame, guilt, frustration, and fear.
4. The **motherhood mandate** refers to the idea that something is wrong with a woman if she has not had children by a certain age.
5. Support groups like RESOLVE can help couples deal with infertility.
6. Infertility can be caused by past sexually transmitted infections or pelvic inflammatory disease.
7. Sometimes changing lifestyle patterns, reducing stress, avoiding rigorous exercise and maintaining a recommended weight may restore fertility.

B. **Options for Infertile Couples**

- Many of the options are time consuming, expensive and do not guarantee success.

1. Fertility Drugs
    a. Fertility drugs are used to treat hormonal irregularities in men but mostly women, and there have been some promising results so far.
    b. These drugs work to increase ova production, resulting in an increased possibility of multiple births.

2. Surgery
    a. Scar tissue, cysts, tumors, adhesions, or blockages inside the Fallopian tubes may be surgically removed.
    b. The use of diagnostic techniques such as **laparoscopy** (a procedure that allows a physician to have direct view of all the pelvic organs and to perform a number of surgical procedures) may be used.
    c. In men, surgery may be used to reverse a prior sterilization procedure, remove any blockage in the vas deferens or epididymis, or to repair a **varicocele**.
3. Artificial Insemination
    a. **Artificial insemination** is the process of introducing sperm into a woman's reproductive tract without sexual intercourse.
    b. Some men who may be undergoing surgery or chemotherapy may choose to store their sperm in a **sperm bank**.
    c. Prior to using these artificial insemination procedures, physicians often prescribe fertility drugs to increase the chances that there will be healthy ova present when the sperm is introduced.
4. In Vitro Fertilization
    a. **In vitro fertilization (IVF)** or **test-tube babies** refers to a procedure where ova are put into a petri dish and mixed with washed sperm from the father and once fertilization has occurred, the zygotes are transferred to a woman's uterus.
    b. Most of the women who use this method have Fallopian tube blockage that does not allow for fertilization.
    c. Fertility drugs are usually used before the procedure to help the ovaries release multiple ova.
    d. It is estimated that only 5% to 30% of zygotes implant in the uterine wall.
5. Gamete Intra-Fallopian Tube Transfer (GIFT)
    a. It is similar to IVF in that ova and sperm are gathered outside of the body, but the difference is the ova and sperm are placed in the Fallopian tube prior to fertilization.
    b. This results in a much higher implantation rate.
6. Zygote Intra-Fallopian Tube Transfer (ZIFT)
    a. Ova and sperm are fertilized outside of the body, and directly following fertilization, the embryo is placed in a woman's Fallopian tube, where travels to the uterus and implants.
    b. This procedure has been found to yield higher implantation rates than IVF but not higher than GIFT.
7. Intravaginal Cultures
    a. Intravaginal cultures are used in conjunction with IVF and ZIFT, using a woman's vagina as an incubator.
    b. The ova and washed sperm are placed in a small container that is sealed and placed in a woman's vagina, held in place by a vaginal diaphragm.
    c. Fertilization can occur in the right temperature and atmosphere, and after fertilization, it is transferred to the uterus.
8. Zonal Dissection
    a. Zonal dissection responds to problems with the enzyme in the head of the sperm so that a microscopic hole is drilled into the ovum to allow fertilization.
    b. One drawback is that several sperm may enter the ovum, which can cause developmental problems.
9. Intracellular Sperm Injection
    a. Intracellular sperm injection (ICSI) was first used in the mid-1990s, and involves injecting a single sperm into the center of an ovum under a microscope.

b. It was developed to help couples who could not use IVF due to low sperm counts or sperm motility.
c. Early research suggests that there may be an increased risk of genetic defects.

10. Oocyte and Embryo Transplants
    a. Oocyte (egg) and embryo donation is used for women who may not be able to produce health ova.
    b. After fertilization, the embryo is transferred from the donor's to the woman's uterus.
11. Surrogate Parenting
    a. Surrogate parenting may be used if a woman cannot carry a pregnancy to term.
    b. Another woman's uterus is used to carry the pregnancy to term.
12. Other Options
    a. Sperm cryopreservation, or freezing sperm for later use, is possible with the frozen sperm stored in liquid nitrogen for years.
    b. Embryo cryopreservation is also possible, but like sperm, not all embryos can survive the freezing and thawing process.
    c. Recently, researchers have been experimenting with ova cryopreservation to help women who choose to delay childbearing until they are older or for women who undergo chemotherapy.

III. A HEALTH PREGNANCY

A. The Prenatal Period: Three Trimesters

- Pregnancy can vary from 38 to 40 weeks in length, and it is divided into 3 three-month periods called **trimesters**.

1. First Trimester
    a. First 3 months of pregnancy
    b. Prenatal Development
    - By the end of the first month of pregnancy, the fetal heart is formed and begins to pump blood with the circulatory system the first organ system to function in the embryo.
    - The beginnings of the brain, spinal cord, nervous system, muscles, arms, legs, eyes, fingers, and toes.

    c. Changes in the Pregnant Mother
    - Specific food cravings are normal as is an increased sensitivity to smells and odors.
    - The fetal heartbeat can usually be heard through ultrasound by the end of the first trimester.
    - After a heartbeat is seen by ultrasound, the probability of miscarriage drops significantly.
    - Ultrasounds help to confirm a pregnancy, rule out abnormalities, indicate gestational age, and confirm multiple pregnancies.
2. Second Trimester
    a. Prenatal Development
    - The fetus grows dramatically during the second trimester and has developed tooth buds and reflexes.
    - During the 2nd trimester, the mother can feel the fetus moving inside.
    - Soft hair, called **lanugo** and a waxy substance called **vernix caseosa** both cover the body of the fetus.

    b. Changes in the Pregnant Mother
    - Nausea and breast sensitivity begin to decrease.

- Fatigue may continue along with increase in appetite, heartburn, edema and a noticeable vaginal discharge.
- Many women report an increased sex drive during the second trimester.

3. Third Trimester

a. Prenatal Development

- By the end of the 7th month, the fetus begins to develop fat deposits and can react to pain, light, and sounds.
- There is often stronger and more frequent fetal movement.

b. Changes in the Pregnant Mother

- Many of the earlier symptoms continue, including backaches, leg cramps, increases in varicose veins, hemorrhoids, sleep problems, shortness of breath. **Braxton-Hicks** contractions also occur, which are intermittent, relatively painless contractions of the uterus.
- A clear liquid called **colostrum** may be secreted from the nipples as the breasts prepare to produce milk for breast-feeding.

B. **The Father's Experience**

1. For men, pregnancy can be a time of joy and anticipation in addition to stress and anxiety.
2. Now in the United States, men are encouraged to play a larger role in pregnancy and childbirth.

IV. HEALTH CARE DURING PREGNANCY

A. **Exercise and Nutrition**

1. Most physicians agree that a woman's exercise routine should not exceed her pre-pregnancy exercise levels.
2. Light exercise has been found to result in a greater sense of well being, shorter labor and fewer obstetric problems.
3. Water exercise may be the best form of exercise for a pregnant woman since it causes fewer maternal and fetal heart rate changes and lowers maternal blood pressure more than land exercise.
4. Physical stresses such as prolonged standing, long work hours, and heavy lifting can also affect a pregnancy by reducing blood flow to the uterus, resulting in lower birth weights and prematurity.
5. Nutritional requirements during pregnancy call for extra protein, iron, calcium, folic acid, and vitamin B6, in addition to increasing overall caloric intake.
6. Iron may be diluted in the blood so many pregnant women are advised to take prenatal vitamins that include iron supplements.
7. Drugs and Alcohol

a. It is recommend that women avoid caffeine, nicotine, alcohol, marijuana, and other drugs.

b. Alcohol has been linked with **fetal alcohol syndrome (FAS)** which leads to low birth weight and mental delays.

c. Cigarette use has been associated with spontaneous abortion, low birth weight, prematurity, and low iron levels.

B. **Pregnancy in Women Over Thirty**

1. Over the last few years it has become common for women to delay having children until they are 35 or older.

2. Delayed pregnancy does carry some risks, which include an increase in spontaneous abortion, first-trimester bleeding, low birth weights, increased labor time and rate of **cesarean section (c-section)**.
3. The likelihood of a chromosomal abnormality increase over the age of 30, and it becomes more difficult to get pregnant due to aging follicles.

C. **Sex During Pregnancy**
1. In the United States many women continue to have satisfying sexual relations during pregnancy.
2. Sexual intercourse during pregnancy is safe for most mothers and the developing child up until the last several weeks of pregnancy.
3. Orgasm during pregnancy is safe, but occasionally it may cause painful uterine contractions.

V. PROBLEMS IN THE PREGNANCY

A. **Ectopic Pregnancies**
1. In an ectopic pregnancy the zygote implants outside of the uterus, with 97% implanting in the Fallopian tube (tubal pregnancies) and the remaining 3% occurring in the abdomen, cervix, or ovaries.
2. Rates of ectopic pregnancies have been increasing due to increases in the incidence of pelvic inflammatory disease caused by chlamydia.
3. It is very dangerous and is the primary cause of maternal mortality in the first trimester of pregnancy, with 30-40 women dying each year in the U.S.
4. Smoking has been linked to ectopic pregnancies based on changes in tubal contractions and muscular tone of the Fallopian tubes due to nicotine.
5. Symptoms include abdominal pain, cramping, pelvic pain, vaginal bleeding, nausea, dizziness, and fainting and may cause a rupture, causing internal hemorrhaging and possibly death.

B. **Spontaneous Abortions**
1. Also known as a miscarriage, which is a natural termination of a pregnancy before the time that the fetus can live on its own.
2. About 10% of all diagnosed pregnancies end in miscarriage, with most occurring during the first trimester.
3. In a significant number of miscarriages, there is some chromosomal abnormality, but other factors may be that the uterus was too small or too weak, or it may be caused by maternal stress, nutritional deficiencies, excessive vitamin A, drug exposure, or pelvic infection.
4. Symptoms include vaginal bleeding, cramps, and lower back pain.

C. **Chromosomal Abnormalities**
1. Once an abnormality is detected by amniocentesis, a woman must decide whether or not to continue the pregnancy, although she will already be four months into the pregnancy.
2. **Chorionic villus sampling (CVS)** can be used between the 8th and 12th week of pregnancy, and involves removing a sliver of tissue from the chorion, the tissue that develops into the placenta.
3. It can result in false positives, and there is an increased risk of miscarriage and limb deformities associated with this test.
4. **Maternal-serum alpha-fetoprotein screening (MSAFP)** is used to detect defects such as **spina bifida** or **anencephaly**, and it is a blood test done between the 16th and 18th weeks of pregnancy.
5. **Down syndrome**, the most common chromosomal abnormality, occurs when an extra chromosome has been added to the 21st chromosome.

6. Down syndrome is associated with older aged mothers.

D. **Rh Incompatibility**

1. If the mother's blood is Rh-negative and the fetal blood is Rh-positive, the mother will begin to manufacture antibodies against the fetal blood during childbirth (not during pregnancy). This will affect future pregnancies and can cause fetal death.
2. After an Rh-negative woman has delivered, she is given Rhogam, which prevents antibodies from forming and ensures that her future pregnancies will be healthy.
3. It's important that an Rh-negative woman is also given **Rhogam** if she has an amniocentesis, miscarriage, or abortion since these may cause her to begin producing the antibodies.

E. **Toxemia**

1. In the last 2 to 3 months of pregnancy, 6% to 7% of women experience **toxemia**, a form of blood poisoning caused by kidney disturbances.
2. Symptoms include rapid weight gain, fluid retention, an increase in blood pressure, and/or protein in the urine.
3. If toxemia is not treated, it can lead to **eclampsia**, which involves convulsions, coma, and in about 15% of cases, death.
4. These cases occur primarily in women who neglect good prenatal care and are relatively rare in women with good medical care.
5. **Preeclampsia** is a condition of hypertension typically accompanied by leg swelling and other symptoms.

VI. CHILDBIRTH

- Only 4% of American babies are born exactly on the due date predicted.
- In the U.S., the drug, pitocin, is used to speed up labor, but in Bolivia nipple stimulation is used because it leads to a release of oxytocin, a natural form of pitocin.

A. **Preparing for Birth**

1. Some couples participate in **Lamaze** classes, which is a prepared childbirth method in which couples are provided information about the birth process and are taught breathing and relaxation exercises to use during labor.
2. A few weeks before delivery, the fetus usually moves into a "head down" position in the uterus that is called **engagement**.
3. Ninety-seven percent of fetuses are in the "head down" position, but if a baby's feet or buttocks are first (**breech birth**), the physician may either try to rotate the baby prior to birth or recommend a cesarean section.

B. **Birthplace Choices**

1. Many people believe that babies can be safely delivered at home with the help of a **midwife**, a trained professional who assists during childbirth.
2. There is criticism of the mainstream hospital births that it has been the medical establishment that has moved delivery into hospitals as a way of making money and controlling women's bodies.
3. The majority of babies in the U.S. are born in hospitals, but there are more birthing centers that provide comfortable rooms that are meant to be cozier.

C. **Inducing the Birth**

1. Inducing the birth involves using techniques to artificially start the birth process, which is usually in the form of drugs given in increasing doses to mimic the natural contractions of labor, although these contractions can be more painful and prolonged than natural labor.
2. Childbirth inductions have increased in recent years.
3. Reasons for inducing labor include being two or more weeks past a due date or to avoid having a large baby that might require a cesarean.

D. **Birthing Positions**

1. Despite criticisms of this position, in the U.S., the majority of hospitals have women in the semi-lying down position with her feet up in stirrups.
2. A woman on her hands and knees or in the squatting position allows her pelvis and cervix to be at its widest, and the force of gravity can be used to help in the birth process.
3. Positions vary from culture to culture and include underwater births.

E. **Stages of Childbirth**

- The beginning of birth is usually marked by an expulsion of the mucus plug from the cervix, which is often combined with blood, giving it the name "**bloody show**."

1. Stage One (**cervical effacement** and **dilation**)
    a. The first stage can last anywhere from 20 minutes to 24 hours.
    b. The cervix begins dilation (the expansion of the opening of the cervix in preparation for birth) and effacement (thinning out) to allow for fetal passage.
    c. The cervix increases from 0 to 10 centimeters, and toward the end the amniotic sac usually ruptures, although this may happen earlier in some women.
    d. When contractions are about 5 minutes apart, the physician will advise women to come to the hospital.
    e. The last phase of stage one is called **transition**, where contractions are the strongest and the periods in between contractions are the shortest.
    f. The woman's body produces pain-reducing hormones called **endorphins**.
    g. In recent years, there has been a movement away from the use of drugs during delivery due to risks to the baby.
    h. Fetal monitoring is done to check for signs of **fetal distress** such as slowed heart rate or lack of oxygen which can be done through a woman's abdomen or by a sensor on the fetus's scalp accessed through the cervix.
2. Stage Two (expulsion of the fetus)
    a. Contractions are somewhat less intense, and an **episiotomy** may be performed to reduce the risk of tearing the tissue between the vaginal opening and the anus.
    b. As the woman pushes during contractions, the top of the head of the baby soon appears at the vagina, which is known as **crowning**.
    c. The umbilical cord is cut (which is painless), and eye drops are put in the baby's eyes to prevent bacterial infection.
    d. An **Apgar** test is perform shortly after birth, which is a system of assessing the general physical condition of a newborn infant based on a rating of 0, 1, or 2 for five criteria: heart rate, respiration, muscle tone, skin color, and response to stimuli. The five scores are added together with a perfect score being 10.
3. Stage Three (expulsion of the placenta or afterbirth)
    a. The placenta is expelled by strong contractions after the baby is born.
    b. The placenta must be checked to make sure all of it has been expelled.

VII. PROBLEMS DURING BIRTHING

A. **Premature Birth: The Hazards of Early Delivery**

1. While the majority of babies are born late rather than early, if a birth takes place before the 37th week of pregnancy, it is considered **premature**. About 8% of births in the U.S. are premature.
2. Prematurity increases the risk of birth-related defects and infant mortality.
3. Reasons for prematurity include early labor, early rupture of the amniotic membranes, or maternal or fetal problems.
4. Twins or other multiple births occur early.
5. Risk factors for prematurity include smoking, alcohol or drug use, inadequate weight gain or nutrition, heavy physical labor, infections, and adolescent pregnancy.

B. **Breech Birth: Feet First into the World**

1. In three to four percent of births, the fetus is in the breech position.
2. Some physicians and midwives will successfully flip the baby to the head down position, but in many cases in the U.S. a cesarean section is performed to ensure a healthy baby.

C. **Stillbirth: Sad Circumstance**

1. When a fetus dies after 20 weeks of pregnancy, it is called a **stillbirth**.
2. Reasons for a stillbirth include umbilical cord accidents, problems with the placenta, birth defects, infections, and maternal diabetes or high blood pressure.
3. The frequency of stillbirths has been decreasing over the last few years due to better monitoring.

D. **Cesarean Section (C-Section) Delivery**

1. A c-section involves the delivery of the fetus through an incision in the abdominal wall.
2. In 1997, about 21% of deliveries were c-sections.
3. The reasons for c-sections include the baby being too large, the woman being unable to push the baby out, the placenta either blocking the cervix (**placenta previa**) or separating from the baby prior to birth, or the baby being in fetal distress.
4. The operation usually lasts between 20 to 90 minutes, and women need more recovery time in the hospital.
5. Women who have had a c-section can deliver their next child vaginally.

VIII. POSTPARTUM PARENTHOOD

A. **More Physical Changes for the Mother**

1. Following delivery, the uterus returns to its original shape in about 6 weeks, but in breast-feeding women, the uterus returns to its original size quicker due to the secretion of oxytocin.
2. A vaginal discharge continues for 10 days to a month.
3. If women had an episiotomy, that can take time to heal and be painful.

B. **Postpartum Psychological Changes**

1. Minor depression occurs in 25% to 67% of women, and in severe cases it is called **postpartum depression**.
2. Physical exhaustion, physiological change, and an increased responsibility of childrearing all contribute to these feelings along with post-partum hormonal changes.
3. Fathers may experience postpartum depression also.
4. In rare case, mental disturbances called **postpartum psychosis** occurs, leading to women who have killed or neglected their babies.

C. **Sexuality for New Parents**

1. Heterosexual couples can resume intercourse two weeks after delivery if it was uncomplicated and without any tears or episiotomy.
2. It's necessary sometimes to wait to prevent infection and make sure the cervix has returned to its original position.
3. If an episiotomy or c-section was performed, it may take longer to resume sexual activity depending upon a woman's level of comfort and/or pain.
4. Soon after delivery, many women report slower and less intense excitement stages and a decrease in vaginal lubrication, but after three months postpartum, most women return to their original levels of desire and excitement.

D. **Breast-Feeding the Baby**

1. Within an hour after birth, the newborn usually begins a rooting reflex that signals hunger, and the baby's sucking triggers the flow of milk from the breast.

2. In the first few days of breast-feeding, the breasts release colostrum, which is important in strengthening the baby's immune system.
3. Women can use a breast pump to express milk from her breasts while she is away.
4. The American Academy of Pediatrics recommends exclusive breast-feeding for 6 months and then continued breastfeeding for a minimum of one year.
5. Breastfeeding advocates argue that due to societal stigma against breast-feeding, women stop too early.
6. Research has found that the natural age of weaning is 2 ½ years, with a maximum of 6-7 years.

TEST YOURSELF

Below you will find fill-in-the-blank and short answer essay questions for topics covered in chapter 12. Check your answers at the end of this chapter.

Fertility and Infertility

1. A woman's sexual desire is usually at its peak during her ________________.
2. During ovulation, a ________________ ________________ in the cervix disappears, making it easier for sperm to enter the uterus.
3. During intercourse, the female's ________________ ________________ attacks the semen immediately after ejaculation, thinking it is unwanted bacteria.
4. Although it is not clear how sperm locates the ovum, preliminary research indicates that the ovum releases ________________ ________________ that indicate its location.
5. ________________ usually occurs in the ampulla, the funnel-shaped open end of the Fallopian tube.
6. The fertilized ovum is referred to as a(n) ________________.
7. Approximately 3-4 days after conception, the blastocyst enters the ________________.
8. From the second through the eighth weeks of pregnancy, the developing human is referred to as a(n) ________________.
9. The umbilical cord connects the fetus to the ________________.
10. During the 16th or 17th week of pregnancy, a procedure called ________________ can determine, among other things, the chromosomal sex of the fetus.
11. ________________ is defined as the inability to conceive (or impregnate) after one year of regular sexual intercourse without the use of any form of birth control.
12. Since all fertility drugs work to increase ova production in women, there is an increased possibility of ________________ ________________ and a condition known as "ovarian hyperstimulation syndrome".

13. Prior to using artificial insemination procedures, doctors often prescribe ________________ ________________ to increase the chances that there will be healthy ova present when the sperm is introduced.

14. ________________ ________________ ________________ is a procedure in which a woman's ova are removed from her body, fertilized with sperm in a laboratory, and then surgically implanted back into her uterus.

15. Although ________________ ________________ ________________ ________________ is similar to IVF in that ova and sperm are mixed in an artificial environment, the main difference is that here both the ova and the sperm are placed in the Fallopian tube *prior* to fertilization.

16. What is a blastocyst?

17. Describe the placenta.

18. What causes identical twins to occur?

19. Describe the procedure called amniocentesis.

20. What is artificial insemination?

Pregnancy

21. Human pregnancy is divided into three three-month periods called ________________.

22. The ________________ system is the first organ system to function in the embryo.

23. A(n) ________________ can capture images of the embryo for measurement as early as five and a half weeks into the pregnancy, while a heartbeat can be seen by six weeks.

24. During the second trimester, soft hair called ________________, and a waxy substance, known as ________________ ________________, cover the body of the fetus.

25. Fetal movement often begins to be felt in the ________________ trimester, sometimes as early as the 16th week.

26. Although it is true that pregnant women are "cardio-vascularly challenged" early in pregnancy, it is a ________________ that too much exercise may cause a miscarriage or harm the developing fetus.

27. It is estimated that a pregnant woman of average size should gain between ________________ and ________________ pounds throughout a pregnancy.

28. Alcohol intake has been linked with ________________ ________________ syndrome, which can produce an infant who is undersized and mentally deficient.

29. Sexual intercourse during pregnancy is safe for most mothers and the developing child up until the last ________________ ________________ of pregnancy.

30. Ninety-seven percent of ectopic pregnancies occur when the fertilized ovum implants in the ________________ ________________.

31. A(n) ________________ ________________, or miscarriage, refers to a natural termination of a pregnancy before the time that the fetus can live on its own.

32. Research indicates that repeat miscarriages may be linked to defective ________________.

33. ________________ ________________ ________________ has been used, which means a sliver of tissue from the chorion is removed and is checked for chromosomal abnormalities.

34. The most common chromosomal abnormality occurs on the twenty-first chromosome and is known as ________________ ________________.

35. In the last two to three months of pregnancy, 6% to 7% of women experience ________________, a form of blood poisoning caused by kidney disturbances.

36. When is it that the most important embryonic development takes place?

37. What are Braxton-Hicks contractions?

38. What are some nutritional requirements during pregnancy?

39. List some problems associated with the pregnancies of women who smoke cigarettes throughout their pregnancies?

40. What is an ectopic pregnancy?

Childbirth and Postpartum

41. The ________________ method of childbirth teaches women and their partners what to expect during labor and delivery and how to control the pain through breathing and massage.

42. A few weeks before delivery, the fetus usually moves into a "head down" position in the uterus. This is referred to as ________________.

43. Many people believe that babies can be safely delivered at home with the help of a ________________.

44. At the beginning of birth, the mucus plug is often combined with blood, giving it the name ________________ ________________.

45. Sometimes women experience ________________ ________________, where contractions are irregular and do not dilate the cervix.

46. The last period in labor in which contractions are strongest and the periods in between contractions are the shortest is known as ________________.

47. Much of the post-birth pain American women experience is due to a procedure called an ________________.

48. As the woman pushes during contractions, the top of the head of the baby soon appears at the vagina, which is know as ________________.

49. If birth takes place before the thirty-seventh week of pregnancy, it is considered ________________.

50. When a fetus dies after 20 weeks of pregnancy it is called a ________________.

51. Following delivery, the ________________ returns to its original shape in about six weeks.

52. In many mammals, it is quite common for the mothers to eat the placenta after delivery, a process known as ________________.

53. In the first few days of breast-feeding, the breasts release a fluid called ________________, which is important in strengthening the baby's immune system.

54. Some women who want to breast-feed but who also wish to return to work use a breast ________________.

55. The American Academy of Pediatrics recommends exclusive breastfeeding (no other fluids or food) for ________________ months.

56. Describe how nipple stimulation may speed up labor.

57. What are the three stages of birth?

58. What are some risk factors that may cause a premature birth?

59. What is a cesarean section?

60. What causes postpartum depression?

POST TEST

Below you will find true/false, multiple-choice and matching quiz items covering the entire chapter. Check your answers at the end of this chapter.

True/False

1. Cervical mucus plays an important role in conception, making it easier for sperm to move through the cervix.

2. The placenta supplies nutrients to the developing fetus, aids in respiratory and excretory functions, and secretes hormones necessary for the continuation of the pregnancy.

3. A higher percentage of couples in the United States have been identified as infertile compared with other developed countries.

4. Surrogate parenting is illegal in the United States.

5. Light exercise has been found to result in a greater sense of well being, shorter labor and fewer obstetric problems.

6. Chlamydia can be linked to an ectopic pregnancy.

7. Inducing labor with drugs usually causes contractions that are more painful and prolonged than natural labor.

8. About 8% of births in the U.S. are premature

9. Heterosexual couples should wait at least six months postpartum to engage in intercourse to prevent infection.

10. Research has found that the natural age of weaning is 2 ½ years.

Multiple-Choice

11. How long can the majority of sperm live in the female reproductive tract?
    a. 6 minutes
    b. 2 hours
    c. 12 hours
    d. 34 hours
    e. 72 hours

12. Where does fertilization take place?
    a. uterus
    b. ovaries
    c. vagina
    d. Fallopian tube
    e. cervix

13. What is the name of the fertilized ovum?
    a. placenta
    b. blastocyst
    c. amnion
    d. zygote
    e. embryo

14. What is the term for when men experience the symptoms of their pregnant partners, including nausea, vomiting, and increased appetite?
    a. couvade
    b. amniocentesis
    c. eclampsia
    d. pseudocyesis
    e. patriocele

15. What is human chorionic gonadotropin (hCG)?
    a. the fluid that is drawn out during an amniocentesis to test for genetic abnormalities
    b. the chemical on the sperm that allows it to break through the ovum
    c. the hormone that causes the nausea in morning sickness
    d. the chemical in a pregnant woman's blood that will tell the sex of the fetus
    e. the hormone that is identified by pregnancy tests

16. What is the name of the standard method used for due date calculation?
    a. RIA Determination
    b. Naegeles Rule
    c. Ovulation Assessment
    d. Apgar Numbering
    e. Braxton-Hicks Method

17. Which procedure below will ensure that a couple will have a boy?
    a. a woman lies on her left side directly after intercourse
    b. a man ties his left testicle
    c. having intercourse close to ovulation
    d. having sexual intercourse in a north wind
    e. None of the above

18. What is term for the inability to conceive (or impregnate) after one year of regular sexual intercourse without the use of any form of birth control?
    a. miscarriage
    b. pseudocyesis
    c. infertility
    d. varicocele
    e. anencephaly

19. What the idea that something is wrong with a woman if she has not had children by a certain age?
    a. infertility
    b. the motherhood mandate
    c. childfree discrimination
    d. couvade encouragement
    e. societal parenting pressure

20. What is IVF?
    a. In vitro fertilization
    b. Intra-vaginal fertility
    c. In vivo fetus
    d. Internal vulvar fatality
    e. Intra-vivo fertile

21. Which fertility procedure used to be referred to as test-tube babies?
    a. gamete intra-Fallopian tube transfer
    b. zonal dissection
    c. in vitro fertilization
    d. oocyte transplants
    e. intracellular sperm injections

22. What is a physiological change of pregnancy?
    a. increased weight
    b. heart pumps more blood
    c. the lungs and digestive system work harder
    d. the thyroid gland grows
    e. All of the above

23. What is the name for the process by which the fetal heartbeat can be heard?
    a. zona reaction
    b. amniocentesis
    c. microsampling
    d. Apgar monitoring
    e. ultrasound

24. What is the name of the waxy substance that coats and protects the fetus in utero?
    a. vernix caseosa
    b. colostrum
    c. lanugo
    d. couvade
    e. varicocele

25. What substance do physicians recommend that pregnant women avoid?
    a. nicotine
    b. alcohol
    c. marijuana
    d. caffeine
    e. All of the above

26. In the U.S., from 1970 to 1990, the proportion of first births increased ______% among women 30-39 years old?
    a. 10%
    b. 27%
    c. 52%
    d. 73%
    e. 100%

27. When it comes to sexual behaviors during pregnancy, which of the following statements is FALSE?
    a. Orgasm during pregnancy may cause a woman to go into premature labor.
    b. Sexual intercourse during pregnancy is safe for most mothers and fetuses up until the last several weeks of pregnancy
    c. Many women continue to have satisfying sexual relations during pregnancy.
    d. All of the above
    e. None of the above

28. What is an ectopic pregnancy?
    a. when the developing fetus has a chromosomal abnormality, usually Down syndrome.
    b. when the mother is producing too much colostrum and needs extra iron supplements.
    c. when the fertilized egg implants outside of the uterus, usually in the Fallopian tube.
    d. when the mother develops high blood pressure and gestational diabetes
    e. when the fetus is growing upside down in the uterus

29. What is eclampsia?
    a. the implantation of the fertilized egg outside the uterus
    b. the stretching and thinning of the cervix in preparation for birth
    c. a condition in which the placenta is abnormally positioned in the uterus so that it covers the opening of the cervix
    d. a progression of toxemia with worsening conditions
    e. a condition in which a woman believes she is pregnant and experiences signs of pregnancy

30. What percent of American babies are born exactly on the due date predicted?
    a. 4%
    b. 16%
    c. 35%
    d. 58%
    e. 79%

31. What is the name for the process of the fetus moving into a "head down" position in the uterus a few weeks before delivery?
    a. transition
    b. effacement
    c. engagement
    d. dilation
    e. crowning

32. What is the term for the thinning out of the cervix that occurs during stage one of childbirth?
    a. breeching
    b. dilation
    c. effacement
    d. bloody show
    e. crowning

33. What happens during the last phase of stage one, called transition?
    a. the top of the head of the baby soon appears at the vagina
    b. there's an expulsion of the mucus plug from the cervix
    c. the fetus usually moves into a "head down" position in the uterus
    d. the cervix begins dilation
    e. the contractions are the strongest and the periods in between contractions are the shortest

34. What does NOT happen during stage two?
    a. the expulsion of the fetus
    b. the expulsion of the placenta
    c. the umbilical cord is cut
    d. crowning
    e. an episiotomy may be performed
35. Which of the following is NOT tested as part of the Apgar test?
    a. response to stimuli
    b. skin color
    c. muscle tone
    d. eyesight
    e. heart rate

36. What happens during stage three?
    a. the umbilical cord is cut
    b. the expulsion of the placenta
    c. eye drops are put into the baby's eyes
    d. the expulsion of the fetus
    e. All of the above

37. What is a reason for prematurity?
    a. multiple births
    b. early labor
    c. early rupture of the amniotic membranes
    d. a fetal problem
    e. All of the above

38. Regarding cesarean sections, which of the following statements below is FALSE?
    a. Women who have had a c-section cannot deliver their next child vaginally.
    b. C-sections are sometimes performed because the placenta is blocking the cervix.
    c. Women who've had c-sections need more recovery time in the hospital.
    d. C-sections usually lasts between 20 to 90 minutes.
    e. None of the above

39. Post-partum hormonal changes, physical exhaustion, physiological changes, and an increased responsibility of childrearing can contribute to what phenomenon?
    a. postpartum diabetes
    b. pseudocyesis
    c. maternal eclampsia
    d. postpartum depression
    e. couvade

40. What is the name of the fluid released from the mother's breasts in the first few days of breast-feeding that strengthens the baby's immune system?
    a. vernix caseosa
    b. colostrum
    c. lanugo
    d. couvade
    e. varicocele

Matching

Column 1		Column 2
A. eclampsia	____	41. the hormone that pregnancy tests detect
B. lanugo	____	42. During the 16th or 17th week of pregnancy a small sample of fluid is drawn out of the uterus through a needle inserted in the abdomen in order to test for genetic abnormalities.
C. human chorionic gonadotropin (hCG)	____	43. when the fetus moves into a "head down" position a few weeks before delivery
D. amniocentesis	____	44. the soft covering of hair over the fetus
E. engagement	____	45. a progression of toxemia with similar but worsening conditions
F. Chorionic villus sampling (CVS)	____	46. the waxy substance that covers the fetus in utero for protection
G. transition	____	47. the sampling and testing of the tissue that develops into the placenta for fetal abnormalities used between the 8th and 12th week of pregnancy
H. placenta	____	48. a type of pregnancy when the implantation of the fertilized egg takes place outside of the uterus, usually in a Fallopian tube
I. ectopic	____	49. the last period in labor in which contractions are strongest and the periods in between contractions are the shortest
J. vernix caseosa	____	50. the structure through which the exchange of materials between fetal maternal circulations occurs

Test Yourself Answer Key

Fertility and Infertility

1. ovulation (p. 350)
2. mucus plug (p. 350)
3. immune system (p. 351)
4. chemical signals (p. 351)
5. Fertilization (p. 351)
6. zygote (p. 351)
7. uterus (p. 353)
8. embryo (p. 354)
9. placenta (p. 354)
10. amniocentesis (p. 357)
11. Infertility (p. 358)
12. multiple births (p. 360)
13. fertility drugs (p. 362)
14. In vitro fertilization (p. 362)
15. gamete intra-fallopian tube transfer (p. 362)
16. A blastocyst is the hollow ball of embryonic cells that enters the uterus from the Fallopian tube and eventually implants. Once implantation has been achieved, pregnancy has begun. (p. 353)
17. It is the structure through which the exchange of materials between fetal and maternal circulations occurs. (p. 354)
18. Identical twins occur when a single zygote completely divides into two separate zygotes. This process produces twins who are genetically identical. (p. 354)
19. A small sample of amniotic fluid is drawn out of the uterus through a needle inserted in the abdomen. The fluid is then analyzed to detect genetic abnormalities in the fetus or to determine the sex of the fetus. (p. 357)
20. It is the process of introducing sperm into a woman's reproductive tract without sexual intercourse. (p. 362)

Pregnancy

21. trimesters (p. 364)
22. circulatory (p. 364)
23. ultrasound (p. 365)
24. lanugo, vernix caseosa (p. 366)
25. second (p. 366)
26. myth (p. 367)
27. 31, 40 (p. 368)
28. fetal alcohol (p. 369)
29. several weeks (p. 370)
30. Fallopian tube (p. 370)
31. spontaneous abortion (p. 371)
32. sperm (p. 371)
33. Chorionic villus sampling (p. 371)
34. Down syndrome (p. 372)
35. toxemia (p. 372)
36. the first trimester (p. 364)
37. They are intermittent, relatively painless, contractions of the uterus that may occur after the third month of pregnancy. (p. 367)
38. extra protein, iron, calcium, folic acid, vitamin B6, and increased caloric intake (p. 368)

39. spontaneous abortion, low birth weight, prematurity, low iron levels, risk of vascular damage to the developing baby's brain (p. 369)
40. The implantation of the fertilized ovum outside the uterus, for example, in the Fallopian tubes of the abdomen. (p. 370)

Childbirth and Postpartum

41. Lamaze (p. 374)
42. engagement (p. 374)
43. midwife (p. 374)
44. bloody show (p. 375)
45. false labor (p. 375)
46. transition (p. 375)
47. episiotomy (p. 377)
48. crowning (p. 377)
49. premature (p. 377)
50. stillbirth (p. 378)
51. uterus (p. 379)
52. placentophagia (p. 379)
53. colostrum (p. 379)
54. pump (p. 380)
55. six (p. 380)
56. In Bolivia, certain groups of people believe that nipple stimulation helps the birth move quicker. So if a birth is moving too slowly, a woman's nipples may be massaged. Biologically, nipple stimulation leads to a release of oxytocin, a natural form of the labor inducing drug pitocin. (p. 373)
57. cervical effacement and dilation, expulsion of the fetus, and expulsion of the placenta. (p. 375)
58. smoking during pregnancy, alcohol or drug use, inadequate weight gain or nutrition, heavy physical labor during the pregnancy, infections, and teenage pregnancy. (p. 377)
59. A cesarean section involves the delivery of the fetus through an incision in the abdominal wall. (p. 378)
60. physical exhaustion, physiological changes, the increased responsibility of childrearing, post-partum hormonal changes (p. 379)

Post Test Answer Key

True/False

1. T (p. 350)
2. T (p. 354)
3. T (p. 358)
4. F (pp. 363-364)
5. T (p. 367)
6. T (p. 371)
7. T (p. 373)
8. T (p. 377)
9. F (p. 379)
10. T (p. 380)

Multiple Choice

11. e (p. 351)
12. d (p. 351)
13. d (p. 351)
14. a (p. 354)
15. e (p. 355)
16. b (p. 356)
17. e (pp. 356-358)
18. c (p. 358)
19. b (p. 358)
20. a (p. 362)
21. c (p. 362)
22. e (p. 364)
23. e (p. 365)
24. a (p. 366)
25. e (p. 368)
26. e (p. 369)
27. a (p. 370)
28. c (p. 370)
29. d (p. 373)
30. a (p. 373)
31. c (p. 374)
32. c (p. 375)
33. e (p. 375)
34. b (p. 377)
35. d (p. 377)
36. b (p. 377)
37. e (p. 377)
38. a (p. 378)
39. d (p. 379)
40. b (p. 379)

Matching

41. C (p. 355)
42. D (p. 357)
43. E (p. 374)
44. B (p. 366)
45. A (p. 373)
46. J (p. 366)
47. F (p. 371)
48. I (p. 356)
49. G (p. 375)
50. H (p. 354)

Chapter 13 – Contraception and Abortion

CHAPTER SUMMARY

The origins of contraception go far back in history to the ancient Greeks and Egyptians. In the early 1950s, a tremendous increase in reproductive and contraceptive research occurred in the United States, resulting in a new selection of birth control methods. Each method has a specific way it works to prevent pregnancy with effectiveness rates and a variety of advantages and disadvantages that potential users need to examine and evaluate with their own needs in mind.

Barrier contraceptives are some of the oldest methods; although more advanced forms are continually being researched. Barrier methods include male and female condoms, the diaphragm, contraceptive sponge, cervical cap, and Lea's Shield. Hormonal methods have become some of the most popular and most effective methods in the United States and typically work by preventing ovulation, thickening the cervical mucus, and making implantation unlikely for a fertilized ovum. They come as a combined-hormone method with both estrogen and progestin or progestin only. Hormonal methods include oral contraceptives, hormonal injectibles, the hormonal ring, and the hormonal patch. Chemical methods of contraception include spermicides such as creams, jellies, foams, suppositories, and films, which deactivate sperm and create a barrier over the cervix. Intrauterine devices have become less popular in the United States but are still used frequently in other countries. They work by creating a low-grade infection in the uterus, which may break apart the fertilized ovum, inhibit implantation, and/or increase prostaglandins. Natural family planning and fertility awareness methods offer a natural way for couples to explore other types of sexual expression while a woman is ovulating. Tubal ligation and vasectomies are two permanent methods of contraception for women and men that are highly effective and popular worldwide. Emergency contraception (EC) is designed to prevent pregnancy after unprotected vaginal intercourse, in case of unanticipated sexual intercourse, contraceptive failure, or sexual assault and involves taking pills that interfere with ovulation and/or fertilization.

Every year there are 50 million abortions done worldwide; 30 million legal, while 20 million are illegal. The abortion debate continues to be very emotional and violent. Historically, there are few large-scale societies historically where abortion has not been practiced, including early Greek, Roman, and Hebrew cultures. In 1973 the U.S. Supreme Court ruled in Roe v. Wade that women have a constitutionally protected right to have an abortion in the early stages of pregnancy.

First-trimester procedures are done before 14 weeks of gestation and involve vacuum aspiration, where dilation rods are used to open the cervix and a cannula is inserted into the cervix and is attached to a vacuum aspirator that empties the contents of the uterus. The risks of first-trimester abortions are much lower compared to second-trimester abortions, which typically involve dilation and evacuation done under general anesthesia. Dilators are used to dilate the cervix 24 hours prior to the abortion. A woman returns for a 15 to 30 minute procedure to empty the uterus. Newer nonsurgical abortion procedures involve pills that allow a woman to have an abortion at home over a period of several days.

LEARNING OBJECTIVES

After studying Chapter 13, you should be able to:

1. Discuss some of the historical practices around birth control.
2. List the factors for consideration when a person is choosing a method of contraception.
3. Explain the functions, advantages, and disadvantages of the different barrier contraceptives.
4. Discuss the history and future plans of oral contraceptives.
5. Explain the function, advantages, and disadvantages of oral contraceptives.
6. Explain the functions, advantages, and disadvantages of the other different types of hormonal methods of contraception.
7. Compare and contrast combined hormone methods with progestin-only methods.
8. Explain the functions, advantages, and disadvantages of spermicides, IUDs, and natural family planning.
9. Explain the functions, advantages, and disadvantages of permanent female and male sterilization.
10. Discuss some of the research about contraceptive methods for men.
11. Identify ineffective methods of contraception.
12. Explain the function and availability of emergency contraception.
13. Identify some worldwide trends around contraception.
14. Discuss the research in contraception in the future.
15. Describe the historical developments of abortion.
16. Identify the procedures involved in first- and second-trimester abortions.
17. Describe the nonsurgical abortion procedures and how they work to terminate a pregnancy.
18. Identify the reasons women cite for having abortions.
19. Discuss the research on the physiological and psychological reactions to having an abortion.
20. Identify the legislation relevant to teenagers and abortion.
21. Describe some of the worldwide trends related to abortion.

CHAPTER OUTLINE

I. THE HISTORY OF CONTRACEPTION

A. The origins of contraception go far back in history to the ancient Greeks and Egyptians.

B. In the early 1800s, several groups in the U.S. were interested in controlling fertility in order to reduce poverty.

C. In 1873 Anthony Comstock worked with Congress to pass the Comstock Laws, which prohibited the distribution of all obscene material including contraceptive information and devices.

II. CHOOSING A METHOD OF CONTRACEPTION

A. A variety of methods of contraception or birth control are currently available, and other methods are currently being evaluated for possible approval by the **Food and Drug Administration (FDA)**.

B. When choosing a method of birth control, a number of factors must be considered.

1. personal health and health risks
2. the number of sexual partners
3. frequency of sexual intercourse
4. risk of acquiring an STI
5. personal responsibility
6. cost of methods
7. advantages and disadvantages of methods

III. BARRIER CONTRACEPTIVES: CONDOMS AND CAPS

A. **Condoms**

1. Penile covering have been used as a method of contraception since the beginnings of recorded history.

2. Rubber male condoms were available in the U.S. in 1850, and female condoms were first available in the U.S. in 1994.

   a. *Reality Vaginal Pouch*, a polyurethane female condom is about 7 inches long and has two flexible rings that help with insertion. It is fairly expensive at $2.00 each, but some people like the polyurethane feel more than latex.

3. How It Works

   a. The male condom is placed on an erect penis prior to vaginal contact, making sure to leave room at the tip for the ejaculatory fluid.

   b. Water-based lubricants can be used with a condom, but oil lubricants will damage the latex.

   c. After ejaculation, the condom must be removed while holding the penis so it doesn't slip off.

   d. The spermicide, nonoxynol-9, was widely recommended for use with condoms, but in recent years it is no longer recommended due to the concern that it irritates mucous membranes and may increase the transmission of HIV or other STIs.

4. Effectiveness

   a. Latex condoms are 88-98% effective, but if used properly with spermicides, effectiveness can be higher.

   b. Female condoms have a 79% effectiveness rate.

   c. Studies have demonstrated that the overall risk of condom breakage, if used correctly, is very low.

   d. It's important that condom users check expiration dates and not store condoms in warm locations such as wallets or cars.

5. Advantages
   a. Barrier methods offer the most protection from STIs, including HIV.
   b. Condoms are relatively inexpensive and can be bought over the counter.
   c. Some people like them for intercourse since they eliminate semen leaking from the vagina (postcoital drip) afterwards.
6. Disadvantages
   a. Condoms can reduce spontaneity.
   b. They may reduce sensation or cause an allergic reaction to latex.
7. Cross-Cultural Use
   a. Condoms are popular outside of the U.S. with 78% of Japanese couples reporting condom use.
   b. The cost varies greatly and may be prohibitive in some countries, such as Brazil.

B. **Diaphragm**

1. Diaphragms are made of latex, come in several different sizes, cost $2- to $30.
2. Current diaphragms are not widely used, but new silicone diaphragms are being developed that may not require a physician's visit.
3. How It Works
   a. Diaphragms work by creating a barrier to the cervical entrance with the latex and a spermicidal jelly.
   b. A visit to a health care provider is necessary in order to make sure it fits properly over the cervix.
   c. Prior to insertion, the diaphragm rim is covered with spermicidal jelly, and one tablespoon of jelly is put into the dome of the diaphragm, and more must to inserted for subsequent acts of intercourse.
   d. After intercourse, the diaphragm must be left in place for at least six to eight hours but not more than 24 hours.
4. Effectiveness
   a. Effectiveness rates range from 82% to 94%.
   b. Research suggests that users who are less than 30 years old or who have intercourse more than 4 times a week have a double risk of failure.
5. Advantages
   a. It can be inserted prior to sexual activity, increasing spontaneity.
   b. The spermicidal jelly provides some protection from STIs and pelvic inflammatory disease.
   c. It doesn't affect hormone levels and is relatively inexpensive.
6. Disadvantages
   a. an office visit for fitting along with learning proper insertion and removal techniques
   b. increase risk of toxic shock syndrome, urinary tract infections, and postcoital drip.
   c. It may move during different sexual positions.
   d. A woman must be comfortable touching her vulva.
7. Cross-Cultural Use
   a. The diaphragm has low usage rates outside of the U.S.
   b. A shortage of physicians may inhibit its use.

C. **Contraceptive Sponge**

1. The *Today* sponge was pulled off the market in 1995 for manufacturing reasons, but it will be returning soon, and is currently sold in Canada.
2. Some women have been using a Canadian-made sponge, *Protectaid*, which costs $3 each.

3. How It Works
    a. Sponges work in three ways: as a barrier, which blocks the entrance to the cervix, through the absorption of sperm, and by deactivating sperm.
    b. Intercourse can occur as many times as desired without adding additional spermicidal jelly or cream, and it must remain in place for six hours after intercourse.
4. Effectiveness
    a. Effectiveness rates for the sponge range from 75-83%.
    b. Like the diaphragm, failure rates are higher in women under the age of 30 or who have frequent sexual intercourse.
5. Advantages
    a. They can be purchased without a prescription or fitting.
    b. It can be inserted prior to sexual intercourse, increasing spontaneity, and no extra spermicide is needed for subsequent acts of intercourse.
    c. They do not affect hormone levels, are disposable, and do not require routine cleaning.
6. Disadvantages
    a. They may increase the risk of toxic shock syndrome and urinary tract infections.
    b. Unlike the diaphragm, the sponge cannot be left in place during a woman's menstrual period.
    c. possible allergic reaction to the spermicide
    d. A woman must be comfortable touching her genitals.
7. Cross-Cultural Use
    a. For many years, women in France have been using vaginal sponges.
    b. Sponges tend to have low usage rates in other cultures.

D. **Prentif Cervical Cap**
1. The cervical cap is a thimble-shaped rubber dome that is placed in the vagina over the cervix and must be fitted by a physician.
2. It is similar to the diaphragm, but it is smaller and can remain in place longer, up to 48 hours.
3. How It Works
    a. It blocks the cervical opening.
    b. It deactivates sperm through the use of spermicidal cream or jelly.
4. Effectiveness
    a. Effectiveness rates for the cervical cap range from 82-94%.
    b. The primary reason for failure is inconsistent or incorrect use.
5. Advantages
    a. It can be left in place for up to 48 hours.
    b. It can be inserted earlier and left in longer than a diaphragm and has higher effectiveness rates than the diaphragm.
    c. It does not affect hormonal levels.
6. Disadvantages
    a. It may increase the risk of toxic shock syndrome, cause abnormal Pap smears, increase the risk of urinary tract infections, and cause possible allergic reactions to the rubber.
    b. It must be fitted by a physician and may dislodge during intercourse.
7. Cross-Cultural Use
    a. It is widely used in England.
    b. It is used infrequently in less-developed countries.

E. **Lea's Shield**

1. In March 2002, the FDA approved the newest female barrier method of contraception.
2. It is a reusable barrier vaginal contraceptive that is made of silicone and used with a spermicidal gel.
3. It has lower slippage rates than the diaphragm and comes in one size only.
4. Although it requires a physician's prescription in the U.S., it is available over the counter in Germany, Austria, Switzerland, and Canada.
5. How It Works
    a. It blocks the entrance to the uterus and deactivates sperm with spermicidal cream or jelly.
    b. It should be left in place for 8 hours after the last intercourse.
6. Effectiveness
    a. Effectiveness rates range from 85-91%.
    b. The primary reason for failure is inconsistent or incorrect use.
7. Advantages
    a. Similar to the cervical cap, Lea's Shield can be inserted earlier and left in longer than a diaphragm.
    b. There is no need to apply more spermicide, and it does not affect hormonal levels.
    c. It has a one-way release valve to reduce the risk of toxic shock syndrome, and there are no latex allergies since it is made of silicone.
8. Disadvantages
    a. It may increase the risk of urinary tract infections.
    b. a physician must fit the device.
9. Cross-Cultural Use
    a. It has been available over-the-counter since 1993 in Germany, Austria, Switzerland, and Canada.

IV. HORMONAL METHODS FOR WOMEN: THE PILL, THE PATCH, AND MORE

- By changing hormonal levels, production of ova can be interrupted and fertilization and implantation can also be prevented.
- Hormonal methods do not protect against STIs.

A. **Combined-Hormone Methods**

1. **Birth Control Pills**
    a. Margaret Sanger was the first to envision **oral contraceptives**, and they received federal approval in 1960.
    b. At first, the pills contained high levels of estrogen that lead to negative side effects. After a reduction in estrogen, current pills have less than half the dose of estrogen than the first pills did.
    c. After 40 years on the market, oral contraceptives still remain the most popular contraceptive method in the U.S. and in the world.
    d. Combination birth control pills that contain synthetic estrogen and a type of progesterone are the most commonly used contraceptive method in the U.S., costing $12-25 per month and requiring a prescription from a physician.
    e. Birth control pills were designed to mimic an average menstrual cycle, but a new pill will be available sometime in 2004 in which women will have only 4 periods a year.
    f. How It Works
        - The increased estrogen and progesterone in the pill prevent the pituitary gland from releasing hormones that cause the ovaries to

begin maturation of an ovum so a woman does not ovulate, the cervical mucus thickens, and the endometrium doesn't build up.

- **Monophasic** pills contain the same amount of hormones in each pill, while **multiphasic** pills vary the hormonal amount.
- Each pill must be taken every day at approximately the same time in order to maintain a certain hormone level in the bloodstream.
- If one pill is forgotten, it should be taken as soon as remembered, and if two pills are forgotten, many physicians recommend taking two pills for the next two days and using a backup method of contraception for the remainder of the cycle.
- Women with a history of circulatory problems, strokes, heart disease, breast or uterine cancer, migraine headaches, hypertension, diabetes, and undiagnosed vaginal bleeding are generally advised not to take oral contraceptives.
- **Triphasil pills** were introduced in the 1990s and have been growing in popularity. They contain three different sets of pills for the month, increasing the hormonal dosage in increments.
- Side effects may include nausea, increase in breast size, breast tenderness, water retention, headaches, increased appetite, fatigue, depression, decreased sexual drive, and high blood pressure.
- **Minipills** or **POPs (progestin-only pills)** contain only progesterone, which reduces side effects since there's no estrogen, although they are less effective.

g. Effectiveness

- Effectiveness rates range from 97-99%.
- To be most effective, they must be taken every day, at the same time of day.

h. Advantages

- If used correctly, they have one of the highest effectiveness rates.
- They don't interfere with spontaneity and reduce the flow of menstruation, menstrual cramps, and premenstrual syndrome.

i. Disadvantages

- They offer no protection from STIs.
- Physical and psychological side effects are common.
- There are increased risks for women who smoke.
- They can be expensive and must be remembered every day.
- There is decreased effectiveness when certain other medications such as antibiotics are used.
- Some women with certain medical conditions cannot use them.

j. Cross-Cultural Use

- They are used around the world more than any other method.
- In 1999, after 35 years of debate, the pill was approved for use in Japan after fears that it would lead to sexual immorality.

2. Hormonal Injectibles

a. *Lunelle* is another hormonal method that is a monthly injection of estrogen and progestin.

b. How It Works

- It works in three ways: preventing ovulation, thickening the cervical mucus, and making implantation unlikely for a fertilized ovum.

c. Effectiveness
- It is 99% effective.

d. Advantages
- It has a high effectiveness rate, does not interfere with spontaneity, and reduces the flow of menstruation, menstrual cramps, and premenstrual syndrome.
- Unlike Depo Provera, it has a fairly quick return to fertility once stopped.

e. Disadvantages
- It offers no protection from STIs.
- It may cause side effects similar to oral contraceptives.
- The effectiveness may be decreased when certain other medications are used.

f. Cross-Cultural Use
- Worldwide scientific studies have found that Lunelle has been safely and effectively used outside the U.S.

3. Hormonal Ring

a. *NuvaRing* is a small plastic ring that it inserted into the vagina once a month (and remains there), releasing a constant does of estrogen and progesterone.

b. How It Works
- It works by inhibiting ovulation, increasing cervical mucus, and rendering implantation impossible.
- The ring is inserted into the vagina, and heat releases the hormones.
- One ring is left in place for three weeks then taken out for one week, when a woman will get her period.

c. Effectiveness
- Less than one woman out of 1000 will become pregnant with perfect use, although effectiveness rates may be lower if the unopened package is exposed to high temperatures or direct sunlight or if they ring is left in the vagina for more than three weeks.

d. Advantages
- similar to other combined-hormone methods of birth control

e. Disadvantages
- no protection from STIs
- side effects, although they may clear up after regular use
- effectiveness is decreased when certain other medications are used.

f. Cross-Cultural Use
- It has been used safely outside the U.S.

4. Hormonal Patch

a. The *Ortho Evra* patch is a thin patch that sticks to the skin and time-releases hormones into the bloodstream.

b. How It Works
- Like other combined-hormone methods it works by inhibiting ovulation, increasing cervical mucus, and rendering implantation impossible.
- It is placed on the buttock, stomach, or upper torso for three weeks, and it will not fall off after swimming or showering.

c. Effectiveness

- It has a 99% effectiveness rate, similar to other combined-hormone methods, although it may be less effective in women who weigh more than 198 pounds.

d. Advantages

- similar to other combined-hormone methods

e. Disadvantages

- similar to other combined-hormone methods
- Some women report that the patch collects fuzz and lint from their clothing.

f. Cross-Cultural Use

- It has been used safely outside the U.S.

B. **Progestin-Only Methods**

- Progestin-only birth control methods (POPs) are those that do not contain estrogen so they can be used by women who cannot take estrogen or who are breastfeeding.
- Minipills are another POP, but they require obsessive regularity in pill-taking and are less likely to be stocked by pharmacies.

1. Subdermal Implants

a. A constant dose of progestin is time-released into the body.

b. Norplant was the first method introduced in the U.S. in 1990, which consisted of silicone cylinders that are implanted in a woman's forearm through a small incision.

c. Other subdermals are in development that will attempt to address the issues of insertion and removal.

d. How It Works

- The matchstick-sized cylinders contain time-released hormones that suppress ovulation, thicken cervical mucus, and render implantation difficult.
- Side effects include irregular bleeding, arm pain, headaches, nausea, etc.

e. Effectiveness

- They are 99.95% effective during the first year of use, but that decreases consistently after the third year,
- Women who weigh more have a greater chance of getting pregnant.

f. Advantages

- Highly effective and long lasting.
- Possibility for women who cannot use oral contraceptives.

g. Disadvantages

- Expensive implantation and lengthy and painful removal process for Norplant.
- No protection from STIs
- Side effects including scars after removal

h. Cross-Cultural Use

- Norplant has high usage rates around the world.

2. Hormonal Injectibles: Depo-medroxyprogesterone acetate (DMOA or Depo-Provera)

a. Depo-Provera was approved for use in the U.S. in 1992 and is the most popular form of non-oral contraceptive. It is injected once every three months.

b. How It Works

- It contains synthetic progesterone, which works by preventing ovulation and thickening cervical mucus.

c. Effectiveness

- 99.7% effectiveness rate

d. Advantages
- It is highly effective, reversible, and does not restrict spontaneity.
- It can be used by women who cannot tolerate estrogen.

e. Disadvantages
- side effects of irregular bleeding with long-term use, resulting in amenorrhea.
- May increase certain cancers and cannot be used by women with a history of liver disease or breast cancer.

f. Cross-Cultural Use
- It has been approved for use in more than 90 countries and is available in 80 countries.

V. CHEMICAL METHODS FOR WOMEN: SPERMICIDES

A. Chemical methods of contraception include **spermicides** such as creams, jellies, foams, suppositories, and films.

1. They are relatively inexpensive, but most of them contain nonoxynol-9, which can be a problem with HIV.
2. Research is underway to develop new **microbicides** that work by inhibiting sperm function, are effective against HIV and STIs, and are not harmful to the vagina or cervical cells.
3. How It Works
   a. Spermicides contain two components: an inert base such as a jelly or cream that holds the spermicide close to the cervix and the spermicide itself.
   b. Most are inserted into the vagina with an applicator.
   c. **Vaginal contraceptive film (VCF)** is now available over the counter and is a thin, small, square sheet that is inserted into the vagina.
4. Effectiveness
   a. Effectiveness rates range from 79-97% with foam being considered the most effective.
   b. They are considerably more effective when used with a diaphragm or condom.
5. Advantages
   a. available over the counter and simple to use
   b. provide more lubrication during intercourse
6. Disadvantages
   a. must be used each time intercourse occurs
   b. postcoital drip and allergies
   c. less effective if used alone
   d. increased risk of HIV if using nonoxynol-9
7. Cross-Cultural Use
   a. Spermicides are not widely used in other countries, probably due to the cost.

VI. INTRAUTERINE METHODS FOR WOMEN: IUDs and IUSs

A. As of 1998, there were only two intrauterine devices (IUDs) approved for use in the U.S., Progestasert and the ParaGard, and 1% of women were using them.
B. In 2000, the FDA approved the *Mirena*, an intrauterine system (IUS), which contains progestin that is time-released into the uterus.
C. The GyneFix is used in Britain and is attached to the uterine wall by a nylon thread and does not cause heavier and more painful periods like other IUDs.
D. The IUD is a method that was more popular in the 1970s and went out of favor due to negative publicity around one type that was linked to pelvic inflammatory disease.

E. IUDs and IUSs

1. How It Works

a. They create a low-grade infection in the uterus, which may either break apart the fertilized ovum, inhibit implantation, and/or increase prostaglandins.

b. They also interfere with sperm mobility and block sperm from passing into the Fallopian tubes.

c. The IUS also releases progesterone, which works like other hormonal methods.

d. A physician must insert IUDs/IUSs, and a string hangs down from the cervix that must be checked monthly.

2. Effectiveness

a. Effectiveness rates range from 97.4-99.2% and depend upon the age of a woman and her past pregnancy history because a woman who has never been pregnant is more likely to expel it through her cervix.

3. Advantages

a. Inexpensive since they can remain in place 5-8 years.

b. Do not interfere with spontaneity.

c. Progestin will decrease menstrual flow.

4. Disadvantages

a. no protection from STIs and may increase transmission

b. risk of uterine perforation, irregular bleeding, and painful insertion and removal

c. can increase menstrual flow and severity of cramps

d. can be expelled from uterus

5 Cross-Cultural Use

a. The IUD is the most frequently used form of contraception in many countries including China, England, and Korea.

VII. NATURAL METHODS FOR WOMEN AND MEN

- Natural methods of contraception do not alter any physiological function.

A. **Natural Family Planning and Fertility Awareness**

1. With **natural family planning (NFP)** (or the **sympto-thermal method**), a woman charts her menstrual periods by taking a daily basal body temperature (BBT) and checking cervical mucus in order to determine when she ovulates. During ovulation, she abstains from sexual intercourse.

2. The **rhythm method** is another term, but this doesn't involve monitoring the signs of ovulation. When charting is used in conjunction with another form of birth control, it is referred to as **fertility awareness**.

3. How It Works

a. A woman takes her BBT every morning before she gets out of bed and records it on a basal body temperature chart because it rises slightly before ovulation.

b. Cervical mucus becomes thin and stretchy during ovulation.

c. After six months of consistent charting, a woman will be able to estimate the approximate time of ovulation, and she can then either abstain from intercourse or use contraception during high-risk times.

4. Effectiveness

a. Effectiveness rates range from 75-99%.

b. The timing of ovulation can be influenced by diet, stress, and alcohol use.

5. Advantages

a. It works for people who can't use other methods due to religious reasons.

b. It is inexpensive and teaches couples about the menstrual cycle and also helps them to communicate about contraception.

c. It offers the opportunity to try other types of sexual expression when intercourse needs to be avoided.

6. Disadvantages
   a. offers no protection from STIs and restricts spontaneity.
   b. low effectiveness and takes time and commitment
   c. difficult for women with irregular cycles
7. Cross-Cultural Use
   a. In places like Peru, Sri Lanka, the Philippines, and Ireland, natural family planning and the rhythm method are the most popular methods of contraception.

B. **Withdrawal**

1. Withdrawal, or coitus interruptus, was the most popular method of birth control in the mid-1800s, but today about 2% of couples use this method.
2. How it Works
   a. Prior to ejaculation, the male withdraws his penis from the vagina so the ejaculate does not enter the uterus.
3. Effectiveness
   a. Effectiveness rates range from 81-96%.
   b. Failures often occur because pre-ejaculatory fluid can contain sperm.
4. Advantages
   a. It may be acceptable for those who cannot use another method for religious reasons.
   b. It's better than no method at all.
5. Disadvantages
   a. no protection from STIs and has low effectiveness rates.
   b. It may contribute to early ejaculation and may be stressful for both partners.
6. Cross-Cultural Use
   a. Some research suggests that it is used by more than 50% of couples in a variety of countries where there are limited contraceptive choices.

C. **Abstinence**

1. Abstinence, or not engaging in sexual intercourse at all, is the only 100% effective method of contraception.
2. Couples who practice abstinence may or may not engage in other sexual behaviors.

VIII. PERMANENT (SURGICAL) METHODS

- Married couples are more likely to use sterilization methods than any other form of birth control, with 48% using female sterilization and 11% using vasectomy.
- It is usually irreversible and requires surgery

A. **Female Sterilization**

1. Female sterilization or tubal ligation is the most widely used method of birth control in the world.
2. A physician will sever or block the Fallopian tubes so that the ovum and sperm cannot meet. It will be done with **cauterization** or **ligation**.
3. The procedure is generally done with the use of a laparoscope through a small incision either under the navel or lower in the abdomen.
4. It has been found to substantially reduce the risk of ovarian cancer.
5. In 2002, *Essure* was approved, which is a tiny, spring-like device that is threaded into the Fallopian tubes where it creates tissue growth around the device, blocking fertilization.

B. **Male Sterilization**

1. Male sterilization, or vasectomy, blocks the flow of sperm through the vas deferens so that a man ejaculates semen that contains no sperm.

2. All functions such as ejaculation, testosterone production, erections, and urination are unaffected by a vasectomy.
3. The surgery is done on an outpatient basis.
4. With a patient under local anesthesia, a physician makes two small incisions in the scrotum and clips the vas deferens.

C. **Effectiveness**
1. Tubal sterilizations are effective immediately, while vasectomies require semen analysis for two months after the procedure to ensure no viable sperm remain.
2. Effectiveness ranges from 99-100%.

D. **Advantages**
1. Sterilization is a highly effective permanent method of contraception.
2. It does not interfere with spontaneity and has no ongoing medical risks.

E. **Disadvantages**
1. no protection from STIs
2. surgery is required and it is irreversible

F. **Cross-Cultural Use**
1. Outside the U.S., female sterilization is the most common contraceptive method.
2. In many countries, it is the only method available and people may receive monetary compensation.

IX.CONTRACEPTION FOR MEN: WHAT'S AHEAD?

A. Women are typically responsible for using birth control and must deal with potential side effects.
B. Some possible male options in the future include implants, hormonal and drug treatments, and gossypol injections, which is an ingredient in cottonseed oil that reduces the quantity of sperm produced.
C. Researchers are also looking into inhibiting the maturing process of sperm in the epididymis.

X. INEFFECTIVE METHODS: DON'T COUNT ON IT!

A. Douching
1. Douching involves using a syringe-type instrument to inject a stream of water in to the vagina, and this was advocated as a method of contraception in the mid-1800s.
2. Now it is known that it is ineffective since sperm move quickly up through the cervix and douching may push them up more quickly.
3. Douching in general is discouraged since it is unnecessary and may lead to an increased risk for STIs and vaginal infections.

B. Breastfeeding is an unreliable method of birth control since a woman may ovulate while she is breastfeeding, even if menstruation doesn't occur.

XI. EMERGENCY CONTRACEPTION

A. Emergency contraception (EC) is designed to prevent pregnancy after unprotected vaginal intercourse, in case of unanticipated sexual intercourse, contraceptive failure, or sexual assault.
B. EC interferes with fertilization and implantation, through either the use of emergency contraceptive pills or copper IUD insertion within five days of unprotected intercourse (although IUDs are not as common).
C. The **Yuzpe Regimen** involved using high doses of combination birth control pills to inhibit pregnancy after unprotected intercourse.
D. Currently, The *Preven* Emergency Contraceptive Kit and *Plan B* are both approved by the FDA with Preven containing both estrogen and progestin and having a 75% effectiveness rate, while Plan B is progestin-only and has a 89% effectiveness rate.

E. Other methods of emergency contraception include early **vacuum aspiration** or **menstrual extraction**, which involves removing the uterine contents prior to a positive pregnancy test.

XII. CONTRACEPTION IN THE FUTURE

A. Many new developments are on the way that includes either completely new methods or improvements to current methods such as cervical caps and spermicidal gels.
B. **Immunocontraceptives** are vaccines that suppress testicular function and would eliminate sperm and testosterone production (which is unacceptable but may be promising for the future).
C. Saliva and urine tests may help natural planning methods.

XIII. THE ABORTION DEBATE

A. Every year there are 50 million abortions done worldwide; 30 million legal, while 20 million are illegal.
B. **Pro-life supporters** believe that abortion should be illegal or strictly regulated by the government.
C. **Pro-choice supporters** believe that the abortion decision should be left up to the woman and not regulated by the government.

XIV. HISTORICAL PERSPECTIVES ON ABORTION

A. There are few large-scale societies historically where abortion has not been practiced, including early Greek, Roman, and Hebrew cultures.
B. For most of Western history, both Christianity and Judaism condemned abortion and punished those who used it, although it was still used.
C. New York became the first state to permit abortion so women flocked to New York or had illegal abortions or, **back-alley abortions** that often resulted in multiple complications and even death.
D. In 1973 the Supreme Court ruled in Roe v. Wade that women have a constitutionally protected right to have an abortion in the early stages of pregnancy.
E. Since Roe v. Wade, restrictions have been passed to limit abortions that include waiting periods, mandatory counseling, parent involvement requirements, and public funding limitations.
F. In 1994 the Supreme Court upheld a decision barring antiabortion demonstrators from getting within 36 feet of an abortion clinic after Dr. David Gunn was shot and killed outside of his clinic.
G. Medical schools have been pressured to eliminate abortion training in residency programs.
H. **Attitudes Toward Abortion**

1. College students have generally been viewed as fairly liberal in their attitudes about abortion, but studies have found a normal distribution of abortion attitudes.
2. No gender differences in attitudes about abortion have been found, but regarding race, African-Americans tend to be more pro-choice than white Americans.

I. **Legal vs. Illegal Abortions**

1. Since 1973, the number of deaths from abortion has declined dramatically since the legalization ensured that sanitary conditions were strictly followed.
2. Pro-life supporters believe that the legalization of abortion has caused women and men to become irresponsible about sexuality and contraceptive use.

XV. TYPES OF ABORTIONS

A. **Abortion Procedures**

1. In an effort to reduce costs, abortions have moved from hospitals to clinics.
2. Like all surgical procedures, the most serious risks of abortion include uterine perforation, cervical laceration, severe hemorrhaging, infection, and anesthesia-related complications.

3. **First-trimester procedures** are done before 14 weeks of gestation and are simpler and safer, while **second-trimester procedures** are done between 14 and 21 weeks.

B. **First Trimester Abortion**

1. A first-trimester abortion is called a **vacuum aspiration**, or suction, abortion and is usually performed on an outpatient basis, using local anesthesia, and is the most common type of abortion procedure in the U.S. today, accounting for 97% of all abortions.
2. **Dilation rods** are used to open the cervix and a **cannula** is inserted into the cervix and is attached to a **vacuum aspirator** that empties the contents of the uterus.
3. It usually takes 4 to 6 minutes and requires a stay at a clinic or doctor's office for a few hours. Once a woman is home, she is advised to rest and avoid sexual intercourse for at least 2 weeks to reduce infections.
4. The risks are much lower compared to second-trimester abortions and include excessive bleeding, possible infection, and uterine perforation.

C. **Second Trimester Abortion**

1. About 11% of abortions occur during the second trimester due to medical complications, fetal deformities, marital problems, miscalculation of the due date, financial or geographic problems, or denial of the pregnancy.
2. Between 13 and 16 weeks, a **dilation and evacuation (D & E)** is the most commonly used procedure. It is done under general anesthesia when dilators such as laminaria (seaweed that expands) are used to dilate the cervix 24 hours prior to the abortion.
3. A woman returns for a 15 to 30 minute procedure to empty the uterus, which includes risks such as increased pain, blood loss, and cervical trauma.
4. In the late part of the second trimester some physicians may use **induced labor procedures**, including **saline** or **prostaglandin**.
5. A **hysterotomy** is a second trimester abortion that may be used if the above methods are not possible or if the woman's life is in danger.

XVI. NEWER (NONSURGICAL) DEVELOPMENTS

A. **RU-486**

1. Although Mifepristone (RU-486) has only been FDA-approved in the U.S. since 2000, it has been used in several European countries for more than a decade.
2. RU-486 is an antiprogestin that blocks the development of progesterone, causing a breakdown in the uterine lining that is needed to sustain a pregnancy.
3. Three pills are taken, then two days later a woman takes an oral dose of prostaglandin, which causes uterine contractions that expel the fertilized ovum.
4. It is between 95-97% effective and can be used up until 7 weeks from the last menstrual period, but does have side effects including nausea, cramping, vomiting, and uterine bleeding.

B. **Methotrexate**

1. The FDA approved Methotrexate in 1953 as a breast cancer drug.
2. When it is given in the form of an injection and used in combination with a prostaglandin, it has been found to cause a drug-induced miscarriage.
3. It works by stopping the development of the developing cells of the zygote and can be used up to 7 weeks past the last menstrual period

XVII. WHY DO WOMEN HAVE ABORTIONS?

A. Many people claim that women have abortions because they do not use contraception; however, studies indicate that 70% of women who undergo an abortion had either been using a method or had discontinued use within three months of conception.

B. Nine percent of women who have abortions never used any contraception.

C. Reasons for abortion are varied.

1. The majority report that a baby would interfere with other responsibilities, such as educational or career goals.
2. an inability to financially provide for a child
3. difficulties in the relationship with a male partner
4. not wanting people to know they are sexually active
5. pressure from their partners or families
6. fetal deformity
7. risks to mother's health
8. having several children already
9. rape or incest

XVIII. HOW DO WOMEN REACT TO ABORTION?

- In the late 1980s, the Surgeon General released a reported that documented that physiological health consequences, including fertility, incompetent cervix, miscarriage, premature birth, and low birth weight, are not more frequent among women who experience abortion than they are among the general population of women.

A. **Physiological Symptoms**

1. After an early abortion, many women report increased cramping, heavy bleeding with possible clots, and nausea.
2. The death rate from an abortion before nine weeks is 1 in 260,000.
3. Despite concerns from pro-life supporters, over thirty published studies have found no relationship between abortion and breast cancer.

B. **Psychological Symptoms**

1. The majority of evidence from scientific studies indicates that most women who undergo abortion have very few psychological side effects later on, with relief being the most prominent response for the majority of women.
2. Some research does suggest that some women experience intense, negative psychological consequences that include guilt, anxiety, depression, and regret.
3. Certain conditions may put a woman more at risk for developing severe psychological symptoms.

a. being young
b. not having family or partner support
c. being persuaded to have any abortion when a woman doesn't want one
d. having a difficult time making the decision
e. having a strong religious background
f. having an abortion for medical or genetic reasons
g. having a history of psychiatric problems before the abortion
h. having a late abortion procedure
i. women who decide to tell no one about their decision

XIX. TEENS AND ABORTION

A. Each year in the U.S. one million adolescents become pregnant, and 85% of these pregnancies are unintended.

B. Many states have passed laws that control and limit teenagers' access to abortion such as parental notification laws, parental consent laws, or judicial bypass options.

C. As of 1999, 32 states require minors to inform at least one parent of their decision to have an abortion, while 21 states require parental consent.

D. Studies show that in states without mandatory parental consent or notice requirements, 75% of minors involve one or both parents.

XX. HOW PARTNERS REACT TO ABORTION

A. Some studies suggest that abortion causes couples to break up, but there is also evidence that if couples can communicate about their thoughts, feelings and fears, an abortion may actually bring them closer.

B. Women whose partners support them and help them through the abortion show more positive responses after an abortion.

C. Men also have a difficult time with the decision to abort, and they often experience sadness, a sense of loss, and fear for their partner's well being

XXI. CROSS-CULTURAL ASPECTS OF ABORTION

A. It is estimated that worldwide, 46 million women have abortions each year, and 78% of these women live in developing countries, while 22% live in developed countries.

B. The highest reported abortion rates are Romania, Cuba, and Vietnam, while the lowest are Belgium, the Netherlands, Germany, and Switzerland.

C. About 20 million unsafe abortions take place each year, which include taking drugs, inserting objects into the vagina, flushing the vagina with certain liquids, or having the abdomen vigorously massaged.

TEST YOURSELF

Below you will find fill-in-the-blank and short answer essay questions for topics covered in chapter 13. Check your answers at the end of this chapter.

Contraception

1. The ancient __________________ used magic, superstition, herbs, and drugs to try and control their fertility.

2. With the advent of __________________ in the 1980s, condoms have become popular not only for their contraceptive abilities, but also for the protection they provide from STIs.

3. __________________ __________________ can be purchased without a prescription, and once inserted, sexual intercourse can take place as many times as desired during a twenty-four-hour period.
4. Combination birth control pills can either be __________________ or __________________.

5. *NuvaRing* is a small plastic ring that is inserted into the __________________ once a month and releases a constant dose of estrogen and progesterone.

6. Like other hormonal methods of birth control, the *Ortho Evra* patch has a __________________ effective rate, does not interfere with __________________, and reduces the flow of menstruation, menstrual cramps, and premenstrual syndrome.

7. Subdermal contraceptive implants time-release a constant dose of __________________.

8. Chemical methods of contraception include __________________ such as creams, jellies, foams, suppositories, and films.

9. IUDs and IUSs have been found to create a low-grade infection in the __________________, which may break apart, the fertilized ovum, inhibit implantation, and/or increase prostaglandins.

10. A woman using the sympto-thermal method charts her menstrual periods by taking a daily basal body ________________ and checking ________________ ________________ in order to determine when she ovulates.

11. Female sterilization, or ________________ ________________, is the most widely used method of birth control in the world.

12. Male sterilization, or vasectomy, blocks the flow of sperm through the ________________ ________________.

13. Two ineffective methods of contraception are ________________ and ________________.

14. ________________ ________________ is designed to prevent pregnancy after unprotected vaginal intercourse, or in case of unanticipated sexual intercourse, contraceptive failure, or sexual assault.

15. ________________ are vaccines that suppress testicular function and would eliminate sperm and testosterone production.

16. Why do diaphragms need to be prescribed by a physician?

17. Why do oral contraceptives need to be taken every day, at approximately the same time?

18. How long does one injection of Depo-Provera last?

19. How effective are male and female sterilization methods?

20. What country passed a law in 2000 that allows high school nurses in public and private schools to provide emergency contraception (EC) to all students?

Abortion

21. Aristotle argued that ________________ was necessary as a backup to contraception.

22. Illegal abortions, known as ________________ ________________, were very dangerous because they were often performed under unsanitary conditions and resulted in multiple complications, sometimes ending in death.

23. Because of the controversy over abortion, ________________ ________________ have been targeted to eliminate abortion training in residency programs.

24. The bitter battle between the ________________ and ________________ factions has resulted in picketing and demonstrations outside abortion clinics.

25. The ________________ of abortion ensured that sanitary conditions were strictly followed, and if infections developed, the patients were treated immediately.

26. A first-trimester abortion, ________________ ________________, is usually performed on an outpatient basis, using local anesthesia.

27. During a first-trimester abortion, dilation rods are used to open the ________________ and usually cause mild cramping of the uterus.

28. Between thirteen and sixteen weeks of pregnancy, a ________________ and ________________ is the most commonly used abortion procedure.

29. Many women perceive ________________ abortion as more acceptable and natural and cite the fact that it allows them to have the abortion in the privacy of their own homes.

30. Although Mifepristone (RU-486) has only been FDA-approved in the U.S. since 2000, it has been used in several European countries for more than a ________________.

31. Methotrexate is often given in the form of an ________________ and when it is used in combination with a prostaglandin, it has been found to cause a drug-induced abortion.

32. Studies indicate that ________ % of women who undergo an abortion had either been using a method birth control or had discontinued use within three months of conception.

33. Over thirty published studies have found no relationship between ________________ and breast cancer.

34. Women who decide to tell no one about their decision to abort, because they encounter a lack of support, were found to have more ________________ psychological symptoms following the procedure.

35. Worldwide, the lifetime average is ________________ abortion per woman.

36. What is the name of the Supreme Court decision from 1973 that stated that women have a constitutionally protected right to have an abortion in the early stages of pregnancy?

37. What are the most serious risks of abortion?

38. Why do most physicians advise women who are considering abortions to have a first-trimester abortion?

39. How does RU-486 work to cause an abortion?

40. What are a few of the unsafe methods women use worldwide to cause abortions?

POST TEST

Below you will find true/false, multiple-choice and matching quiz items covering the entire chapter. Check your answers at the end of this chapter.

True/False

1. Ancient Egyptians inserted a mixture of crocodile feces, sour milk, and honey into the vagina as a form of birth control.

2. Users of the contraceptive sponge need to add additional spermicidal jelly or cream for subsequent acts of intercourse.
3. Taking antibiotics can reduce the effectiveness of oral contraceptives.

4. Withdrawal or coitus interruptus is usually highly effective at preventing pregnancy.

5. Tubal ligation procedures are usually reversible.

6. Due to the abortion controversy, medical schools have been pressured to eliminate abortion training in residency programs.

7. The risks of a first-trimester abortion are much lower compared to second-trimester abortions.

8. The FDA approved RU-486 in 1953 as a breast cancer drug.

9. Women who decide to tell no one about their decision to have an abortion are at more risk for developing serious psychological reactions.

10. A high number of unsafe abortions continue to take place worldwide.

Multiple-Choice

11. Which statement below regarding the female condom is FALSE?
    a. It is 7 inches long.
    b. It has two flexible rings that help with insertion.
    c. It is made of latex.
    d. It is fairly expensive at $2.00 each.
    e. None of the above

12. Which statement below regarding male condoms is FALSE?
    a. Latex condoms are between 88-98% effective, but if used properly with spermicides, effectiveness can be higher.
    b. Condoms are relatively expensive, and a person must be 18 to buy them.
    c. Studies have demonstrated that the overall risk of condom breakage, if used correctly, is very low.
    d. Water-based lubricants can be used with a condom, but oil lubricants will damage the latex.
    e. None of the above

13. What contraceptive method works by creating a barrier to the cervical entrance with latex and spermicidal jelly?
    a. female condom
    b. Depo-Provera
    c. Hormonal Ring
    d. diaphragm
    e. contraceptive sponge

14. What is NOT an advantage of the contraceptive sponge?
    a. They can remain in place for up to one week.
    b. They can be purchased without a prescription or fitting.
    c. They can be inserted prior to sexual intercourse, increasing spontaneity
    d. No extra spermicide is needed for subsequent acts of intercourse.
    e. They are disposable and do not require routine cleaning.

15. What birth control method is widely used in England?
    a. Hormonal Ring
    b. Depo-Provera
    c. cervical cap
    d. Lea's shield
    e. contraceptive sponge

16. What barrier contraceptive has a one-way release valve to reduce the risk of toxic shock syndrome?
    a. diaphragm
    b. contraceptive sponge
    c. Depo-Provera
    d. Lea's shield
    a. cervical cap

17. What symptom below should an oral contraceptive user report to a physician immediately?
    a. chest or abdominal pain
    b. severe headaches
    c. severe leg or calf pain
    d. vision or eye problems
    e. All of the above

18. What statement below regarding oral contraceptives is FALSE?
    a. To be most effective, they must be taken once a week, on the same day each week.
    b. They offer no protection from sexually transmitted infections.
    c. There are increased risks for women who smoke.
    d. They can be expensive.
    e. There is decreased effectiveness when certain other medications are used such as antibiotics.

19. What kind of method is Depo Provera?
    a. spermicide
    b. oral contraceptive
    c. hormonal injectible
    d. barrier contraceptive
    e. subdermal implant

20. What is an advantage of spermicides?
    a. They decrease menstrual flow.
    b. They do not require a prescription.
    c. They last for three months.
    d. They are highly effective if used alone.
    e. All of the above

21. What is the name of the small plastic device that is inserted into the uterus to prevent pregnancy?
    a. cervical cap
    b. Depo-Provera
    c. Norplant
    d. Lea's Shield
    e. IUD

22. What activities may be done as part of natural family planning?
    a. checking the consistency of cervical mucus
    b. charting the menstrual cycle on a calendar
    c. taking a daily basal body temperature
    d. All of the above
    e. None of the above

23. When was the withdrawal or coitus interruptus popular?
    a. in the 1980s
    b. in the 1920s
    c. in the late 12th century
    d. in the mid-1800s
    e. It's always been popular.

24. What is the most widely used method of birth control in the world?
    a. tubal ligation
    b. oral contraceptives
    c. intrauterine devices
    d. condoms
    e. natural family planning

25. What part of the male anatomy is clipped during a vasectomy?
    a. vas deferens
    b. testicles
    c. urethra
    d. epididymis
    e. Cowper's glands

26. Even though it is an unreliable method, why has breastfeeding been considered a form of birth control over the years?
    a. Most women do not engage in intercourse while breastfeeding.
    b. The hormones released during breastfeeding make implantation of a fertilized ovum unlikely.
    c. The hormones released during breastfeeding create an acidity in the vagina that deactivates sperm.
    d. While women are breastfeeding, ovulation is less likely to occur.
    e. All of the above

27. What is the name for birth control that is designed to prevent pregnancy after unprotected vaginal intercourse, or in case of unanticipated sexual intercourse, contraceptive failure, or sexual assault?
    a. gossypol injections
    b. emergency contraception
    c. contraceptive sponges
    d. Depo-Provera
    e. tubal ligation

28. What is the name for vaccines in the developmental stage that suppress testicular function and would eliminate sperm and testosterone production?
    a. cannula injections
    b. DES acetate
    c. Mifepristone
    d. laminaria vaccines
    e. immunocontraceptives

29. When it comes to the history of abortion, which of the following statements is TRUE?
    a. There are few large-scale societies historically where abortion has not been practiced.
    b. Aristotle argued that abortion was necessary as a backup to contraception.
    c. Before the common era, pagans, Jews, and Christians used abortion, usually to hide sexual activity.
    d. In early Roman society, abortions were allowed, but husbands had the power to decide if their wives would undergo the procedure.
    e. All of the above

30. What 1973 Supreme Court ruling stated that women have a constitutionally protected right to have an abortion?
    a. Planned Parenthood v. Casey
    b. Madison v. Women's Health Center
    c. Roe v. Wade
    d. Gunn v. Louisiana
    e. NARAL v. Parsons

31. What type of abortion would a woman have before 14 weeks of gestation?
    a. second-trimester abortion
    b. hysterotomy
    c. first-trimester abortion
    d. induced labor procedures
    e. All of the above

32. In a first-trimester abortion, what is used to open the cervix?
    a. dilation rods
    b. intrauterine devices
    c. gossypol injections
    d. curettes
    e. microbicides

33. What is a hysterotomy?
    a. a surgical instrument which is used to empty the uterus during an abortion procedure
    b. a second trimester abortion that may be used if the woman's life is in danger or other methods aren't possible
    c. a type of first-trimester abortion
    d. the insertion of an IUD after pregnancy
    e. when a woman's uterus is removed as part of an abortion procedure

34. What is another term for RU-486?
    a. RU-486
    b. Lunelle
    c. Depo-Provera
    d. Mifepristone
    e. gossypol

35. What abortion method stops the development of the developing cells of the zygote?
    a. inhibin
    b. RU-486
    c. Mifepristone
    d. prostaglandins
    e. Methotrexate

36. What percent of women who get an abortion have never used contraception in their lives?
    a. 9%
    b. 26%
    c. 52%
    d. 77%
    e. 96%

37. What is a short-term physiological reaction reported by many women after an early abortion?
    a. increased cramping
    b. nausea
    c. heavy bleeding
    d. All of the above
    e. None of the above

38. According to research, what is the most common psychological response women report after having an abortion?
    a. indifference
    b. depression
    c. relief
    d. All of the above
    e. None of the above

39. What is the name of the abortion legislation that exists in 32 states that requires the notification of the parents of a minor prior to an abortion procedure?
    a. parental notification
    b. adolescent access restraints
    c. teen abortion restriction
    d. parent approval rule
    e. judicial bypass option

40. What is a reaction men report when it comes to their partners having an abortion?
    a. Counseling services for men are often not available at abortion clinics.
    b. They often do not discuss the pregnancy with anyone other than their partner.
    c. Men may try to bury their feelings by becoming rational and intellectual.
    d. All of the above
    e. None of the above

Matching

Column 1		Column 2
A. mifepristone	____	41. a thimble-shaped rubber dome that is placed in the vagina over the cervix, blocking sperm from moving through the uterus
B. tubal ligation	____	42. a surgical procedure in which the vas deferens is cut, tied, or cauterized for permanent contraception.
C. emergency contraception	____	43. a type of natural family planning that involves monitoring basal body temperature and cervical mucus
D. triphasil	____	44. process where the Fallopian tubes are cut or blocked, preventing the joining of sperm and ova.
E. vasectomy	____	45. chemical name for RU486 which has been used to induce early abortion
F. Depo Provera	____	46. a pump that is used during abortion procedures
G. cervical cap	____	47. a popular injectible hormonal contraception that is injected every 3 months
H. nonoxynol-9	____	48. a spermicide that deactivates sperm but has been found to irritate mucous membranes and may increase the transmission of HIV or other STIs
I. sympto-thermal method	____	49. contraception that is designed to prevent pregnancy after unprotected vaginal intercourse, or in case of unanticipated sexual intercourse, contraceptive failure or sexual assault.
J. vacuum aspirator	____	50. a type of multiphasic oral contraceptives that contain three different types of pills will different levels of hormones

Test Yourself Answer Key

Contraception

1. Greeks (p. 385)
2. AIDS (p. 388)
3. Contraceptive sponges (p. 394)
4. monophasic, multiphasic (p. 397)
5. vagina (p. 401)
6. high, spontaneity (p. 402)
7. progestin (p. 402)
8. spermicides (p. 404)
9. uterus (p. 406)
10. temperature, cervical mucus (p. 407)
11. tubal sterilization (p. 409)
12. vas deferens (p. 410)
13. douching, breastfeeding (p. 411)
14. Emergency contraception (p. 412)
15. Immunocontraceptives (p. 413)
16. They come in a variety of different sizes, and a health care provider needs to ensure a proper fit over the cervix. (p. 392)
17. in order to maintain a certain hormonal level in the bloodstream so that ovulation does not occur. (p. 397)
18. three months (p. 404)
19. Effectiveness for both procedures ranges from 99% to 100%. (p. 410)
20. France (p. 412)

Abortion

21. abortion (p. 414)
22. back-alley abortions (p. 414)
23. medical schools (p. 414)
24. pro-choice, pro-life (p. 415)
25. legalization (p. 415)
26. vacuum aspiration (p. 416)
27. cervix (p. 416)
28. dilation, evacuation (p. 417)
29. nonsurgical (p. 418)
30. decade (p. 418)
31. injection (p. 418)
32. 70 (p. 418)
33. abortion (p. 419)
34. negative (p. 421)
35. one (p. 422)
36. Roe v. Wade (p. 414)
37. uterine perforation, cervical laceration, severe hemorrhaging, infection, and anesthesia-related complications (p. 416)
38. The risks are much lower in a first-trimester abortion, partly due to the use of local anesthesia versus general anesthesia. (p. 417)
39. It is an antiprogestin that blocks the development of progesterone, causing a breakdown of the uterine lining, which is needed to sustain a pregnancy (p. 418).
40. taking drugs, inserting objects into the vagina, flushing the vagina with certain liquids or having the abdomen vigorously massaged (p. 422)

Post Test Answer Key

True/False

1. T (p. 385)
2. F (p. 394)
3. T (p. 400)
4. F (p. 408)
5. F (p. 409)
6. T (p. 414)
7. T (p. 417)
8. F (p. 418)
9. T (p. 421)
10. T (p. 422)

Multiple Choice

11. c (p. 388)
12. b (pp. 388-391)
13. d (p. 391)
14. a (pp. 394-395)
15. c (p. 395)
16. d (p. 396)
17. e (p. 399)
18. a (pp. 399-400)
19. c (p. 400)
20. b (p. 405)
21. e (p. 405)
22. d (p. 407)
23. d (p. 408)
24. a (p. 409)
25. a (p. 410)
26. d (p. 411)
27. b (p. 412)
28. e (p. 413)
29. e (p. 414)
30. c (p. 414)
31. c (p. 416)
32. a (p. 416)
33. b (p. 417)
34. d (p. 418)
35. e (p. 418)
36. a (p. 419)
37. d (p. 419)
38. c (p. 420)
39. a (p. 421)
40. d (pp. 421-422)

Matching

41. G (p. 395)
42. E (pp. 409-410)
43. I (p. 407)
44. B (p. 409)
45. A (p. 418)
46. J (p. 417)
47. F (p. 404)
48. H (pp. 388-390)
49. C (p. 412)
50. D (p. 399)

Chapter 14 – Challenges to Sexual Functioning

CHAPTER SUMMARY

The Diagnostic and Statistical Manual (DSM) classify sexual dysfunctions. A sex therapist's first task is to ascertain whether a problem is psychological, physiological, or mixed. A primary sexual dysfunction is one that has always existed, while a secondary sexual dysfunction is one in which a problem developed after a period of adequate functioning. A situational sexual dysfunction is a difficulty that occurs during certain sexual activities or with certain partners, while a global sexual dysfunction is a problem that occurs in every situation, during every type of sexual activity, and with every sexual partner.

Sexual desire disorders include hypoactive sexual desire and sexual aversion and are considered by many therapists to be the most complicated sexual dysfunction to treat. Sexual arousal disorders include female sexual arousal disorder and male erectile disorder. A tremendous amount of research has been dedicated to finding causes and treatment options for erectile disorder. Treatment options include drugs such as Viagra, hormonal treatments, intracavernous injections, vacuum constriction devices, and surgical treatments. Orgasm disorders include female and male orgasmic disorder, premature ejaculation, and retarded ejaculation. Female orgasmic disorder and premature ejaculation are more common, while male orgasmic disorder and retarded ejaculation are less common. The recommended treatment for premature ejaculation includes techniques done by a man (or his partner) during sexual activity called the squeeze technique or the stop-start technique. Pain disorders include vaginismus and dyspareunia, which is often caused by a type of vulvodynia called vulvar vestibulitis syndrome. Other sexual problems that aren't necessarily classified as dysfunctions but may be linked to dysfunctions include faking orgasms, sleep sex, and Peyronie's disease.

Treatments for the various dysfunctions depend on the causes and may be more medical, such as medication or surgery, or they may be more psychological, such as counseling or bibliotherapy. Bibliotherapy refers to reading sexuality books or educational material and has been found to be particularly helpful in reducing certain sexual problems such as orgasmic dysfunction.

Sexual functioning involves a complex physiological process that can be impaired by pain, immobility, changes in bodily functions, or medications. Individuals with a variety of medical conditions may have to deal with temporary or chronic sexual dysfunctions. People with cardiovascular problems often fear that their damaged heart is not up to the strain of intercourse or orgasm, but research has found that the risk of sexual activity triggering a heart attack is extremely low. Cancer can affect sexual functioning in a variety of ways including physical scars, the loss of limbs or body parts, changes in skin texture due to radiation, the loss of hair, nausea, bloatedness, weight gain or loss, and ostomies. Different cancers that affect men and women's reproductive system may have more impact on their sexual functioning, such as cancer of the breast, prostate, cervix, uterus, or testes. Sexuality is affected by other chronic conditions such as chronic obstructive pulmonary disease, diabetes, multiple sclerosis, alcoholism, HIV/AIDS, and mental illness. People with spinal cord injuries may have to develop new and broader ways to explore sexual expression.

LEARNING OBJECTIVES

After studying Chapter 14, you should be able to:

1. Identify psychological factors related to sexual dysfunctions.
2. Identify physical factors related to sexual dysfunctions.
3. Distinguish between the four general categories of sexual dysfunctions: primary, secondary, situational, and global.
4. List and define sexual desire disorders.
5. Describe treatment options for sexual desire disorders.
6. List and define sexual arousal disorders.
7. Describe treatment options for sexual arousal disorders.
8. Compare and contrast treatment options for erectile dysfunction.
9. List and define orgasm disorders.
10. Describe treatment options for orgasm disorders.
11. Explain the squeeze and stop-start techniques.
12. List and define pain disorders.
13. Describe treatment options for pain disorders.
14. Define bibliotherapy.
15. Describe the sexual functioning issues relevant to a variety of illness and/or chronic conditions: cardiovascular problems, cancer, chronic pain, chronic obstructive pulmonary disease, diabetes, and multiple sclerosis.
16. Discuss the short- and long-term effects of alcohol.
17. Explain some specific considerations for the sexuality of disabled individuals.
18. Discuss the sexuality of people living with HIV/AIDS.
19. Discuss the challenges faced by people diagnosed with mental illness when it comes to sexuality.
20. Discuss the challenges faced by people with developmental delays when it comes to sexuality.

CHAPTER OUTLINE

I. WHAT IS SEXUAL DYSFUNCTION?

A. Even with the increased interest in female sexuality recently, the research still focuses largely on male sexual dysfunction.

B. Sexual dysfunctions are classified by the **Diagnostic and Statistical Manual (DSM)**, which is the major diagnostic system used in U.S. research and therapy. However, this system remains heterosexist, classifying dysfunctions in terms of sexual intercourse.

C. **Are All Sexual Problems Sexual Dysfunctions?**

1. Many men and women experience sexual problems, but sexual problems aren't the same as sexual dysfunctions.
2. Even though couples with sexual dysfunctions often have lower frequencies of sexual intercourse, they report feeling very positive about their sexual relationships.
3. A sex therapist's first task is to ascertain whether a problem is psychological, physiological, or mixed.

D. **Psychological Factors in Sexual Dysfunction**

1. Some psychological causes include unconscious fears, anxiety, stress, depression, guilt, anger, and fear of infidelity, partner conflict, fear of intimacy, dependency, abandonment, or loss of control.
2. Both **performance fears** and an excessive need to please a partner interfere with sexual functioning.
3. Sex therapy usually begins by overcoming performance fears, feelings of sexual inadequacy, and other anxieties.
4. Distractions, shifts in attention, or preoccupation during sexual arousal may interfere with the ability to become aroused, as can **spectatoring** (acting as an observer or judge of one's own sexual performance).

E. **Physical Factors in Sexual Dysfunction**

1. Physical causes such as disease, disability, illness, and many commonly used drugs can interfere at any point in the sexual response cycle in both men and women.
2. A person with poor health, who smokes, or takes medications or illegal drugs experiences more sexual dysfunctions.
3. Sexual problems for women, with the exception of lubrication, have been found to decrease with age, while erection problems in men tend to increase with age.

II. SEXUAL DYSFUNCTIONS: WHAT THEY ARE AND HOW WE TREAT THEM

A. **Categorizing the Dysfunctions**

1. A **primary sexual dysfunction** is one that has always existed, while a **secondary sexual dysfunction** is one in which a problem developed after a period of adequate functioning.
2. A **situational sexual dysfunction** is a difficulty that occurs during certain sexual activities or with certain partners.
3. A **global sexual dysfunction** is a problem that occurs in every situation, during every type of sexual activity, and with every sexual partner.
4. The type affects treatment strategies since primary problems tend to have more biological or physiological causes, while secondary problems tend to have more psychological causes.

B. **Treating Dysfunctions**

1. Treatment of most sexual dysfunctions begins with a medical history and an evaluation of any past sexual trauma or abuse.
2. Treatment of sexual dysfunctions may be **multimodal**, involving more than one type of therapy.

3. Success rates of treatment vary for different types of dysfunctions.

III. SEXUAL DESIRE DISORDERS

A. There appear to be fewer cases of male hypoactive sexual desire than female, which may be because men feel less comfortable discussing the problem.

B. Sexual desire disorders are considered by many therapists to be the most complicated sexual dysfunction to treat.

C. **Hypoactive Sexual Desire**

1. When someone has hypoactive sexual desire (HSD), there is a low or absent desire for sexual activity, even though a person can still function normally.
2. There may be a lack of sexual fantasies, a reduction of or absence in initiating sexual activity, a lack of physiological response when sexually stimulated, or a decrease in self-stimulation.
3. Psychological causes for HSD include a lack of attraction to one's partner, fear of intimacy and/or pregnancy, marital or relationship conflicts, religious concerns, depression, and other psychological disorders.
3. HSD can also result from negative messages about female sexuality, anorexia, or sexual coercion.
4. HSD in both men and women may also be due to biological factors such as hormonal problems, medication side effects, and illness, along with chronic use of alcohol, marijuana, and/or cocaine.

D. **Treating Sexual Desire**

1. Cognitive-behavioral therapy is most promising for women, as it emphasizes the importance of how a person thinks and the effects these thoughts have on a person's feelings and behaviors.
2. Testosterone has been used for both men and women; however, the majority of men who experience low sexual desire have normal levels of testosterone.
3. The **discrepancy in desire** between the partners can create a problem for a couple.

E. **Sexual Aversion**

1. A person with sexual aversion reacts with strong disgust or fear to a sexual interaction, blocking the ability to become sexually aroused.
2. Sexual abuse and anorexia have been found to be associated with sexual aversion.

F. **Treating Sexual Aversion**

1. The most common treatment for sexual aversion involves discovering and resolving the underlying conflict that is contributing to the sexual aversion, often using cognitive-behavioral therapy.
2. Treatment often includes goal setting and the completion of homework assignments. However, successful treatment varies, and there have been few studies that have documented success in men and women.

IV. SEXUAL AROUSAL DISORDERS

A. Sexual arousal disorders occur even when the client reports adequate focus, intensity, and duration of sexual stimulation and may be primary, or more commonly, secondary in that it only occurs with a certain partner or specific sexual behavior.

B. **Female Sexual Arousal Disorder (FSAD)**

1. FSAD is an inability to either obtain or maintain adequate lubrication in response to sexual excitement.
2. Physiological factors include decreased blood flow to the vulva, while psychological factors include fear, guilt, anxiety, and depression.
3. **Persistent sexual arousal syndrome** is the opposite of FSAD, where a woman experiences excessive and unremitting arousal.

C. **Treating Female Sexual Arousal Disorder**

1. Clinical trials of Viagra to treat FSAD have shown that although it can increase vasocongestion and lubrication, it provides little overall benefit in the treatment of FSAD.
2. The most effective treatment may be a combination of drugs and psychological therapy.
3. The FDA approver the EROS Clitoral Therapy Device in 2000 for the treatment of FSAD.

D. **Male Erectile Disorder (ED)**

1. ED is defined as the persistent inability to obtain or maintain an erection sufficient for satisfactory sexual performance.
2. Psychological factors (including fear of failure and performance anxiety) may also affect erectile functioning.
3. The majority of cases of ED are caused by a combination of physiological and psychological factors, with older men's problems more linked to physiological issues.
4. To diagnose causes of erectile disorder, sex therapists use the **nocturnal penile tumescence test (NPT)**, which tests whether or not a man experiences erections while sleeping, which will indicate whether the problem is primarily psychological or physiological.

E. **Treating Male Erectile Disorder**

1. A tremendous amount of research has been dedicated to finding causes and treatment options for ED.
2. The primary psychological treatments for ED include systematic desensitization and sex therapy that includes education, **sensate focus** (a series of touching experiences that are assigned to couples in sex therapy), and communication training.
3. In 1998, the FDA approved the use of Viagra, the first oral medication for ED, which helps to produce muscle relaxation in the penis, dilation of the arteries supplying the penis, and an inflow of blood, which can lead to erection. (However, it will not increase sexual desire.)
4. **Hormonal Treatments**
    a. Hormonal injections may help improve erections in men with hormonal problems.
    b. They will improve erectile functioning only in men who have low levels of testosterone to begin with.
5. **Intracavernous Injections**
    a. Men and their partners are taught to self-inject vasodilator preparations directly into the corpora cavernosa while the penis is gently stretched out.
    b. **Priapism** is a possible side effect, which occurs in 4% to 8% of men who use these injections.
6. **Vacuum Constriction Devices**
    a. These use suction to induce erections and have become more popular in the last several years because they are less invasive and safer than injections.
    b. These devices can be expensive, bulky, noisy, and require use before having an erection, which some men find unappealing.
7. **Surgical Treatments**
    a. Surgical intervention has increased, with some physicians performing revascularization or in other cases, prosthesis implantation may be recommended.
    b. Today there are two types of implants: semirigid rods which provide a permanent state of erection but can be bent up and down and inflatable devices that become firm when the man pumps them up.

8. **What the Future Holds**
    a. Currently there are several creams, pills, and even nasal sprays in production for the treatment of ED.
    b. Despite the popularity of Viagra, it is ineffective for about one third of the men taking it.
    c. Several other drugs are close to being approved by the FDA and may work faster and have fewer side effects than Viagra.

V. ORGASM DISORDERS

A. Every individual reaches orgasm differently and has different wants and needs to build sexual excitement.

B. **Female Orgasmic Disorder**
1. Female orgasmic disorder is defined as a delay or absence of orgasm following a normal phase of sexual excitement.
2. Studies have found that this is a common complaint with 24% of women reporting with this disorder.
3. Women with orgasmic disorders have more negative attitudes about masturbation, believe more myths about sexuality, and possess greater degrees of sex guilt.
4. Physical factors can also be a cause, including diabetes, neurological problems, hormonal deficiencies, alcoholism, and medications like antidepressants.
5. **Treating Female Orgasmic Disorder**
    a. The majority of treatment programs involve a combination of different treatment approaches, including homework assignments, sex education, communication skills, training, cognitive restructuring, and desensitization, with treatments being more successful for primary rather than secondary disorders.
    b. The most effective treatment involves teaching a woman to masturbate to orgasm.
    c. **Systematic desensitization** is helpful when there is a great deal of sexual anxiety. A relaxation technique is used to reduce the anxiety.

C. **Male Orgasmic Disorder**
1. Male orgasmic disorder is relatively rare, with 8% of men reporting problems reaching orgasm.
2. It is defined as a delay or absence of orgasm following a normal phase of sexual excitement.
3. **Treating Male Orgasmic Disorder**
    a. It is uncommon and rarely treated by sex therapists.
    b. Treatment options include psychotherapy and, if necessary, changing medications.

D. **Premature Ejaculation (PE)**
1. Defining premature ejaculation has always been difficult for professionals, but it usually refers to a man reaching orgasm just prior to, or directly following, penetration.
2. According to research, estimates are that close to 30% of men reported experiencing PE in the last year.
3. PE has been associated with anxiety, depression, drug and alcohol abuse, and personality disorders.
4. **Treating Premature Ejaculation**
    a. **Squeeze Technique**
    - Sexual intercourse or masturbation is engaged in just short of orgasm and then stimulation is stopped.
    - The man or his partner applies pressure to the penis for several seconds until the urge to ejaculate subsides.

b. **Stop-Start Technique**

- Stimulation is stopped until the ejaculatory urge subsides.
- Stimulation is repeated and the process is repeated over and over.

c. These techniques are used for 6 to 12 months to help a man gain some control over his erection.
d. These techniques may help a man get in touch with his arousal levels and sensations.

E. **Retarded Ejaculation** (inhibited ejaculation)

1. It refers to a situation in which a man may be entirely unable to reach orgasm during certain sexual activities or may only be able to ejaculate after prolonged intercourse (30-45 minutes).
2. It may the result of physical and/or psychological factors and may be situational.
3. Psychological factors include a strict religious upbringing, unique or atypical masturbation patterns, or ambivalence over sexual orientation, while physical factors include diseases, injuries, or medications.
4. **Treatment for Retarded Ejaculation**

a. Psychological factors are the most common cause so in many cases psychotherapy is used to help work through issues.
b. Retarded ejaculation can be very difficult to treat.

VI. PAIN DISORDERS

- Genital pain disorders can occur at any stage of the sexual response cycle, and although they are more frequent in women, they also occur in men.

A. **Vaginismus**

1. Vaginismus involves involuntary contractions of the pubbococcygeus muscle that surrounds the entrance to the vagina and controls the vaginal opening. Penetration during sexual intercourse can be impossible and may cause pain.
2. It may be situation-specific so that a woman may experience penetration with fingers or during a pelvic exam but not vaginal-penile intercourse.
3. It is common in women who have been sexually abused or raped.
4. **Treating Vaginismus**

a. Women often don't seek treatment, but a physical examination is a preliminary check for any physical problems.
b. After a confirmed diagnosis, one of the most effective treatments is the use of dilators.

- The size of the dilator is slowly increased.
- These dilators help to open and relax the vaginal muscles.

c. It is also helpful for couples to become more informed about vaginismus and sexuality.

B. **Dyspareunia and Vulvodynia**

1. Dyspareunia is defined as painful intercourse and may involve only slight pain or may be extreme, making intercourse very difficult or impossible.
2. Physical problems, allergies, or infections, along with psychological problems can cause dyspareunia.
3. Vulvodynia can be a cause of dyspareunia, and **vulvar vestibulitis syndrome (VVS)** is a subtype of vulvodynia that is considered one of the most common causes of dyspareunia today.
4. Men can also experience dyspareunia, which may cause pain in the testes or penis, either during or after sexual intercourse.
5. **Treating Dyspareunia and Vulvodynia**

a. Dyspareunia needs to be evaluated medically to see if there is a physical problem such as an infection.
b. Treatments for VVS include biofeedback, psychotherapy, and surgery, which have reported significant reduction.

VII. OTHER SEXUAL PROBLEMS

A. **Faking Orgasms** often occurs as a result of a dysfunction.

B. **Sleep Sex**

1. Researchers have newly discovered a condition known as **sleep sex**, which causes people to commit sexual acts in their sleep.
2. Treatment involves medication, such as Valium, combined with psychotherapy.

C. **Peyronie's Disease**

1. Peyronie's disease is a disorder where the penis develops nodules that can be painful and lead to a sexual dysfunction.
2. It usually lasts about 2 years and goes away, and is often treated with medication or surgery.

VII. OTHER TREATMENTS FOR SEXUAL DYSFUNCTION

A. **Bibliotherapy**

1. Bibliotherapy, or reading self-help materials, has been found to be particularly helpful in reducing certain sexual problems such as orgasmic dysfunction.
2. It has been found to be less successful in cases of hypoactive sexual desire, erectile disorder, and dyspareunia.

B. **Hypnosis**

1. Hypnotherapists use relaxation and direct suggestion to help improve sexual functioning.
2. At this time, little clinical evidence exists on the success of hypnosis in the treatment of sexual dysfunction.

C. **Drugs**

1. Much of the current clinical research today focuses on developing new drugs to treat dysfunctions, even though a number of dysfunctions may be caused or worsened by other medications.
2. There is also a big business in health supplements that aid in sexual functioning, including aphrodisiacs.

VIII. ILLNESS, DISABILITY, AND SEXUAL FUNCTIONING

A. Sexual functioning involves a complex physiological process that can be impaired by pain, immobility, changes in bodily functions, or medications.

B. A majority of the research on the sexuality of the disabled has been done on men, and physicians are more likely to talk to their male patients about sexual issues than their female ones.

C. **Cardiovascular Problems: Heart Disease and Stroke**

1. A person with heart disease, even a person who has had a heart transplant, can return to a normal sex life shortly after recovery.
2. Many patients (or their partners) fear that their damaged (or new) heart is not up to the strain of intercourse of orgasm, but research has found that although sexual activity can trigger a heart attack, this risk is extremely low.
3. Some forms of heart disease can result in erectile difficulties, and some heart medications can dampen desire or cause erectile or lubrication problems.
4. In most cases of stroke, sexual functioning itself is not damaged, and many people go on to resume sexual activity.

5. Stroke may result in physiological changes such as jerking motions, paralysis, or aphasia that can affect sexual communication or functioning.
6. Stroke patients may go through periods of disinhibition where they exhibit behavior that they would have suppressed before the stroke, which may include hypersexuality.

D. **Cancer: "The Big C"**

1. Cancer can affect sexual functioning in a variety of ways including physical scars, the loss of limbs or body parts, changes in skin texture due to radiation, the loss of hair, nausea, bloatedness, weight gain or loss, and ostomies.
2. The psychological issues can lead to depression.
3. **Breast Cancer**
    a. The numbers of mastectomies have decreased today as many women opt for lumpectomies.
    b. The most important factor in resuming a normal sex life is the encouragement and acceptance from the woman's sexual partner, assuring her that she is still sexually attractive and desirable.
4. **Pelvic Cancer and Hysterectomies**
    a. Cancer can affect a woman's vagina, uterus, cervix or ovaries, affecting her physically and/or psychologically.
    b. An oophorectomy (removal of the ovaries) results in hormonal imbalances since the ovaries produce most of a woman's estrogen and progesterone.
    c. Depression and the disruption of intimate relationships are common after a hysterectomy, depending upon why it occurred and how a woman perceives it.
5. **Prostate Cancer**
    a. In the past, a prostatectomy involved cutting the nerves necessary for erection, resulting in erectile dysfunction. However, newer techniques allow for more careful surgery without these problems.
    b. One result of prostatectomy may be incontinence, which may necessitate the need for a catheter, which can in turn affect body image.
6. **Testicular Cancer**
    a. Cancer of the penis or scrotum is rare, and cancer of the testes is only slightly more common.
    b. Research has found that although sexual problems are common after treatment for testicular cancer, there is considerable improvement one year after diagnosis.
    c. Some men experience psychological distress due to body image concerns.

E. **Chronic Illness and Chronic Pain**

1. Many people with chronic illnesses learn to make adjustments in many parts of their lives, including their sexual behaviors.
2. Chronic pain from arthritis, migraine headaches, and lower back pain can make intercourse difficult or impossible at times.

F. **Chronic Obstructive Pulmonary Disease**

1. Another group of conditions that affect sexual functioning are the chronic obstructive pulmonary diseases (COPD), which include asthma, emphysema, tuberculosis, and chronic bronchitis.
2. These diseases can make physical exertion difficult, along with potentially impairing perceptual and motor skills.

G. **Diabetes**

1. Sexual problems (especially difficulty in getting an erection and vaginitis) may be one of the first signs of diabetes.
2. A large number of men in the later stages of diabetes have penile prostheses implanted.
3. Type I diabetic women, aside from some problems with vaginal lubrication, do not seem to have significantly more problems than unaffected women, while Type II diabetic

women show loss of desire, difficulties in lubrication, less satisfaction in sex, and difficulty reaching orgasm.

H. **Multiple Sclerosis (MS)**

1. Multiple Sclerosis involves a breakdown of the myelin sheath that protects all nerve fibers, and it can be manifested in a variety of symptoms, such as dizziness, weakness, blurred or double vision, muscle spasms, spasticity, and loss of control of limbs and muscles.
2. Between 60-80% of men with MS experience ED problems, while women may have a lack of vaginal lubrication, altered feelings during orgasm, or difficulty experiencing orgasm.
3. Both men and women may become hypersensitive to touch, experiencing even light caresses as painful or unpleasant.

I. **Alcoholism**

1. Alcoholism can impair spinal reflexes and decrease serum testosterone levels, which can lead to erectile dysfunction.
2. Long-term alcohol abuse can lead to hyperestrogenemia, which may cause feminization, gynecomastia, testicular atrophy, sterility, ED, and decreased sexual desire.
3. In women, liver disease can lead to decreased or absent menstrual flow, ovarian atrophy, loss of vaginal membranes, infertility, and miscarriages.
4. Problem drinking may lead to guilt and low self-esteem.

J. **Spinal Cord Injuries**

1. If the spinal cord injury is above a certain vertebra and the cord is not completely severed, a man may still be able to have an erection, although most of them are not able to ejaculate.
2. Women with spinal cord injury can still get pregnant and must use contraception, but they may lose sensation in their genitals and the ability to lubricate.
3. Skin sensation in the areas unaffected by the injury can become greater, and new erogenous zones can appear.
4. Men and women with spinal cord injuries report decreased sexual desire and frequency.
5. Treatments for erectile dysfunction are possible, and many couples learn to shift the focus of sexual interactions from erections and penetration to other pleasurable contact.
6. Part of the problem is that individuals with spinal cord injury do not receive adequate education and information about their sexuality, leaving them to feel asexual.

K. **AIDS and HIV**

1. People living with advanced HIV may feel tired, lose their appetite, lose weight, have a fever, diarrhea, night sweats, and swollen glands.
2. Sexual dysfunctions are common in people with HIV due to both physical and psychological factors.
3. People with HIV need to avoid the exchange of bodily fluids but can continue to have sexual relations.

L. **Mental Illness and Retardation: Special Issues**

1. People with psychiatric disorders have historically been treated as asexual, or their sexuality has been viewed as illegitimate, warped, or needing external control.
2. Drugs given for schizophrenia can cause a host of sexual side effects.
3. Educators have designed special sexuality education programs for mentally retarded and developmentally disabled to make sure that they can express their sexuality in a socially approved manner and not be exploited sexually.

IX. GETTING THE HELP YOU NEED

A. Physicians and other health care professionals have historically been uncomfortable learning about the sexual needs of the disabled and discussing these needs with their patients.
B. Now sexuality counseling is becoming a more common part of the recuperation from many diseases and injuries in many hospitals.

TEST YOURSELF

Below you will find fill-in-the-blank and short answer essay questions for topics covered in chapter 14. Check your answers at the end of this chapter.

Sexual Dysfunctions

1. The classification system for sexual dysfunctions appears rather ________________ in that to be diagnosed, a person needs to be experiencing problems with sexual intercourse.

2. Both ________________ ________________ and an excessive need to please a partner may interfere with sexual functioning.

3. When ________________ levels are high, physiological arousal may be impossible.

4. A person with poor health, who smokes, or takes medications experiences more ________________ ________________.

5. Sexual problems for women, with the exception of lubrication, have been found to ________________ with age, while erection problems in men tend to ________________ with age.

6. A ________________ sexual dysfunction is one that has always existed.

7. A ________________ sexual dysfunction is one in which a problem developed after a period of adequate functioning.

8. A ________________ sexual dysfunction is a difficulty that occurs during certain sexual activities or with certain partners.

9. A ________________ sexual dysfunction is a problem that occurs in every situation, during every type of sexual activity, and with every sexual partner.

10. When someone has ________________ ________________ ________________, there is a low or absent desire for sexual activity.

11. A person with a ________________ ________________ reacts with strong disgust or fear to a sexual interaction.

12. Female sexual arousal disorder is an inability to either obtain or maintain adequate ________________ in response to sexual excitement.

13. ________________ ________________ is defined as the persistent inability to obtain or maintain an erection sufficient for satisfactory sexual performance.

14. To diagnose the causes of erectile disorder, sex therapists use tests such as the ________________ ________________ ________________ test.

15. The DSM-IV has two categories of pain disorders, ________________, which occur in women, and ________________, which can affect both men and women.

16. What are some common sexual problems as distinguished from sexual dysfunctions?

17. What are some psychological causes of sexual dysfunction?

18. Name the four basic ways that sexual dysfunctions are categorized.

19. What are the causes of Female Sexual Arousal Disorder?

20. What are some of the differences between women with orgasmic disorders and orgasmic women?

Illness, Disability, and Sexual Functioning

21. A majority of the research on the sexuality of the disabled has been done on ________________.

22. Physicians are more likely to talk to their ________________ patients about sexual issues than their ________________ ones.

23. Research has found that although sexual activity can trigger a myocardial infarction, this risk is ________________.

24. Some forms of heart disease can result in ________________ difficulties.

25. Some heart ________________ can dampen desire or cause erectile problems, or less often, women may experience a decrease in lubrication.

26. Some men find that after a stroke their erections are __________________ because the nerves controlling the erectile tissue on one side of the penis are affected.

27. Some stroke victims go through periods of __________________, where they exhibit behavior that, before the stroke, they would have been able to suppress.

28. There is no reason that a mastectomy should interfere with normal sexual functioning. The most important factor in resuming a normal sexual life is the __________________ and __________________ from the woman's sexual partner.

29. Cancer of the reproductive organs in women may result in a __________________.

30. Almost all men will experience a normal enlargement of the __________________ gland.

31. In some rare cases, cancer of the penis may necessitate a partial or total __________________.

32. Sexual problems (especially difficulty in getting an erection for men and vaginitis or yeast infections in women) may be one of the first signs of __________________ .

33. __________________ __________________ can affect sexual functioning in many ways. For example, both men and women may become hypersensitive to touch, experiencing even light caresses as painful or unpleasant.

34. In males, __________________ can result from the liver damage due to alcoholism, which, combined with lower testosterone levels, may cause feminization, gynecomastia, testicular atrophy, sterility, ED, and decreased libido.

35. Some women and men with spinal cord injuries report experiencing __________________ __________________, a psychic sensation of having an orgasm without the corresponding physical reactions.

36. What is a major reason that sexual activity decreases so much after cardiac incidents?

37. What are some psychological issues associated with heart problems that can affect sexual activity?

38. What is hypersexuality?

39. What are some ways that cancer can affect sexual functioning?

40. How can a hysterectomy affect sexual functioning and pleasure?

POST TEST

Below you will find true/false, multiple-choice and matching quiz items covering the entire chapter. Check your answers at the end of this chapter.

True/False

1. Sexual problems for women have been found to decrease with age, while erection problems in men tend to increase with age.
2. Treatment of most sexual dysfunctions begins with a medical history and an evaluation of any past sexual trauma or abuse.
3. Anorexia has been associated with the diagnosis of sexual aversion.
4. Clinical trials of Viagra to treat female sexual arousal disorder have shown that it provides significant benefit.
5. Retarded ejaculation refers to a situation in which a man may be entirely unable to reach orgasm during certain sexual activities or may only be able to ejaculate after prolonged intercourse.
6. Allergies can cause dyspareunia.
7. Faking orgasms often occurs as a result of a dysfunction.
8. People who've had heart attacks should avoid strenuous sexual activity, especially orgasm, since it could trigger another one.
9. Removal of the prostate always results in erectile dysfunction due to damage to the nerves.
10. A man with a spinal cord injury may still be able to have an erection.

Multiple-Choice

11. What is a concern the authors discuss with respect to research and knowledge around sexual dysfunction?
    a. It still focuses largely on female sexual dysfunction with much less exploration of male dysfunction.
    b. It fails to examine issues of sexual desire but focuses solely on visible signs of functioning.
    c. It still focuses largely on male sexual dysfunction with much less exploration of female dysfunction.
    d. It only explores physiological factors, ignoring psychological issues.
    e. All of the above

12. What is a psychological factor that might interfere with sexual functioning?
    a. performance fears
    b. partner conflict
    c. fear of intimacy
    d. fear of infidelity
    e. All of the above

13. If a sex therapist recommends that a client should try to avoid "spectatoring," what should the client do?
    a. avoid accusing a partner of cheating
    b. stop engaging in aggressive sexual behavior in public
    c. avoid acting as an observer or judge of his or her own sexual performance
    d. avoid thinking about engaging in sexual activity all the time
    e. None of the above

14. What problem below could be classified as a secondary sexual dysfunction?
    a. Sheila has never experienced any type of sexual desire despite having several partners throughout her life.
    b. Joseph started having difficulty obtaining an erection about three weeks ago after he lost his job.
    c. Karla has always experienced pain during intercourse ever since her first sexual experience when she was 15.
    d. As long as he can remember, Juan has ejaculated moments after becoming aroused, and it has recently become a serious problem in his relationship.
    e. All of the above

15. What is a global sexual dysfunction?
    a. a problem that occurs in every situation, during every type of sexual activity, and with every sexual partner.
    b. a problem that occurs only when a couple is away from a typical routine
    c. a difficult that occurs only in sexual relationships that have moved to a level of commitment
    d. a difficulty that occurs during certain sexual activities or with certain partners.
    e. None of the above

16. If a woman comes to sex therapy and she reports never being able to achieve an orgasm, what type of sexual dysfunction is this?
    a. secondary sexual dysfunction
    b. premature sexual dysfunction
    c. primary sexual dysfunction
    d. multimodal sexual dysfunction
    e. affective sexual dysfunction

17. When determining a plan of treatment for a sexual dysfunction, a sex therapist may choose _______________ treatment, involving more than one type of therapy?
    a. pandemic
    b. aphasic
    c. triphasil
    d. global
    e. multimodal

18. Which of the following is NOT a potential cause of hypoactive sexual desire?
    a. chronic use of alcohol
    b. fear of intimacy
    c. medication side effects
    d. depression
    e. None of the above

19. Which of the following statements regarding sexual aversion is FALSE?
    a. People with sexual aversion are unable to become sexually aroused.
    b. There have been few studies that have documented success in treating sexual aversion.
    c. Sexual aversion affects more men than women.
    d. All of the above
    e. None of the above

20. If woman reports that she is experiencing an inability to either obtain or maintain an adequate lubrication in response to sexual excitement, what diagnosis will a sex therapist likely offer?
    a. persistent sexual arousal syndrome
    b. sexual aversion
    c. vaginismus
    d. female sexual arousal disorder
    e. vulvodynia

21. The nocturnal penile tumescence test is used to diagnose what sexual dysfunction?
    a. erectile disorder
    b. dyspareunia
    c. hypoactive sexual desire
    d. retarded ejaculation
    e. premature ejaculation

22. What treatment has been used for female orgasmic disorder that involves neutralizing the anxiety-producing aspects of sexual situations and behavior by a process of gradual exposure?
    a. situational relaxation
    b. affective bibliotherapy
    c. multimodal sensate focus
    d. psychogenic spectatoring
    e. systematic desensitization

23. What male sexual dysfunction is uncommon and rarely treated by sex therapists?
    a. premature ejaculation
    b. retarded ejaculation
    c. male orgasmic disorder
    d. vulvodynia
    e. male erectile disorder

24. According to research, estimates are that close to _____ % of men report experiencing premature ejaculation in the last year?
    a. 8
    b. 16
    c. 30
    d. 77
    e. 90

25. What is the most effective treatment for vaginismus?
    a. applying testosterone cream to the clitoris
    b. teaching a woman to masturbate to orgasm
    c. Viagra
    d. using dilators to help open and relax the vaginal muscles
    e. cognitive-behavioral therapy

26. William complains to his doctor that for the last month he has been experiencing moderate pain in his testicles after intercourse. What is the term for this condition?
    a. erectile disorder
    b. psychogenic aphasia
    c. vulvar vestibulitis syndrome
    d. dyspareunia
    e. hyperestrogenemia

27. How do sex therapists typically treat sleep sex, which causes people to commit sexual acts in their sleep?
    a. dilators
    b. medication and psychotherapy
    c. clitoral/penile therapy devices
    d. hypnosis and bibliotherapy
    e. hormonal injections

28. What is Peyronie's disease?
    a. a disorder where the penis develops nodules that can be painful and lead to a sexual dysfunction
    b. a complication of erectile disorder when the corpora cavernosa becomes infected
    c. having an excessive amount of estrogens in the blood
    d. a condition in which erections are long lasting and often painful
    e. a syndrome that causes pain in the scrotum when it is touched

29. If a sex therapist suggests that a client should try "bibliotherapy," what is the therapist suggesting?
    a. using relaxation and direct suggestion to help improve sexual functioning
    b. the use of an herbal substance that increases a person's sexual desire
    c. a series of touching experiences that are assigned to teach nonverbal communication and reduce anxiety
    d. using books and educational materials for the treatment of sexual dysfunction
    e. gradually exposing a client to the anxiety-producing aspects of sexual situations

30. According to the textbook, why does sexual activity decrease so much after cardiac incidents?
    a. feelings of depression
    b. fear
    c. heart medications
    d. erectile difficulties
    e. All of the above

31. What organs are typically removed during a total hysterectomy?
    a. uterus and cervix
    b. uterus and vagina
    c. ovaries and Fallopian tubes
    d. Fallopian tubes and cervix
    e. cervix and vagina

32. Which of the following statements regarding testicular cancer is FALSE?
    a. Sexual problems are common after treatment for testicular cancer.
    b. Some men experience psychological difficulties after testicular cancer treatment due to body image concerns.
    c. Testicular cancer is most common in men who are over the age of 60.
    d. The removal of a testicle does not affect the ability to reproduce.
    e. None of the above

33. What chronic illness often causes severe pain during sexual activities?
    a. diabetes
    b. arthritis
    c. aphasia
    d. alcoholism
    e. All of the above

34. What is a possible sexual functioning problem common to chronic obstructive pulmonary diseases, which include asthma, emphysema, tuberculosis, and chronic bronchitis?
    a. hypersensitivity to touch
    b. difficulty with physical exertion
    c. difficulty in getting an erection
    d. problems with vaginal lubrication
    e. muscle spasms

35. What sexual problem may be one of the first signs of diabetes in women?
    a. female orgasmic disorder
    b. vulvodynia
    c. persistent sexual arousal syndrome
    d. vaginismus
    e. yeast infections

36. What is a fairly common problem of sexual functioning experienced by men with multiple sclerosis?
    a. erectile disorder
    b. hypoactive sexual desire
    c. dyspareunia
    d. persistent sexual arousal syndrome
    e. All of the above

37. Hyperestrogenemia is the result of what chronic condition?
    a. diabetes
    b. emphysema
    c. alcoholism
    d. schizophrenia
    a. multiple sclerosis

38. How might a person's sexual functioning adapt to a spinal cord injury?
    a. Sexual behaviors become broader.
    b. New erogenous zones can appear as skin sensations can become greater.
    c. Some people report experiencing a psychic sensation of having an orgasm without the corresponding physical reactions.
    d. All of the above
    e. None of the above

39. Swollen glands, loss of appetite and weight, fatigue, fever, diarrhea, and night sweats can be advanced symptoms of what chronic condition?
    a. alcoholism
    b. diabetes
    c. asthma
    d. HIV/AIDS
    e. hypertension

40. Which statement below regarding sexuality and schizophrenia is TRUE?
    a. A third of men with schizophrenia reported no sexual activity or feelings at all.
    b. Drugs given for schizophrenia can cause a host of sexual side effects.
    c. People with schizophrenia have historically been treated as though their sexuality needs to be controlled.
    d. All of the above
    e. None of the above

Matching

Column 1		Column 2
A. persistent sexual arousal syndrome	____	41. a man ejaculating just prior to, or directly following, penetration
B. female sexual arousal disorder	____	42. an inability to either obtain or maintain adequate lubrication in response to sexual excitement
C. hypoactive sexual desire	____	43. involuntary contractions of the pubbococcygeus muscle which surrounds the entrance to the vagina and controls the vaginal opening
D. male erectile disorder	____	44. a woman experiences excessive and unremitting arousal
E. retarded ejaculation	____	45. the persistent inability to obtain or maintain an erection sufficient for satisfactory sexual performance
F. female orgasmic disorder	____	46. strong disgust or fear to a sexual interaction, blocking the ability to become sexually aroused
G. sexual aversion	____	47. a man may be entirely unable to reach orgasm during certain sexual activities or may only be able to ejaculate after prolonged intercourse
H. premature ejaculation	____	48. painful intercourse
I. vaginismus	____	49. a low or absent desire for sexual activity, even though a person can still function normally
J. dyspareunia	____	50. a delay or absence of orgasm following a normal phase of sexual excitement

Test Yourself Answer Key

Sexual Dysfunctions

1. heterosexist (p. 429)
2. performance fears (p. 431)
3. anxiety (p. 431)
4. sexual dysfunctions (p. 431)
5. decrease, increase (p. 431)
6. primary (p. 431)
7. secondary (p. 431)
8. situational (p. 431)
9. global (p. 431)
10. hypoactive sexual desire (p. 434)
11. sexual aversion (p. 435)
12. lubrication (p. 437)
13. Erectile disorder (p. 437)
14. nocturnal penile tumescence (p. 438)
15. vaginismus, dyspareunia (p. 444)
16. insufficient foreplay, lack of enthusiasm for sex, inability to relax (p. 430)
17. unconscious fears, ongoing stress, anxiety, depression, guilt, anger, fear of infidelity, partner conflict, fear of intimacy, dependency, abandonment, or loss of control. (pp. 430-431)
18. primary or secondary, and situational or global (p. 431)
19. Physiological factors include decreased blood flow to the vulva; psychological factors include fear, guilt, anxiety, and depression. (p. 437)
20. Women with orgasmic disorders have more negative attitudes about masturbation, believe more myths about sexuality, and possess greater degrees of sex guilt. (p. 441)

Illness, Disability, and Sexual Functioning

21. men (p. 447)
22. male, female (p. 447)
23. low (p. 449)
24. erectile (p. 449)
25. medication (p. 449)
26. crooked (p. 450)
27. disinhibition (p. 450)
28. encouragement, acceptance (p. 451)
29. hysterectomy (p. 451)
30. prostate (p. 452)
31. penectomy (p. 452)
32. diabetes (p. 453)
33. Multiple sclerosis (p. 453)
34. hyperestrogenemia (p. 454)
35. phantom orgasm (p. 454)
36. fear (p. 449)
37. feelings of depression, inadequacy (especially among men), or loss of attractiveness (especially among women) (p. 449)
38. It is abnormally expressive or aggressive sexual behavior, often in public; the term usually refers to behavior due to some disturbance of the brain. (p. 450)
39. Physical scars, the loss of limbs or body parts, changes in skin texture, the loss of hair, nausea, bloatedness, weight gain or loss, and acne are some of the ways that cancer and its treatment can affect the body and one's body image. (p. 450)

40. The resulting hormonal imbalances can lead to reduced vaginal lubrication and mood swings. Also, many women find the uterine contractions of orgasm very pleasurable, and when the uterus is removed, they lose that aspect of orgasm. (p. 451)

Post Test Answer Key

True/False

1. T (p. 431)
2. T (p. 432)
3. T (p. 435)
4. F (p. 437)
5. T (p. 443)
6. T (p. 445)
7. T (p. 446)
8. F (p. 449)
9. F (p. 452)
10. T (p. 454)

Multiple Choice

11. c (p. 428)
12. e (pp. 430-431)
13. c (p. 431)
14. b (p. 431)
15. a (p. 431)
16. c (p. 431)
17. e (p. 432)
18. e (p. 434)
19. c (p. 435)
20. d (p. 437)
21. a (p. 438)
22. e (p. 442)
23. c (p. 442)
24. c (p. 442)
25. d (p. 444)
26. d (p. 445)
27. b (p. 446)
28. a (p. 446)
29. d (p. 446)
30. e (p. 449)
31. a (p. 451)
32. c (p. 452)
33. b (p. 452)
34. b (p. 453)
35. e (p. 453)
36. a (p. 453)
37. c (p. 454)
38. d (p. 454)
39. d (p. 455)
40. d (p. 455)

Matching

41. H (p. 442)
42. B (p. 437)
43. I (p. 444)
44. A (p. 437)
45. D (p. 437)
46. G (p. 435)
47. E (p. 443)
48. J (p. 445)
49. C (p. 434)
50. F (p. 441)

Chapter 15 – Sexually Transmitted Infections and HIV/AIDS

CHAPTER SUMMARY

Young adults in the United States are disproportionately affected by sexually transmitted infections with the incidence continuing to grow. STIs have historically been stigmatized and viewed as symbols of corrupt sexuality, a view that can persist today, which interferes with the act of getting tested. STIs can be caused by ectoparasitic, bacterial, or viral infections, which affect the treatment options available.

Ectoparasitic infections include pubic lice and scabies. Bacterial infections include gonorrhea, syphilis, chlamydia, chancroid, and vaginal infections. Many people, especially women, are asymptomatic for gonorrhea and chlamydia, and untreated infections can cause serious complications such as infertility. Symptoms of syphilis include the appearance of small, painless, red-brown sores, chancres that appear on the genitals. Bacterial infections can be treated and cured by antibiotics, although certain antibiotics have become resistant to these infections. Viral infections include herpes, human papillomavirus, and hepatitis. Once a person is infected with a virus, he or she will have it for the rest of his or her life, even though he or she may not experience any symptoms. Herpes (HSV I & II) comes in two forms that favor either the mouth or genital area and are characterized by small, painful blisters, which appear externally on the vulva or penis. Antiviral drugs that shorten the duration of outbreaks and prevent complications can treat herpes. Human papillomavirus (HPV) includes over 100 types; of which about 30 are transmitted sexually and may cause genital warts. Other forms may cause abnormal pap smears in women and have been linked to cervical cancer. Viral hepatitis is an infection that causes impaired liver function and includes three major types: A, B, and C, with B being the type that is most linked to high-risk sexual behavior. Drugs are available to help manage the virus and reduce viral load. Pelvic inflammatory disease (PID) is an infection of the female genital tract, which can be caused by STIs, particularly gonorrhea and chlamydia infections that have not been treated.

AIDS is caused by a viral infection with the human immunodeficiency virus (HIV), a virus primarily transmitted through body fluids, including semen, vaginal secretions, and blood. HIV attacks the T-lymphocytes in the blood; leaving fewer of them to fight off infection so many opportunistic diseases infect people with HIV. HIV remains in a person's body for life. In 2000, it was estimated there were 800,000 to 900,000 people living in the U.S. with HIV and 42 million people worldwide. Women are the fastest growing group with AIDS in the U.S. as heterosexual transmission has passed injection drug use as the most common way a woman is infected with HIV. Tests for HIV can either identify the virus in the blood, or more commonly, detect whether the person's body has developed antibodies to fight HIV. Since 1995, there has been a tremendous decrease in HIV and AIDS-related deaths primarily due to the development of highly active antiretroviral therapy. The United Nations Program on HIV/AIDS estimates that from 2000 to 2020, 68 million people in the 45 most heavily affected countries will die of AIDS unless improvements are made in treatment and prevention. Seven of every 10 people living with HIV in the world are living in sub-Saharan Africa with only a small percent of them receiving antiretroviral therapy.

LEARNING OBJECTIVES

After studying Chapter 15, you should be able to:

1. Describe the historical stigma that persists around people with sexually transmitted infections.
2. Identify the different types of infections that can cause sexually transmitted infections.
3. Describe the incidence, symptoms, and treatments for pubic lice and scabies.
4. Describe the incidence, symptoms, and treatments for the bacterial infections gonorrhea, syphilis, chlamydia, nongonococcal urethritis, and chancroid.
5. Identify the possible complications of an untreated chlamydia infection.
6. Describe the most common vaginal infections, their causes, and treatments.
7. Describe the incidence, symptoms, and treatments for the viral infections herpes, human papillomavirus, and hepatitis.
8. Compare and contrast bacterial and viral infections.
9. Distinguish between HSV I and HSV II.
10. Identify the possible complications of human papillomavirus.
11. Describe the incidence, symptoms, and treatments for pelvic inflammatory disease (PID).
12. Describe the incidence and symptoms of HIV/AIDS in the United States.
13. Explain the HIV testing process.
14. Discuss the pros and cons of the current treatment regimen for HIV/AIDS.
15. Identify some of the issues faced by families directly affected by HIV/AIDS.
16. Discuss the cross-cultural incidence of HIV/AIDS.
17. Identify some strategies other countries have used to combat HIV/AIDS.
18. Explain some of the challenges around HIV/AIDS faced in sub-Saharan Africa.
19. Compare and contrast sub-Saharan Africa and Latin America when it comes to responses to HIV/AIDS.

CHAPTER OUTLINE

I. ATTITUDES AND SEXUALLY TRANSMITTED INFECTIONS

A. STIs have historically been viewed as symbols of corrupt sexuality, reflecting the **punishment concept**: the idea that people who had become infected with certain diseases, especially STIs, had done something wrong and were being punished.
B. The self-stigmatization of STIs can interfere with the act of getting tested.
C. Young adults are disproportionately affected by STIs, with the incidence continuing to grow.
D. Research has found that the cervix of a teenage girl is more vulnerable to certain STIs than the cervix of an adult woman.

II. SEXUALLY TRANSMITTED INFECTIONS

A. Research suggests that women are at more risk from STIs because the tissue of the vagina is more fragile, and women are more likely to be asymptomatic, meaning they show no symptoms of certain STIs.
B. STIs can affect a developing fetus, causing complications during pregnancy.
C. STIs can be caused by ectoparasitic, bacterial, or viral infections, which affects the treatment options available
D. **Ectoparasitic Infections: Pubic Lice and Scabies**

1. Pubic Lice (also known as "crabs")
    a. They are a parasitic form of an STI—small, wingless insects that can attach to pubic hair and feed off the tiny blood vessels just beneath the skin.
    b. Pubic lice can be spread through casual contact like sharing towels or bedding.
    c. Incidence: Pubic lice are common and regularly seen by health clinics, though there are no mandated reporting laws.
    d. Symptoms: Mild itching is the most common symptom.
    e. Diagnosis: Since pubic lice can be seen with the naked eye, a diagnosis is made fairly quickly and easily.
    f. Treatment
    - Health care providers can prescribe Kwell ointment, which comes in a shampoo or cream that is applied directly to the pubic hair.
    - Sheets and clothing must be dry cleaned, boiled, or washed in very hot water.
2. Scabies
    a. It is ectoparasitic infection of the skin caused by the mite sarcoptes scabei that is spread from skin-to-skin contact during sexual and nonsexual contact. They can live for up to 48 hours on bed sheets and clothing but are impossible to see with the naked eye.
    b. Incidence: Scabies affects millions of people worldwide, but there are no mandated reporting laws.
    c. Symptoms: A rash and intense itching are the first symptoms, which take longer to develop (4-6 weeks) if a person has never been infected before.
    d. Diagnosis: A diagnosis can usually be made through a visual inspection of the rash or scraping of the rash.
    e. Treatment: Topical creams are available to treat scabies, and all bedding and clothes must be washed in hot water.

E. **Bacterial Infections: Gonorrhea, Syphilis, Chlamydia, and More**

1. Gonorrhea (also known as "the clap" or "drip")
    a. Gonorrhea is caused by bacteria that can survive only in the mucus membranes of the body, which limits the transmission to sexual contact (except in the case of newborns who are infected during birth).

b. Incidence: Gonorrhea rates have been declining in the U.S., although the highest rates are among those 25 years old and younger.

c. Symptoms

- The majority of women with gonorrhea is asymptomatic but can still transmit it to their partners.
- Women who do have symptoms may have a discharge from the cervix, frequent urination, abnormal uterine bleeding, and if left untreated, may develop PID.
- The majority of men with gonorrhea experience epididymitis (an inflammation of the epididymis), urethra, discharge, and painful and frequent urination.
- Symptoms usually appear 2 to 6 days after exposure.

d. Diagnosis: A sample of discharge from the cervix, urethra or other affected area is tested for the presence of **gonococcus bacterium**.

e. Treatment: Gonorrhea can be treated effectively with antibiotics, although because certain antibiotics have become resistant, it is important that a person be retested to make sure the infection is gone.

2. Syphilis

a. Syphilis is caused by a bacterium that can survive only in the mucus membranes of the body, making it transmitted during sexual contact, usually infecting the cervix, penis, anus, lips, or nipples, and it can also infect a newborn during birth.

b. Incidence: The overall rates of syphilis have decreased over the last few years in the U.S., although there has been an increase in certain populations such as crack cocaine users.

c. Symptoms

- The first of three stages occurs about 10 to 90 days after infection. Several small, painless, red-brown sores, **chancres**, will appear on the genitals.
- Once the chancre disappears, if untreated, the second stage begins about 6 weeks after infection, when the bacteria invades the central nervous system, and a person develops a rash and possibly fatigue or fevers.
- During the third stage, the symptoms disappear, but if left untreated, it can cause neurological, sensory, muscular, and psychological difficulties and is eventually fatal.

d. Diagnosis: Tests can be done on the chancres or blood tests can test for the presence of antibodies.

e. Treatment: Antibiotics are the choice of treatment.

3. Chlamydia and Nongonococcal Urethritis

a. Chlamydia can be transmitted through sexual contact, including oral sex.

b. The bacterium that causes chlamydia can also cause epididymitis and **nongonococcal urethritis (NGU)** in men.

c. Incidence

- It is the most frequently reported infectious disease in the U.S. today and is highest among those under 25.
- By the age of 30, at least half of all sexually active women have been infected with chlamydia at some point in their lives.

d. Symptoms
- Approximately 75% of women and 50% of men are asymptomatic for chlamydia.
- Female symptoms can include burning during urination, pain during intercourse, pain in the lower abdomen, and cervical bleeding or spotting.
- Male symptoms include a discharge from the penis, burning during urination, and pain or swelling in the testicles.
- Chlamydia is thought to be one of the agents more responsible for pelvic inflammatory disease (PID).

e. Diagnosis: A culture of cervical discharge or a blood test will screen for chlamydia in women or a urine test for men.
f. Treatment: Chlamydia can be treated effectively with antibiotics, although because certain antibiotics have become resistant, it is important that a person be retested to make sure the infection is gone.

4. Chancroid
a. Chancroid is a bacterial sexually transmitted infection that may look similar to syphilis, but the sores have soft edges, and they eventually rupture, forming a painful ulcer that may last for weeks or months.
b. Incidence: It is relatively rare in the U.S., but is one of the most prevalent STI in poor areas in Africa and Asia and has been associated with HIV transmission.
c. Symptoms: Women are often asymptomatic, while men get small lesions on their penis that ruptures to form painful ulcers.
d. Diagnosis: Diagnosis can be difficult, but a fluid sample from an ulcer can be examined for the bacteria.
e. Treatment: Chancroids are treated with antibiotics.

5. Vaginal Infections
a. **Trichomoniasis**
- a form of vaginitis that may cause discharge or burning in women but may be asymptomatic in men.
- The oral medication, Flagyl, is the most common treatment.

b. **Hemophilus** is a vaginal infection that may occur with other vaginal infections and may be asymptomatic or cause discharge or soreness.
c. **Bacterial vaginosis (BV)**
- It is the most common cause of vaginal discharge and odor, although about half of infected women are asymptomatic.
- It occurs when there is an overabundance of certain types of bacteria that are normally present in the vagina.

d. **Vulvovaginal candiasis** (yeast infections)
- These infections are caused by a fungus that is normally present in the vagina but multiplies when the pH balance of the vagina is disrupted by antibiotics, douching, pregnancy, oral contraceptives, or diabetes.
- Symptoms include discharge and pain, and treatment involves over-the-counter drugs.
- It is not sexually transmitted, yet women with multiple infections may want their partners to be evaluated.

F. **Viral Infections: Herpes, Human Papillomavirus, and Hepatitis**
1. Once a person is infected with a virus, he or she will have it for the rest of his or her life, even though he or she may not experience any symptoms.

2. **Herpes (HSV)**
    a. Herpes is caused by an infection with the herpes simplex virus (HSV), with **herpes simplex I** preferring to infect the mouth and face, and **herpes simplex II** preferring the genitals. Once infected, they can overlap.
    b. HSV-I and HSV-2 are contained in the sores that the virus causes, but they are also released between outbreaks from the infected skin called **viral shedding**.
    c. The majority of genital herpes cases are caused by an infection with HSV-2, although more and more are being caused by HSV-1.
    d. After a certain time, herpes will seem to disappear, but it can lie dormant in the body so the person can still transmit the virus.
    e. Incidence: The herpes simplex virus-2 is one of the most common STIs in the U.S. today. with more than 1 in 5 Americans infected.
    f . Symptoms
    - The most common sites for HSV infection are the skin and mucus membranes, with the first symptoms usually appearing with 4 days after exposure; however, the majority of those infected do not experience any noticeable symptoms.
    - The first occurrence of the HSV sores is usually the most painful, with women having more severe symptoms.
    - At the onset, there's usually a tingling or burning feeling in the affected area, which can grow into itching and a red, swollen appearance of the genitals (**prodromal phase**).
    - Small blisters, which range from mild to severe pain, appear externally on the vulva or penis.
    - Other symptoms include fever, headaches, pain, itching, vaginal or urethral discharge, and general fatigue.

    g. Diagnosis: The presence of blisters is often enough for diagnosis, although a scraping of the blisters can also be taken.
    h. Treatment: Although there is no cure, standard therapy consists of antiviral drugs that shorten the duration of outbreaks, prevent complications, and reduce viral shedding.
3. **Human Papillomavirus (HPV)**
    a. There are over 100 types of the human papillomavirus (HPV), and 30 of these are sexually transmitted through different types of sexual contact.
    b. Some of these are called "high risk" because they can cause abnormal Pap smears and increase the potential for certain kinds of cancers; it has been suggested that almost all cervical disease can be attributed to HPV infection.
    c. Other types are called "low-risk" and can cause genital warts.
    d. Incidence: Between 50-75% of sexually active men and women will become infected with HPV at some point in their lives.
    e. Symptoms
    - Many people infected with HPV are asymptomatic, while others develop symptoms as late as 6 weeks to 9 months after exposure.
    - About 10% of HPV infections lead to genital warts, which are usually flesh-colored and may have a bumpy surface, appearing on the vulva, cervix, penis, or scrotum.

    f. Diagnosis: Health care providers may be able to identify genital warts in women during a routine pelvic exam.
    g. Treatment: Treatment alternatives include chemical topical solutions, cryotherapy, electrosurgical interventions, and laser surgery.

4. Viral Hepatitis
   a. Viral hepatitis is an infection that causes impaired liver function. Three main types are prevalent.
      - Hepatitis A is transmitted through fecal-oral contact and is often spread by food handlers.
      - Hepatitis B is known to be transmitted through high-risk sexual contact.
      - Hepatitis C can be spread through sexual behavior and is caused most by illegal drug use injection.

   a. Incidence: A high number of individuals in the U.S. are infected with the various strains of hepatitis.
   b. Symptoms
      - Hepatitis A: fatigue, abdominal pain, loss of appetite, and diarrhea with no chronic long-term infection.
      - Hepatitis B: mostly asymptomatic, although possible symptoms may include nausea, vomiting, jaundice, headaches, fever, and fatigue, with 15-25% of those infected dying from chronic liver disease.
      - Hepatitis C: asymptomatic or mild illness with 75-85% developing a chronic liver infection.

   c. Diagnosis: Blood tests are used to identify viral hepatitis.
   d. Treatment: Drugs are available to reduce viral load.

III. PELVIC INFLAMMATORY DISEASE (PID)

A. Pelvic inflammatory disease is an infection of the female genital tract, including the endometrium, Fallopian tubes, and the lining of the pelvic area.
B. It is estimated that 10% to 20% of women with gonorrhea or chlamydia will develop PID.
C. Twenty percent of women with PID become infertile, 20% develop chronic pelvic pain, and 10% who conceive have an ectopic pregnancy.
D. Although PID can be asymptomatic, symptoms of PID include acute pelvic pain, fever of 101 degrees or higher, and an abnormal vaginal discharge.
E. Treatment includes antibiotics for 14 days.

IV. THE HUMAN IMMUNODEFICIENCY VIRUS (HIV) AND ACQUIRED IMMUNE DEFICIENCY SYNDROME (AIDS)

A. AIDS was first identified among men who have sex with men, and because of this early identification, the disease was linked with "socially marginal" groups in the population.
B. AIDS is caused by a viral infection with the human immunodeficiency virus (HIV), a virus primarily transmitted through body fluids including semen, vaginal secretions, and blood.
C. HIV remains in the body for the rest of a person's life and is often fatal.
D. HIV attacks the **T-lymphocytes (T-helper cells)** in the blood, leaving fewer of them to fight off infection, so many **opportunistic diseases** infect people with HIV that a healthy person could easily fight off.
E. **Incidence**
   1. In 2000, it was estimated there were 800,000 to 900,000 people living in the U.S. with HIV and 42 million people worldwide.
   2. Women became the fastest growing group with AIDS in the U.S. when heterosexual transmission passed injection drug use as the most common way a woman is infected with HIV.
   3. Infants can acquire HIV through their mothers during birth or breastfeeding.

4. HIV/AIDS continues to disproportionately affect minority communities throughout the world due to late identification of HIV in African-Americans, less access to health care and HIV therapy, and a lack of health insurance.

F. **Knowledge and Attitudes about AIDS**

1. College students tend to overestimate how knowledgeable they actually are about AIDS, but higher knowledge levels about AIDS have not been found to be consistently correlated with behavior changes or the practice of safer sex.
2. Three aspects of AIDS make it different from other diseases: fear of transmission, the social worth of the individuals who have been diagnosed with the disease, and the inability of society to comprehend the magnitude of this disease.
3. Although fewer people hold negative attitudes about people with AIDS today, research suggests that AIDS remains a stigmatized condition in the U.S.

G. **Symptoms**

1. Flu-like symptoms such as fever, sore throat, chronic swollen lymph nodes in the neck or armpits, headaches, and fatigue may appear.
2. Later symptoms may include significant weight loss, severe diarrhea, night sweats, oral candiasis, gingivitis, oral ulcers, and persistent fever.
3. Opportunistic diseases include pneumocystis carinii pneumonia (PCP), toxoplasmosis, cryptococcosis, cytomegalovirus, and Kaposi's sarcoma.
4. Other STIs may appear or progress quickly and may be resistant to treatment.

H. **Diagnosis**

1. Tests for HIV can either identify the virus in the blood, or more commonly, detect whether the person's body has developed antibodies to fight HIV.
2. An ELISA test is the first test, and if it's positive, a Western Blot will be used to determine the presence or absence of HIV antibodies. However, there's a period of time where a person might be infected with HIV, but their body hasn't started to produce the antibodies so he or she will falsely test negative.
3. The above tests can take as long as two weeks before a result, so the OraQuick Rapid HIV-1 Antibody test was the first FDA-approved, non-invasive HIV antibody test that can be used at home after a simple blood test.
4. Effectiveness rates for saliva-based tests are lower than for blood tests, but research continues to evaluate them.

I. **Treatment**

1. Since 1995, there has been a tremendous decrease in HIV and AIDS-related deaths primarily due to the development of **highly active antiretroviral therapy (HAART)**.
2. HAART is the combination of three of more HIV drugs, often referred to as a "drug cocktail" which, in conjunction with the development of HIV RNA testing (that allows health care providers to monitor the amount of virus in the bloodstream) has allowed for better control of HIV and has slowed the disease progression.
3. A viral load test and CD4+ T Cell Count can determine how much HIV is in a person's system and estimate the T-helper white blood cell count.
4. HAART therapy for HIV is complicated and can involve taking 25 or more pills a day at various times of the day. Dosages that are missed can cause drug resistance.
5. There are serious side effects to the medication, and they are very expensive.

J. **Prevention**

1. Many programs have been started to try to change people's behaviors to prevent the increase of HIV infection, including educational programs, advertising, and mailings.
2. HIV-negative men who have sex with men have begun engaging in more unprotected anal sex and worrying less about contracting HIV since the introduction of HAART therapy.

3. In 2003, AIDSVAX, the first AIDS vaccine proved to be an overall failure after a five-year clinical trial.

K. **Families and AIDS**

1. Families and friends of people with HIV/AIDS often do not receive the same social support as do families and friends of people diagnosed with other diseases such as Alzheimer's disease due to the continued social stigma.
2. Mothers are more likely to disclose their HIV status earlier than fathers, and they disclose more often to their daughters than to their sons.
3. Adolescents who were told their parents had HIV/AIDS engaged in more high-risk sexual behaviors and had more emotional distress than adolescents who were uninformed.

V. CROSS-CULTURAL ASPECTS OF AIDS

A. The United Nations Program on HIV/AIDS estimates that from 2000 to 2020, 68 million people in the 45 most heavily affected countries will die of AIDS unless improvements are made in treatment and prevention.

B. The number of children that have been orphaned throughout the world due to HIV/AIDS is equivalent to the total number of children under the age of 5 living in the U.S., with over 90% of the world's HIV/AIDS orphans living in sub-Saharan Africa.

C. **Asia and the Pacific**

1. Much of the increase in people living with HIV in Asia and the Pacific are due to the rising numbers in India and China.
2. Cambodia has been hard hit by HIV/AIDS with estimates suggesting that close to 3% of the entire adult population is infected with HIV.
3. New HIV infections in Thailand have been decreasing over the last 10 years partly due to a 100% condom use program targeted at the prostitution industry.

D. **Eastern Europe and Central Asia**

1. Eastern Europe and Central Asia have the world's fastest growing HIV and AIDS cases, with a total of 1.2 million people living with HIV/AIDS in 2002.
2. An estimated 90% of HIV infections in the Russian Federation are due to intravenous drug use.

E. **Sub-Saharan Africa**

1. Seven of every 10 people living with HIV in the world live in sub-Saharan Africa.
2. In 2002, South Africa's Sesame Street introduced an HIV-positive muppet who is a 5 year-old girl who was orphaned when her parents died of AIDS.
3. One of the biggest obstacles in many parts of Africa is that because of the cost, only a small percent of the people with HIV are receiving HAART therapy.
4. A myth exists in some parts of South Africa that having sex with a young virgin will cure them, so men with HIV have raped dozens of babies.

F. **Latin America and the Caribbean**

1. There are 1.9 million adults and children living with HIV/AIDS in Latin America, making it the second most infected area in the world, behind sub-Saharan Africa.
2. Several countries (Argentina, Costa Rica, Cuba, and Uruguay) guarantee free and universal access to antiretroviral therapy. Honduras and Panama offer price reductions.
3. Beginning in 1997, people with HIV in Brazil could get HIV medications for free because the Brazilian government manufactures generic copies of the expensive drugs.

G. **The Middle East and North Africa**

1. There are a total of 550,000 people estimated to be living with HIV/AIDS in the Middle East and North Africa,
2. It has been difficult to collect actual numbers.

H. **Other Issues**

1. Some argue that people with HIV/AIDS should have access to all drugs, regardless of their ability to pay, while others argue that the money is more wisely spent on education and prevention.
2. Recent research suggests that in developing countries counseling is increasingly recognized as an important part of the care for people with HIV/AIDS and their families.
3. In some countries home-based health care is also being established to remove some of the burden from the hospitals, increase quality health care, and reduce costs.

VI. PREVENTING STIs AND AIDS

TEST YOURSELF

Below you will find fill-in-the-blank and short answer essay questions for topics covered in chapter 15. Check your answers at the end of this chapter.

Sexually Transmitted Infections

1. ____________________ are the most effective contraceptive method for reducing the risk of acquiring a STI.

2. The most effective way of avoiding STI transmission is to abstain from ____________________, vaginal, and anal sex or to be in a long-term, mutually monogamous relationship with someone who is free from STIs.

3. ____________________ infections are those that are caused by parasites that live on the skin surface.

4. ____________________ ____________________, or crabs, are a parasitic STI and are very small, wingless insects that can attach themselves to pubic hair with their claws.

5. Gonorrhea may also be referred to as the ____________________ or drip.

6. In women, ____________________ has been found to cause approximately 40% of the cases of pelvic inflammatory disease.

7. The majority of men infected with ____________________ is symptomatic and may experience symptoms of epididymitis, urethral discharge, painful urination, and an increase in the frequency and urgency of urination.

8. In the first stage of syphilis, the ____________________, which is a round sore with a hard raised edge and a sunken center, is usually painless and does not itch.

9. For syphilis, ____________________ are the treatment of choice.

10. ____________________ is so common in young women today, that by the age of 30, at least half of all sexually active women have been infected with it at some point in their lives.

11. ____________________ ____________________ is the most common cause of vaginal discharge and odor.

12. Vulvovaginal candiasis is also known as a ____________________ infection.

13. HSV-1 and HSV-2 are contained in the sores that the virus causes but they are also released between outbreaks from the infected skin. This is called ____________________ ____________________.

14. At the onset of a(n) ____________________ sore, there is there is usually a tingling or burning feeling in the affected area.

15. Viral ____________________ is an infection that causes impaired liver function.

16. What are some ways that STIs can adversely affect pregnancy?

17. Name the two ectoparasitic infections that are sexually transmitted.

18. Name at least three sexually transmitted infections that are caused by bacteria.

19. What are several common vaginal infections that may also be associated with sexual intercourse?

20. Why are men less likely than women to have problems with yeast infections?

HIV/AIDS

21. AIDS is cause by a viral infection with the ____________________ ____________________ virus.

22. The attack on the T-helper cells causes the immune system to be less effective in its ability to fight disease, and so many ____________________ diseases infect people with AIDS that a healthy person could easily fight off.

23. All 50 states require that the names of people with ____________________ be reported to local or state health departments.

24. ____________________ transmission has passed intravenous drug use as the most common way a woman is infected with HIV.

25. Infection with HIV results in a gradual deterioration of the immune system through the destruction of ____________________ lymphocytes.

26. The average HIV-positive person, who is not on any type of treatment, will develop AIDS within ________ to ________ years.

27. Tests for HIV can either identify the virus in the blood, or more commonly, detect whether the person's body has developed ____________________ to fight HIV.

28. Public service announcements about AIDS were increased on ____________________ stations and many ____________________ programs agreed to address HIV/AIDS in upcoming episodes.

29. Because the virus can remain in the body for several years before the onset of symptoms, some people may not know that they have the virus and are capable of ____________________ others.

30. One of the major contributors to the constant increase in HIV infection in China has been that the supply of ____________________ falls short of the need for them.

31. Eastern ____________________ and Central Asia have the world's fastest growing HIV and AIDS cases.

32. The majority of the 42 million HIV-positive people live in ____________________, where in certain countries as many as 1 in 4 are infected.

33. In 2002, South Africa's *Sesame Street* unveiled an____________________ muppet character who is a 5-year-old girl who was orphaned when her parents died of AIDS.

34. The majority of HIV infected people in the hardest hit areas outside the U.S. do not have access to ____________________ therapy because of the expense or other reasons.

35. Recent research suggests that, in developing countries, ____________________ is being increasingly recognized as an important part of care for people with AIDS and their families.

36. What are three aspects of AIDS that make it different from other diseases?

37. What are some of the early symptoms of HIV?

38. Name some of the opportunistic diseases that may develop as a result of HIV/AIDS?

39. What is highly active antiretroviral therapy (HAART)?

40. What are some of the side effects of HAART therapy?

POST TEST

Below you will find true/false, multiple-choice and matching quiz items covering the entire chapter. Check your answers at the end of this chapter.

True/False

1. Pubic lice can be spread through casual contact like sharing towels or bedding.
2. Approximately 50% of men are asymptomatic for chlamydia.
3. Herpes can infect both the genitals and the mouth.
4. Human papillomavirus has been linked to cervical cancer.
5. Chronic liver disease is a possible complication of pelvic inflammatory disease (PID).
6. For the most part, HIV/AIDS is no longer a stigmatized condition in the U.S. today.
7. Missing a dose of HIV medications can cause drug resistance.
8. The number of children that have been orphaned throughout the world due to HIV/AIDS is equivalent to the total number of children under the age of 5 living in the U.S.
9. Employers in South Africa automatically fire employees if they learn they are living with HIV/AIDS.
10. Some people argue that money provided to poor countries is more wisely spent on education and prevention rather than on antiretroviral medications.

Multiple-Choice

11. If someone advocates the idea that people who become infected with sexually transmitted infections have done something wrong, what is the term for this perspective?
    a. the justice theory
    b. the punishment concept
    c. the revenge model
    d. the superiority mentality
    e. the moralizing perspective

12. What sexually transmitted infections are ectoparasitic infections?
    a. chlamydia and nongonococcal urethritis
    b. herpes and hepatitis
    c. trichomoniasis and candiasis
    d. scabies and pubic lice
    e. human papillomavirus and syphilis

13. How is scabies transmitted?
    a. through saliva
    b. through the blood stream
    c. skin-to-skin contact
    d. any bodily fluids
    e. mucus membrane contact

14. The majority of women infected with gonorrhea have what symptoms?
    a. fatigue and abdominal pain
    b. severe itching in the pubic area
    c. bumpy warts on the vulva
    d. painful blisters on the genitals
    e. The majority of women with gonorrhea are asymptomatic.

15. What is the treatment for gonorrhea?
    a. shampoo or cream to be applied to the pubic hair
    b. laser therapy around the affected area
    c. ice applied to the pubic area
    d. antiviral drugs
    e. antibiotics

16. What is the first symptom of syphilis?
    a. painful red blisters with grayish centers
    b. small, painless, red-brown sores, called chancres
    c. yellowish discharge from the urethra in men and the vagina in women
    d. a reddish rash on the palms of the hands
    e. a tingling or itchy feeling in the infected area

17. The bacterium that causes chlamydia can also cause what sexually transmitted infection in men?
    a. prodromal trichomoniasis
    b. nongonococcal urethritis (NGU)
    c. hemophilus
    d. syphilis
    e. treponema pallidum

18. If a man is diagnosed with chlamydia, what is the most likely treatment his physician will prescribe?
    a. antiviral drugs
    b. cryotherapy to the affected area
    c. antibiotics
    d. shampoo or cream to be applied to the pubic hair
    e. ice applied to the pubic area

19. Which below is a vaginal infection caused by an organism and may result in pain, discharge, and/or inflammation?
    a. trichomoniasis
    b. herpes
    c. nongonococcal urethritis
    d. human papillomavirus
    e. All of the above

20. Tanya goes to her doctor complaining of small, moderately painful blisters that appeared on her vulva. Once the blisters burst, they have a yellowish discharge. What is a possible diagnosis of her condition?
    a. chlamydia
    b. scabies
    c. syphilis
    d. herpes
    e. HIV

21. What virus causes genital warts, which are usually flesh-colored and may have a bumpy surface, appearing on the vulva, cervix, penis, or scrotum?
    a. hepatitis
    b. chlamydia
    c. HIV
    d. human papillomavirus
    e. herpes simplex II

22. What is the predominant source for hepatitis B infection among adults in the U.S.?
    a. illegal drug use injection
    b. high-risk sexual behaviors
    c. unscreened blood transfusions
    d. fecal-oral contact spread by food handlers
    e. breastfeeding

23. What is the most serious complication that can result from infection with hepatitis B or C?
    a. tuberculosis
    b. heart disease
    c. cervical or penile cancer
    d. kidney failure
    e. chronic liver disease

24. If a woman is diagnosed with pelvic inflammatory disease (PID), how is it treated?
    a. It is untreatable, and a woman must wait 1-2 months for the symptoms to fade.
    b. cryotherapy on the cervix
    c. antibiotics for 14 days
    d. cream applied to the vaginal area
    e. antiviral drugs

25. Why are racial minorities at an increased risk for acquiring HIV?
    a. less access to HIV therapy
    b. lack of health insurance
    c. late identification of HIV/AIDS
    d. less access to health care
    e. All of the above

26. Public attitudes about AIDS may be a mixture of the fear of casual contact and ________________.
    a. religiosity
    b. IQ level
    c. knowledge
    d. homophobia
    e. All of the above

27. What is a possible symptom of HIV infection?
    a. persistent fever
    b. night sweats
    c. gingivitis
    d. severe diarrhea
    e. All of the above

28. What type of HIV test is the most accurate?
    a. blood
    b. urine
    c. saliva
    d. skin
    e. All of the above are about the same

29. Which of the following statements is FALSE?
    a. Tests for HIV can detect whether the person's body has developed antibodies to fight HIV.
    b. If a person tests negative for HIV with an antibody test, they do not have HIV.
    c. The ELISA is the most widely used HIV test.
    d. There is an HIV antibody test which can be used at home.
    e. Some HIV tests take as much as two weeks before a result is provided.

30. What has allowed for better control of HIV and has slowed the disease progression?
    a. ELISA tests
    b. spermicides
    c. highly active antiretroviral therapy
    d. Western Blot tests
    e. All of the above

31. What is a possible side effect of HIV medications?
    a. diabetes
    b. diarrhea
    c. fatigue
    d. nightmares
    e. All of the above

32. What HIV prevention strategy proved to be an overall failure after a five-year clinical trial?
    a. HAART therapy
    b. the law mandating that people with HIV tell their past sexual partners
    c. an advertising campaign championed by Nickelodeon
    d. AIDSVAX, the first AIDS vaccine
    e. the home HIV testing kit

33. Which of the following statements regarding HIV/AIDS and families is FALSE?
    a. Adolescents who were told their parents had HIV/AIDS engaged in more high-risk sexual behaviors
    b. Mothers are more likely to disclose their HIV status earlier than fathers
    c. Families and friends of people with HIV/AIDS often do not receive the same social support as do families and friends of people diagnosed with other diseases due to the continued social stigma.
    d. The majority of parents living HIV/AIDS have chosen not to discuss their illness with their family members.
    e. All of the above

34. What countries in Asia and the Pacific have experienced rising numbers of HIV/AIDS cases?
    a. India and China
    b. Thailand and Japan
    c. Cambodia and South Korea
    d. Vietnam and China
    e. North Korea and Thailand

35. Why is it difficult to know the true number of HIV/AIDS cases in the Russian Federation?
    a. They do not keep track of HIV/AIDS cases.
    b. They define AIDS differently than the rest of the world.
    c. They deny the existence of AIDS.
    d. There is a lot underreporting of HIV/AIDS cases.
    e. None of the above

36. What is the likelihood that a 15-year old in Botswana will die from AIDS?
    a. 8%
    b. 21%
    c. 49%
    d. 80%
    e. 92%

37. Why are only a small percent of the many people with HIV receiving HIV medications in many parts of Africa?
    a. They go against tradition.
    b. People don't trust that they are safe.
    c. They are expensive.
    d. Few people know they exist.
    e. All of the above

38. What is the second most infected area in the world, behind sub-Saharan Africa?
    a. Latin America
    b. Eastern Europe
    c. Central Asia
    d. Asia and the Pacific
    e. North America

39. How has the Brazilian government responded to the HIV/AIDS epidemic?
    a. They instituted a 100 percent condom campaign for adolescents.
    b. They began importing HIV medications from the U.S. at a reduced rate.
    c. They began imprisoning intravenous drug users.
    d. They began manufacturing generic copies of the expensive drugs.
    e. All of the above

40. What is being established in some countries to remove some of the burden from the hospitals, increase quality health care, and reduce costs?
    a. HIV/AIDS clinics
    b. the transferring of people with HIV to more developed countries
    c. home-based health care
    d. All of the above
    e. None of the above

Matching

Column 1		Column 2
A. herpes	____	41. It can cause genital warts, which are usually flesh-colored and may have a bumpy surface, appearing on the vulva, cervix, penis, or scrotum.
B. gonorrhea	____	42. The majority of men with this sexually transmitted infection experience epididymitis, urethral discharge, and painful and frequent urination.
C. scabies	____	43. It is the most frequently reported infectious disease in the U.S. today and is highest among those under 25, with approximately 75% of women and 50% of men asymptomatic.
D. nongonococcal urethritis	____	44. sores develop into painful ulcers that may last for weeks or months
E. chlamydia	____	45. develops in men from the bacterium that causes chlamydia
F. syphilis	____	46. a form of vaginitis that may cause discharge or burning in women but may be asymptomatic in men
G. viral hepatitis	____	47. A tingling or burning feeling in the affected area is followed by the appearance of small blisters that range from mild to severe pain.
H. trichomoniasis	____	48. an infection that causes impaired liver function
I. chancroid	____	49. ectoparasitic infection of the skin spread from skin-to-skin contact during sexual and nonsexual contact
J. human papillomavirus	____	50. small, painless, red-brown sores, chancres, appear on the genitals

Test Yourself Answer Key

Sexually Transmitted Infections

1. Condoms (p. 464)
2. oral (p. 464)
3. Ectoparasitic (p. 465)
4. Pubic lice (p. 465)
5. clap (p. 466)
6. gonorrhea (p. 467)
7. gonorrhea (p. 467)
8. chancre (p. 468)
9. antibiotics (p. 470)
10. Chlamydia (p. 471)

11. Bacterial vaginosis (p. 474)
12. yeast (p. 475)
13. viral shedding (p. 476)
14. herpes (or HSV) (p. 476)
15. hepatitis (p. 480)
16. They can cause problems such as miscarriage, stillbirth, early onset of labor, premature rupture of the amniotic sac, mental retardation, and fetal or uterine infection. (p. 463)
17. pubic lice and scabies (p. 465)
18. gonorrhea, syphilis, chlamydia, chancroid (p. 466)
19. trichomoniasis, hemophilus, bacterial vaginosis, and candiasis (p. 474)
20. Men are less likely to have problems with yeast infections because the penis does not provide the warm and moist environment that the vagina does. (p. 475)

HIV/AIDS

21. human immunodeficiency (p. 481)
22. opportunistic (p. 482)
23. AIDS (p. 483)
24. Heterosexual (p. 484)
25. T-helper (p. 486)
26. 8, 10 (p. 486)
27. antibodies (p. 487)
28. radio, television (p. 486)
29. infecting (p. 490)
30. condoms (p. 491)
31. Europe (p. 492)
32. Africa (p. 492)
33. HIV-positive (p. 492)
34. HAART (p. 493)
35. counseling (p. 493)
36. the fear of transmission, the issue of the social worth of the individuals who have been diagnosed with the disease, and the inability of society to comprehend the magnitude of this illness (p. 485)
37. Flu-like symptoms such as fever, sore throat, chronic swollen lymph nodes in the neck or armpits, headaches, and fatigue may appear. (p. 486)
38. pneumocystis carinii pneumonia (PCP), toxoplasmosis, cryptococcosis, cytomegalovirus, Kaposi's sarcoma (KS) (p. 486)
39. HAART is the combination of three or more HIV drugs, often referred to as a "drug cocktail." (p. 487)
40. fatigue, nausea, fever, nightmares, headaches, diarrhea, changes in a person's fat distribution, elevated cholesterol levels, the development of diabetes, decreased bone density, liver problems, and skin rashes (p. 488)

Post Test Answer Key

True/False

1. T (p. 465)
2. T (p. 471)
3. T (p. 475)
4. T (p. 478)
5. F (p. 481)
6. F (p. 485)
7. T (p. 488)
8. T (p. 491)
9. F (p. 492)
10. T (p. 493)

Multiple Choice

11. b (p. 461)
12. d (p. 465)
13. c (p. 466)
14. e (p. 467)
15. e (p. 468)
16. b (p. 468)
17. b (p. 471)
18. c (p. 472)
19. a (p. 474)
20. d (p. 476)
21. d (p. 478)
22. b (p. 480)
23. e (p. 480)
24. c (p. 481)
25. e (p. 484)
26. d (p. 485)
27. e (p. 486)
28. a (p. 487)
29. b (p. 487)
30. c (p. 487)
31. e (p. 488)
32. d (p. 490)
33. d (pp. 490-491)
34. a (p. 491)
35. d (p. 492)
36. d (p. 492)
37. c (p. 492)
38. a (p. 492)
39. d (p. 493)
40. c (p. 493)

Matching

41. J (p. 478)
42. B (p. 466)
43. E (p. 470)
44. I (p. 472)
45. D (p. 471)
46. H (p. 474)
47. A (p. 475)
48. G (p. 480)
49. C (p. 466)
50. F (p. 468)

Chapter 16 – Varieties of Sexual Expression

CHAPTER SUMMARY

Sexual behaviors increase and decrease in popularity as exemplified by the changing historical views of masturbation and oral sex. Paraphilias are sexual behaviors that involve a craving for an erotic object that is "unusual." For some paraphilias, the fantasy or presence of the object of their desire is necessary for arousal and orgasm, while in others, the desire occurs periodically, or exists separately from their other sexual relationships. Many people live comfortably with paraphilias, leading to a controversy where some theorists suggest that the term describes a society's value judgments about sexuality and not a psychiatric or clinical category.

There is no consensus as to how paraphilias develop, but they are undoubtedly complex behavior patterns, which, in different cases, may have biological, psychological, or social origins—or aspects of all three. Biological, psychoanalytic, developmental, behavioral, and sociological theories have been used to explain how paraphilias begin.

Some paraphilias are more common than others, such as fetishism, sadomasochism, exhibitionism, voyeurism, and transvestism. A fetish is an inanimate object or a body part not usually associated with the sex act that becomes the primary or exclusive focus of sexual arousal and orgasm in an individual. Common fetish items include women's underwear, bras, shoes, boots, and objects made of rubber or leather. Sadism and masochism both associate sexuality and pain, and most people who practice one are also involved with the other, which is referred to as sadomasochism or S&M. Much of S&M is centered on playing roles, usually with appropriate attitude, costuming, and scripted talk, and people who practice S&M may or may not see it as a preferred or an exclusive means of sexual arousal and orgasm. The person who becomes sexually aroused primarily from displaying his or her genitals, nudity, or sexuality to strangers is an exhibitionist. The person whose primary mode of sexual stimulation is to watch others naked or engaging in sex is called a voyeur. Fetishistic transvestism refers to a transvestite who obtains sexual pleasure from dressing up in the clothing of the other sex. Transvestism is usually so fixed in a man's personality that eradication is neither possible nor desirable, and treatment typically focuses on coping with anxieties, guilt, and relationships with family and friends.

Pedophilia is sex with children as a preferred or exclusive mode of sexual interaction. It is typically viewed as one of the most serious and most common paraphilias and is most likely to be seen in treatment. Being a victim of sexual abuse in childhood is the most frequently reported risk factor for becoming a pedophile.

Although the majority of paraphiliacs do not seek treatment and are content with balancing the pleasure and guilt of their paraphilias, others find them to be an unwanted disruption in their lives, particularly those with illegal and/or harmful paraphilias. Treatment for paraphilias is generally multifaceted and may include group, individual, and family therapy, medication, education, and/or self-help groups. Shame aversion, systematic desensitization, orgasmic reconditioning, and satiation therapy are all techniques used to treat various paraphilias.

LEARNING OBJECTIVES

After studying Chapter 16, you should be able to:

1. Describe the historical evolution of abnormal sexual behaviors.
2. Define paraphilia.
3. Identify the demographics of people with paraphilias.
4. Describe some of the controversy around the research and treatment of paraphilias.
5. List the different theories used to explain where paraphilias begin.
6. Compare and contrast the different theories used to explain where paraphilias begin.
7. Explain John Money's theory of lovemaps and how it is used to explain paraphilias.
8. List some of the most common paraphilias.
9. Define fetishism.
10. Identify some characteristics of fetishism.
11. Define sadism and masochism.
12. Identify some characteristics of sadism and masochism.
13. Identify some characteristics of exhibitionism and voyeurism.
14. Identify some characteristics of transvestic fetishism.
15. Distinguish between transvestism and transsexualism.
16. Define pedophilia.
17. Identify major traits of pedophiles.
18. List the major treatment techniques used for illegal paraphilias.
19. Define hypersexuality.
20. Define hyposexuality.

CHAPTER OUTLINE

I. WHAT IS "TYPICAL" SEXUAL EXPRESSION?

A. Sexual behaviors increase and decrease in popularity as exemplified by the changing historical views of masturbation and oral sex.

B. In 1906, Krafft-Ebing defined sexual deviance as "every expression of (the sexual instinct) that does not correspond with the purpose of nature—i.e. propagation."

C. Freud stated that the criterion of normalcy was love and that defenses against "perversion" were the bedrock of civilization because perversion trivializes or degrades love.

II. PARAPHILIAS: MOVING FROM EXOTIC TO DISORDERED

A. Paraphilias are sexual behaviors that involve a craving for an erotic object that is "unusual."

1. This behavior causes significant distress and interferes with a person's ability to work, interact with friends, and other important areas.
2. To be diagnosed with a paraphilia, a person must experience symptoms for six months or more.
3. For some paraphilias, the fantasy or presence of the object of their desire is necessary for arousal and orgasm, while in others, the desire occurs periodically, or exists separately from their other sexual relationships.

B. Paraphiliacs come from every socioeconomic and intelligence bracket, every ethnic and racial group, and from every sexual orientation, although the majority are men.

C. Some have suggested that the defining characteristic of paraphilia is that it replaces a whole with a part, that it allows the person to distance himself or herself from complex human sexual contact and replace it with the undemanding sexuality of an inanimate object, a scene, or a single action.

D. Research on paraphilias has been drawn mostly from clinical and incarcerated samples, and the number of people who live comfortably with uncommon sexual habits is hard to determine.

E. Many people live comfortably with paraphilias, leading to a controversy where some theorists suggest that the term describes a society's value judgments about sexuality and not a psychiatric or clinical category.

F. **Theories About Where Paraphilias Begin**

- There's no consensus as to how paraphilias develop, but they are undoubtedly complex behavior patterns, which, in different cases, may have biological, psychological, or social origins—or aspects of all three.

1. Biological Theories
   a. Men without previous paraphilias have begun to display paraphiliac behavior when they developed temporal lobe epilepsy, brain tumors, and disturbances or certain areas of the brain.
   b. Researchers have found that some paraphiliacs have differences in brain structure and brain chemistry and possible lesions in certain parts of the brain.
   c. At most these factors may lead some people to be more likely to develop a paraphilia, and they do not explain the majority of paraphiliac behaviors.
2. Psychoanalytic Theory
   a. Psychoanalytic thought suggests that paraphilias can be traced back to the difficult time the infant has in negotiating his way through the Oedipal crisis and castration anxiety.
   b. Psychoanalytic theory can explain why paraphilias are more common among men in that boys must painfully separate from their mothers to establish a male identity.
   c. Louise Kaplan, a psychoanalyst, suggests that every paraphilia involves issues of masculinity or femininity.

3. Developmental Theories
   a. We learn at an early age what sexual objects society deems as appropriate for us to desire, but advertising sexualizes its products, which, for example, may turn shoes into a fetish object.
   b. John Money's lovemap theory suggests that the auditory, tactile, and especially the visual stimuli we experienced during childhood sex play forms a template in our brain that defines our ideal lover and ideal sexual situation.
      - The theory suggests that if the child is punished for normal sexual curiosity or if there are traumas during this stage such as sexual abuse, the development of a lovemap can be disrupted.
      - In **hypophilia** negative stimuli prevent the development of certain aspects of sexuality, resulting in the inability to orgasm, vaginal pain, etc.
      - In **hyperphilia** a person becomes overly sexually active.
      - The third way a love map is disrupted is when a paraphilia is developed by the substitution of new elements into the lovemap.
      - Once this lovemap is set, it becomes very stable, explaining why changing it is so difficult.
   c. The idea of **courtship disorders** suggests that organizing the paraphilias into "courtship" stages suggests that they paraphiliac's behavior becomes fixed at a preliminary stage of mating that would normally lead to sexual intercourse.
4. Behavioral Theories: Behaviorists suggest that paraphilias develop because some behavior becomes associated with sexual pleasure through **conditioning**.
5. Sociological Theories
   a. Another way of looking at the causes of paraphilias is to look at the ways society encourages certain behaviors.
   b. Because American society is ruled by images from television, movies, etc., it leads to a world where the image takes the place of the reality, where it becomes common to substitute fantasies for reality.
   c. Surrounded by media, the society experiences things vicariously, through reading about it or seeing it rather than actually doing it.

G. **Describing the Paraphilias**
1. Fetishism
   a. A **fetish** is an inanimate object or a body part not usually associated with the sex act that becomes the primary or exclusive focus of sexual arousal and orgasm in an individual.
   b. Common fetish items include women's underwear, bras, shoes, boots, and objects made of rubber or leather.
   c. The strength of the preference for the object varies from thinking about or holding the object to a need to use it during all sexual acts.
   d. Many people enjoy using lingerie or other fabrics as part of their lovemaking without becoming dependent on them for their arousal like someone with a fetish.
   e. For the fetishist, the object can stand for pure eroticism without the complication of having to deal with another person's feelings, wants, and needs.
2. Sadism and Masochism
   a. Sadism refers to the intentional infliction of physical or psychological pain on another person in order to achieve sexual excitement.
      - Sadistic fantasies or acts may include restraint, blindfolding, strangulation, spanking, whipping, pinching, etc.

- The term is named after the novelist Marquis de Sade.

b. Masochism is the achievement of sexual pleasure through one's own physical pain or psychological humiliation.

- It was named after the novelist Sacher-Masoch.
- Masochism involves the act of being humiliated, beaten, bound, or made to suffer.

c. Sadism and masochism both associate sexuality and pain, and most people who practice one are also involved with the other, which is referred to as **sadomasochism** or S&M.

d. In most S&M encounters, one partner plays the **dominant** role and the other the **submissive**, with bondage and restraint as the most common expressions of S&M.

e. A **dominatrix** is a woman, often a prostitute, who humiliates and dominates submissive men.

f. Much of S&M is about playing roles, usually with appropriate attitude, costuming, and scripted talk.

3. Exhibitionism and Voyeurism

a. The person who becomes sexually aroused primarily from displaying his or her genitals, nudity, or sexuality to strangers is an **exhibitionist**.

b. The person whose primary mode of sexual stimulation is to watch others naked or engaging in sex is called a **voyeur** or **scopophiliac**.

c. Exhibitionism involves exposing the genitals to a stranger. An exhibitionist is usually a man who achieves sexual gratification from exposing his genitals in public or to unsuspecting women.

d. The behavior is compulsive and very difficult to stop.

e. Obscene Telephone Callers

- **Scatolophilia** is the technical name for obscene telephone calling and is a form of exhibitionism where a person calls women and becomes excited as the victims react to his obscene suggestions.
- Most masturbate during the call.

f. The typical voyeur is a heterosexual male who begins his voyeuristic behaviors before the age of 15. Voyeurism is rarely the only paraphilia.

g. Although it technically refers to a single couple copulating in front of others, **troilism** has come to mean any sex sessions involving multiple partners.

4. Transvestic Fetishism

a. Fetishistic transvestism refers to a transvestite who obtains sexual pleasure from dressing up in the clothing of the other sex.

b. Many theorists believe that transvestism evolves from an early childhood experience, such as a male masturbating with or in some item of female clothing, and some may move beyond the sexual arousal and begin to feel less anxious and stressed when around the particular item of clothing.

c. One research study with the wives of transvestites suggested that of this sample, most had learned of their husbands' habit early in the relationship and tolerated or even supported it to some degree.

d. Transvestism is usually so fixed in a man's personality that eradication is neither possible nor desirable, and treatment typically focuses on coping with anxieties, guilt, and relationships with family and friends.

5. Pedophilia

a. Pedophilia is defined as sex with children as a preferred or exclusive mode of sexual interaction in an adult and is one of the most common paraphilias.

b. Adult-child sexual interactions have been viewed differently in different cultures and different periods of history.
c. Attraction to postpubertal boys and girls is called ebhebephilia, but it is usually not considered pathological.
d. Pedophiles are usually obsessed with their fantasies, and these fantasies tend to dominate their lives.
e. A child sexual abuser may or may not be a pedophile.
f. Girls are twice as likely as boys to be victims of pedophiliac behavior.

6. Other Paraphilias

a. **Frotteurism**
- This involves a man rubbing his genitals against a woman's thighs or buttocks in a crowded place where he can claim it was an accident and get away quickly.
- This is similar to **toucheurism** where the compulsive desire is to rub strangers with one's hands.

b. **Zoophilia**
- This is also referred to as **bestiality** or sexual contact with animals and is rare.
- Contact between people and animals has been both practiced and condemned since earliest times.
- Research has found that a male dog is the most popular animal sex partner for both men and women who engage in zoophilia.

c. **Necrophilia**
- This is having sex with corpses and has been found even in ancient civilizations.
- Research has identified three types of necrophilia: necrophiliac fantasy, regular necrophilia, and necrophiliac homicide

H. **Assessing and Treating Paraphilias**
- Although the majority of paraphiliacs do not seek treatment and are content with balancing the pleasure and guilt of their paraphilias, others find them to be an unwanted disruption in their lives.

1. Assessment

a. The first step in treating a person with a paraphilia is to assess the nature and scope of the problem through self-report, behavioral observation, or by physiological tests or personality inventories.
b. The most reliable technique for men is **penile plethysmography**, which is often used with male sex offenders.
c. Personality inventories such as the **Minnesota Multiphasic Personality Inventory** can help establish personality patterns and determine whether there are additional psychological disorders.

2. Treatment Options

a. Treatment for paraphilias is generally multifaceted and may include group, individual, and family therapy, medication, education, and/or self-help groups.
b. Behavioral techniques have been the most commonly used and most successful treatments for paraphilias.
c. In aversion therapy the undesirable behavior is linked with an unpleasant stimulus, whereas in **shame aversion**, the unpleasant stimulus is shame, like an exhibitionist exposing himself in front of an audience.

d. In **systematic desensitization** the person is taken through more and more anxiety-provoking or arousing situations and is taught to deeply relax at each step.
e. In **orgasmic reconditioning**, a person masturbates until he feels orgasm is inevitable, and then he switches his fantasy to a more desired one.
f. In **satiation therapy** a person masturbates to a conventional fantasy and then masturbates again to an undesirable fantasy in order to connect the lower desire with the undesirable fantasy.
g. Some treatments involve pharmacological and surgical means.
h. Some therapists treat paraphilias with the same drugs used to treat obsessive-compulsive disorder (OCD) due to the similarities between the two.

III. VARIATIONS IN SEXUAL FREQUENCY

A. Hypersexuality: Does Obsession Imply Addiction?

1. Recently, "sexual addiction," also called compulsive sexual behavior, sexual compulsivity, sexual dependency, sexual impulsivity, and hypersexuality, has become popular and controversial.
2. According to one researcher, Patrick Carnes, it is estimated that 80% of sexual addicts have concurrent addictions including drugs, alcohol, gambling, food, or shopping.
3. Carnes suggests that a sexual addict goes through four cycles repeatedly: a preoccupation with thoughts of sex, ritualization of preparation for sex, compulsive sexual behavior, and despair afterward.
4. Some critics suggest that there is an attempt to return to a pathological model of sexuality, using the concept of addiction.

B. Hyposexuality: Lacking Desire and Avoiding Sex

1. People with hyposexuality have no sexual fantasies or desire for sexual activity.
2. Carnes discusses sexual anorexia, an obsessive condition in which a person spends much of his or her energy avoiding sex.
3. A sexual anorexic suppresses his or her interest in sexuality, and engaging in sex is seen as a weakness and a failure of self-discipline.
4. A sexual anorexic often experiences the following symptoms: fear and dread of sexual pleasure and contact, obsession about sexual matters, a preoccupation with the sexuality of others, judgmental attitudes about sexuality, an avoidance of intimacy, and self-destructive behaviors that are engaged in to limit or avoid sex.

IV. VARIATIONS, DEVIATIONS, AND WHO GETS TO DECIDE?

TEST YOURSELF

Below you will find fill-in-the-blank and short answer essay questions for topics covered in chapter 16. Check your answers at the end of this chapter.

Paraphilias and Variations in Sexual Frequency

1. ______________________________ are sexual behaviors that involve a craving for an erotic object that is "unusual."

2. Many people live comfortably with their ______________________________. A man who has a fetish for lingerie, for example, may find a partner who very much enjoys wearing it for him.

3. ________________________ suggest that paraphilias develop because some behavior becomes associated with sexual pleasure through conditioning.

4. A(n) ________________________ is an inanimate object or a body part not usually associated with the sex act that becomes the primary or exclusive focus of sexual arousal and orgasm in an individual.

5. Sadism refers to the intentional infliction of physical or psychological ________________________ on another person in order to achieve sexual excitement.

6. ________________________ involves exposing the genitals to a stranger.

7. The technical name for obscene telephone calling is ________________________.

8. Although it technically refers to a single couple copulating in front of others, ________________________ has come to mean any sex sessions involving multiple partners.

9. People can be sexually attracted to almost ________________________.

10. ________________________ involves a man rubbing his genitals against a woman's thighs or buttocks in a crowded place, where he can claim it was an accident and get away quickly.

11. ________________________ is also referred to as bestiality.

12. Tales of necrophilia, or having sex with ________________________, have been found even in ancient civilizations.

13. The terms asphyxiophilia or hypoxyphilia refer to the act of partly ________________________ oneself to enhance orgasm.

14. In ________________________ ________________________, a paraphiliac masturbates, and just as he feels orgasm is inevitable, he switches his fantasy to a more desired one, hoping thereby to increasingly associate orgasm with the desirable stimulus.

15. ________________________, like drugs, alcohol, gambling, and all other behaviors that bring a sense of excitement and pleasure, should involve some degree of moderation.

16. Name some of the more common types of paraphilias.

17. What is masochism?

18. What is an exhibitionist?

19. What is the difference between transvestites and transsexuals?

20. Name some treatment options for people who are uncomfortable with their paraphilias.

POST TEST

Below you will find true/false, multiple-choice and matching quiz items covering the entire chapter. Check your answers at the end of this chapter.

True/False

1. Research on paraphilias has been drawn mostly from college student populations.
2. Birching is part of the development of lovemaps.
3. Psychoanalysts believe that paraphilias develop because some behavior becomes associated with sexual pleasure.
4. A large percentage of people into sadomasochism switch between the dominant and submissive roles.
5. The primary goal of therapy with transvestites is stop the cross-dressing.
6. Attraction to postpubertal boys and girls is called satyriasis.
7. Records of contacts between humans and animals can be traced to Greek mythology.
8. Psychoanalytic techniques have been the most commonly used and most successful treatments for paraphilias.
9. Women don't identify as having sexual addictions.
10. Some sexologists object to the concept of sexual addiction.

Multiple-Choice

11. What affects how "appropriate" or "inappropriate" objects of sexual attention differ?
    a. different cultures
    b. different times
    c. different people
    d. All of the above
    e. None of the above

12. To be diagnosed with a paraphilia, how long must a person experience symptoms?
    a. at least 1 month
    b. for 3 months
    c. for 6 months or more
    d. at least a year or more
    e. at least 2 years or more

13. What is the controversy among sexologists related to paraphilias?
    a. Therapists disagree about how long a person should exhibit the desire or behavior in order to be classified as a paraphiliac.
    b. Some researchers dispute the notion that paraphiliacs are mostly men, believing that women are ignored.
    c. Some theorists suggest that paraphilias reflect a society's value judgments about sexuality and are not a psychiatric or clinical category.
    d. Some therapists believe that the paraphilia category should be expanded to include people who prefer positions for sexual intercourse beside male-on-top.
    e. All of the above

14. What below is an example of a biological theory of where paraphilias begin?
    a. At birth we have a general erotic potential that can be attached to almost anything.
    b. Some people with paraphilias have differences in brain structure.
    c. Paraphilias can be traced back to the difficult time the infant has in negotiating his way through the Oedipal crisis.
    d. Paraphilias develop because some behavior becomes associated with sexual pleasure through conditioning.
    e. Society encourages certain behaviors like treating women as sexual objects.

15. The view that an exhibitionist is "masquerading" as a man to cover up feelings of non-masculinity reflects what theoretical perspective of where paraphilias begin?
    a. behavioral theory
    b. sociological theory
    c. developmental theory
    d. psychoanalytic theory
    e. biological theory

16. Reflecting a psychoanalytic perspective of paraphilias, what psychoanalyst suggests that every paraphilia involves issues of masculinity and femininity?
    a. Louise Kaplan
    b. John Money
    c. Alfred Kinsey
    d. Anna Sacher-Masoch
    e. Vern Bullogh

17. Sexualized advertising reflects what theoretical perspective of where paraphilias begin?
    a. developmental theory
    b. biological theory
    c. psychoanalytic theory
    d. humanistic theory
    e. behavioral theory

18. What are lovemaps?
    a. the experience of being stuck in different stages of the normal progression of courtship
    b. recurrent and persistent thoughts about falling in love to the point of an obsession
    c. the physical reaction to falling in love at first sight
    d. the visual stimuli we experienced in childhood sex play that forms a template that defines our ideal lover
    e. therapeutic term used to describe the process of coming to terms with paraphilias through diagrams

19. What is the theory of courtship disorders?
    a. a theory of paraphilias that suggests they develop when a person first falls in love
    b. the idea that girls develops sexual dysfunctions during preadolescence when the process of flirting begins
    c. the development of sadomasochism
    d. a theory of paraphilias that links them to being stuck in different stages of the typical progression of dating and mating
    e. None of the above

20. Which process below is associated with behaviorism?
    a. conditioning
    b. birching
    c. nesting
    d. nocturnal emissions
    e. constancy

21. What statement below reflects a sociological view of paraphilias?
    a. Paraphilias may develop after a person develops a lesion in the brain.
    b. Men who expose themselves in public may be coping with castration anxiety by evoking a reaction to his penis from women.
    c. Representations of eroticism may be easily substituted for sex itself and so paraphilias become common.
    d. Normal sexual curiosity has been made painful, redirecting erotic energy toward other objects.
    e. A boy masturbates on his leather couch and begins to associate leather with sexual excitement.

22. If Chris can only become aroused while he is holding leather, what type of paraphilia is this?
    a. ebhebephilia
    b. fetishism
    c. hypophilia
    d. sadism
    e. satyriasis

23. What is the name for an inanimate object or a body part not usually associated with the sex act that becomes the primary or exclusive focus of sexual arousal and orgasm in an individual?
    a. paraphile
    b. sadist
    c. frottage
    d. fetish
    e. satyrim

24. What is the term for the intentional infliction of physical or psychological pain on another person in order to achieve sexual excitement?
    a. masochism
    b. satyriasis
    c. recidivism
    d. sadism
    e. frotteurism

25. Someone who is repeatedly arrested for being a "peeping Tom" is likely a
    ___________________________.
    a. hebephiliac
    b. nymphomaniac
    c. voyeur
    d. frotteurist
    e. satyriasist

26. What sexual experience may involve aspects of both voyeurism and exhibitionism?
    a. aversionism
    b. bestiality
    c. troilism
    d. flagellism
    e. recidivism

27. What is fetishistic transvestism?
    a. when a person's preferred or exclusive method of sexual arousal or orgasm is through wearing the clothing of the other sex
    b. when a person's exclusive method of sexual arousal is through hiring prostitutes
    c. when a person engages in humiliating sexual interactions in order to become aroused
    d. when a person has to fantasize about fictional women from movies in order to become aroused
    e. None of the above

28. What is another term for a transvestite?
    a. transsexual
    b. mother hen
    c. drag queen
    d. dominatrix
    e. None of the above

29. A pedophile is often _______-years old or more and is at least five years older than his victim.
    a. 12
    b. 16
    c. 21
    d. 30
    e. 35

30. What characteristic below describes pedophiles as reported by research?
    a. poor social relations with adults
    b. low self-esteem
    c. arrested psychological development
    d. childish emotional needs
    e. All of the above

31. What paraphilia is similar to toucheurism?
    a. narratophilia
    b. hyphenism
    c. submissim
    d. frotteurism
    e. scopophilia

32. What is the term for the form of bestiality that involves an emotional and/or sexual attachment or attraction to an animal?
    a. zoophilia
    b. scatolophilia
    c. gerontophilia
    d. animaphilia
    e. stigmatophilia

33. Who prohibited the corpses of the wives of important men from being delivered immediately to the embalmers for fear that the embalmers would violate them?
    a. early American colonists
    b. ancient Greeks
    c. royalty in Renaissance England
    d. ancient Egyptians
    e. French rulers in the 1700s

34. How does the Minnesota Multiphasic Personality Inventory help when assessing and treating paraphilias?
    a. It can rule out the need for physiological intervention.
    b. It can illustrate the types of fantasies a person is having.
    c. It can determine whether there are additional psychological disorders.
    d. All of the above
    e. None of the above

35. A sex therapy patient is being treated for his exhibitionism by being instructed to switch fantasies just at the moment of masturbatory orgasm in order to try and condition himself to become excited by more conventional fantasies. What is this technique?
    a. stimuli desensitization
    b. muted plethysmography
    c. orgasmic reconditioning
    d. secondary reactivism
    e. learned breakthrough

36. What physiological method has been used to treat paraphilias?
    a. chemotherapy
    b. antidepressants
    c. castration
    d. testosterone-suppressing drugs
    e. All of the above

37. Historically, men who had repeated and fleeting sexual encounters which were said to have __________________________.
    a. infantilism
    b. hyposexuality
    c. scatophagia
    d. satyriasis
    e. desensitization

38. What is the name of the researcher who wrote the controversial book on Understanding Sexual Addiction?
    a. Willy Sacher
    b. Patrick Carnes
    c. Vern Bullogh
    d. John Money
    e. Leonard McGrath

39. The condition of having no sexual fantasies or desire for sexual activity can be called __________________________.
    a. pedosexuality
    b. hyposexuality
    c. primsexuality
    d. masosexuality
    e. hypersexuality

40. What is sexual anorexia?
    a. an obsessive condition in which a person spends much of his or her energy engaging in sex.
    b. an obsessive condition in which a person spends much of his or her energy thinking about food during sex.
    c. an obsessive condition in which a person spends much of his or her energy avoiding sex.
    d. an obsessive condition in which a person spends much of his or her energy encouraging others to be sexual.
    e. an obsessive condition in which a person spends much of his or her energy engaging in group sex.

Matching

Column 1		Column 2
A. necrophilia	____	41. sex with children as a preferred or exclusive means of sexual arousal and orgasm
B. frotteurism	____	42. attraction to children who have just passed puberty
C. pedophilia	____	43. focus on receiving pain and/or humiliation as a preferred or exclusive means of sexual arousal and orgasm
D. ebhebephilia	____	44. the sexual attraction to dead bodies in fantasy or through sexual contact as a preferred or exclusive means of sexual arousal and orgasm
E. masochism	____	45. treating the submissive partner as a baby, including dressing the person in diapers
F. scatolophilia	____	46. sexual arousal from making obscene telephone calls
G. fetishism	____	47. focus on administering pain and/or humiliation as a preferred or exclusive means of sexual arousal and orgasm
H. infantilism	____	48. the act of compulsively rubbing against strangers for sexual arousal
I. sadism	____	49. focusing intensely on an inanimate object of body part as a preferred or exclusive means of sexual arousal and orgasm
J. voyeurism	____	50. observing people undressing or engaging in sex without their consent as a preferred or exclusive means of sexual arousal and orgasm

Test Yourself Answer Key

Paraphilias and Variations in Sexual Frequency

1. Paraphilias (p. 501)
2. paraphilias (p. 503)
3. Behaviorists (p. 504)
4. fetish (p. 505)
5. pain (p. 506)
6. Exhibitionism (p. 509)
7. scatolophilia (p. 509)
8. troilism (p. 511)
9. anything (p. 518)
10. Frotteurism (p. 518)
11. Zoophilia (p. 518)
12. corpses (p. 518)
13. strangulating (p. 519)
14. orgasmic reconditioning (p. 520)
15. Sexuality (p. 522)
16. fetishism, sadism, masochism, exhibitionism, voyeurism (pp. 505-511)
17. It is the achievement of sexual pleasure through one's own physical pain or psychological humiliation. (p. 507)
18. An exhibitionist is a person who becomes sexually aroused primarily by displaying his or her genitals, nudity, or sexuality to strangers. (p. 509)
19. Transsexuals feel that they are trapped in the body of the wrong sex, while transvestites obtain sexual pleasure from dressing up in the clothing of the other sex and do not desire to change their biological sex. (pp. 511-512)
20. shame aversion therapy, systematic desensitization, orgasmic reconditioning (pp. 520-521)

Post Test Answer Key

True/False

1. F (p. 502)
2. F (p. 504)
3. F (p. 503)
4. T (p. 508)
5. F (p. 514)
6. F (p. 514)
7. T (p. 518)
8. F (p. 520)
9. F (p. 522)
10. T (p. 522)

Multiple Choice

11. d (p. 500)
12. c (p. 501)
13. c (p. 503)
14. b (p. 503)
15. d (p. 503)
16. a (p. 503)
17. a (p. 504)
18. d (p. 504)
19. d (p. 504)
20. a (p. 505)
21. c (p. 505)
22. b (p. 505)
23. d (p. 505)
24. d (p. 506)
25. c (p. 511)
26. c (p. 511)
27. a (p. 511)
28. e (pp. 511-514)
29. b (p. 514)
30. e (p. 515)
31. d (p. 518)
32. a (p. 518)
33. d (p. 518)
34. c (p. 520)
35. c (p. 520)
36. e (p. 521)
37. d (p. 522)
38. b (p. 522)
39. b (p. 523)
40. c (p. 523)

Matching

41. C (p. 514)
42. D (p. 514)
43. E (p. 507)
44. A (p. 518)
45. H (p. 507)
46. F (p. 509)
47. I (p. 506)
48. B (p. 518)
49. G (p. 505)
50. J (p. 511)

CHAPTER OUTLINE

I. RAPE AND SEXUAL ASSAULT

- Physically or psychologically forcing sexual relations on another person is usually referred to as **rape**.

A. **Defining Rape and Sexual Assault**

1. Every state has its own definitions of sexual assault and rape so there is no single definition of rape.
2. Elements that are generally included in most definitions are lack of consent, force or threat of force, and vaginal penetration (of any kind).
3. A nonpenile sexual attack has also been referred to as sexual assault and is defined as the unwanted touching of an intimate part of another person, including the genitals, buttocks, and/or breasts, for sexual arousal.
4. Society has begun to use the word "survivor" in place of "victim" since it is not a passive term and implies that the person has within themselves the strength to overcome and to survive the rape.

B. **Incidence and Reporting of Rape**

1. It has been estimated that 15% of adult women in the U.S. have been raped at some point in their lives, and another 3% have been victims of an attempted rape.
2. In men the estimates for rape and attempted rape are 2% and 1% respectively.
3. On college campuses it is estimated that about 3% of women experience a completed and/or attempted rape during a typical college year.
4. Nine in 10 of the offenders were known to the women, and 60% of the rape/attempts took place in the women's residence.
5. Forty-four percent of female college students had engaged in unwanted sexual intercourse because of verbal pressure from their partners.
6. It is difficult to assess the actual incidence of rape because forcible rape is one of the most underreported crimes in the U.S., with one study finding that only 21% of stranger rapes and 2% of acquaintance rapes were reported.
7. Forty-four percent of one sample of adult women had been victims of either an attempted or completed rape, yet only 8% reported it to the police.

II. THEORIES AND ATTITUDES ABOUT RAPE

A. **Rapist Psychopathology: A Disease Model**

1. This theory of **rapist psychopathology** suggests that either disease or intoxication forces men to rape and that if they did not have these problems, they would not rape.
2. According to this theory, the rape rate can be reduced by finding these sick individuals and rehabilitating them, which makes people feel safer because it suggests that only sick individuals rape, not "normal" people.
3. Research consistently fails to identify any significant distinguishing characteristics of rapists.

B. **Victim Precipitation Theory: Blaming the Victim**

1. The victim precipitation theory explores the ways that victims make themselves vulnerable to rape, such as how they dress, act, or where they walk.
2. By focusing on the victim and ignoring the motivations of the attacker, many have labeled this a "blame the victim" theory.

C. **Feminist Theory: Keeping Women in Their Place**

1. Feminists contend that rape and the threat of rape are tools used in our society to keep women in their place, keeping women in traditional sex roles that are subordinate to men's.

2. Feminist theorists believe that the social, economic, and political separation of the genders has encouraged rape, which is viewed as an act of domination of men over women.
3. Sex-role stereotyping reinforces the idea that men are supposed to be strong, aggressive, and assertive, while women are expected to be slim, weak, and passive, encouraging rape in our culture.

D. **Sociological Theory: Balance of Power**
1. Sociological theory and feminist theory are related, with sociologists suggesting that rape is an expression of power differentials in society.
2. When men feel disempowered by society, by changing sex roles, or by their jobs, overpowering women with the symbol of their masculinity reinforces men's control over the world.

E. **Measuring Attitudes About Rape:** Researchers have used many techniques to measure attitudes about rape and rape victims including written vignettes, mock trials, videotaped scenarios, still photography, and newspaper reports.

F. **Gender Differences in Attitudes About Rape**
1. Men have been found to be less empathetic and sensitive toward rape than women and to attribute more responsibility to the victim than women, especially men who often watch pornography.
2. Some women believe that if they "led a man on," they gave up their right to refuse sex.

III. RAPE IN DIFFERENT CULTURES

A. The United States has the highest number and rate of reported rapes in the world, though the incidence of rape varies depending on how each culture defines rape.
B. In some cultures, rape has been used as a punishment for women, a sign of masculinity, or for initiation purposes.
C. During times of war, rape is used as a weapon.
D. Research has found that one million women and children are raped in South Africa each year.
E. Research by Peggy Sanday indicates that the primary cultural factors that affect the incidence of rape in a society include relations between the sexes, the status of women, and male attitudes in the society.
F. Societies that promote male violence have higher incidences of rape because men are socialized to be aggressive, dominating, and to use force to get what they want.

IV. RAPE ON CAMPUS

A. **Alcohol and Rape**
1. On college campuses, alcohol use is one of the strongest predictors of acquaintance rape. Half of all rape cases are estimated to involve alcohol consumption by the rapist, victim, or both.
2. One study found that 75% of college men reported giving women alcohol or drugs in an attempt to obtain sex.
3. When a woman experiences a rape while drunk, she is more likely to blame herself and often will not label the attack as a rape even when it clearly was.
4. A man who is drunk and is accused of rape is seen as less responsible, whereas a woman who has been drinking is seen as more responsible for being raped.

B. **Fraternities and Rape**
1. It is estimated that 10% of college rapes happen in fraternities.
2. Many fraternities revolve around an ethic of masculinity where there is a considerable pressure to be sexually successful.
3. Other factors that contribute to rape in fraternities are the inherent secrecy and protection of the group.

4. Peggy Sanday, an expert on fraternity rape, discusses **pulling train**, which refers to the gang rape of a woman as a succession of boxcars on a train.
5. Some fraternities have begun to institute education programs and invite guest speakers from Rape Crisis Centers to discuss the problem of date rape.

C. **Athletes and Rape**

1. Male athletes have been found to be disproportionately over-represented as assailants of rape by women surveyed.
2. Although less than 2% of the men on college campuses are athletes, they represent 23% of the men who are accused of rape on college campuses.
3. Many athletes have been found to view the world in a way that helps to legitimize rape, and many feel a sense of privilege.

V. EFFECTS OF RAPE

A. **Rape Trauma Syndrome**

1. Rape Trauma Syndrome is a two-stage stress response pattern characterized by physical, psychological, behavioral, and/or sexual problems, and it occurs after forced, non-consent sexual activity.
2. The **Acute Phase** is the first stage when most victims feel a fear of being alone, fear of strangers, and it may also include anger, anxiety, depression, confusion, shock, disbelief, incoherence, guilt, humiliation, shame, and self-blame.
3. Research indicates that about 60% of women will eventually talk to someone about the experience, with younger victims more likely to talk.
4. Women who report being raped by strangers experience more anxiety, fear, and startle responses, whereas women who report being raped by acquaintances usually report more depression, guilt, and a decrease in self-confidence.
5. **Long-term Reorganization** is stage two and involves restoring order and control in the victim's lifestyle.
6. The majority of victims report experiencing sexual problems after a rape.

B. **Silent Rape Reaction**

1. Some victims never discuss the rape with anyone and carry the burden of the assault alone within themselves, which as been termed the Silent Rape Reaction and has many similarities with Rape Trauma Syndrome.
2. Those who take longer to confide in someone usually suffer a longer recovery period.
3. The Silent Rape Reaction occurs because some victims deny and repress the incident until a time when they feel stronger emotionally.

C. **Rape of Spouses and Other Special Populations**

1. **Marital Rape**

a. As of 1993, marital rape is considered a crime in all 50 states, although in 32 states there are ways for husbands to be exempted from prosecution for rape, and a handful of states believe that marital rape is a lesser crime than non-marital or stranger rape.
b. It has been estimated that 10% to 14% of all married women are raped by their husbands.
c. There is often little social support for wives who are raped, and those who stay with their husbands often endure repeated attacks.

2. **Lesbians:** For some lesbians it is difficult to assimilate the experience of rape into their own self-image.
3. **Older Women:** Older women are likely to be even more traumatized by rape than younger women because many have conservative attitudes about sexuality, have undergone physical changes that can increase the severity of physical injury, and they have less social support after a rape.

4. **Women with Disabilities**
   a. Women with disabilities are assaulted, raped, and abused at a rate two times greater than women without disabilities.
   b. The impact of a rape may be very intense for women with mental disabilities because of a lack of knowledge about sexuality, the loss of a sense of trust, and the lack of knowledgeable staff who can effectively work with these victims.
5. **Prostitutes:** Because of the general disapproval of prostitution, a prostitute who reports rape is often treated with disdain, so believing and trusting her experience is imperative.

D. **How Partners React to Rape**

1. When a man or woman's partner is raped, the partner often feels anger, frustration, and intense feelings of revenge.
2. In cases of acquaintance rape, men may lose their trust in their partner, feeling that because the partner knew the assailant, she may have expressed sexual interest in him.
3. Couples often avoid dealing with the rape entirely.
4. Even though dealing with a rape in a relationship can be traumatic, it has been found that women who have a stable and supportive partner recover more quickly.

VI. WHEN MEN ARE RAPE VICTIMS

- Each year in the U.S., more than 14,000 men are victims of rape or attempted rape. However, male rape is even more underreported than female rape.

A. **Rape of Men by Women**

1. The myth that a woman cannot rape a man serves to make male rape more painful for many men.
2. When anxious, embarrassed, or terrorized, men can get an erection, and dildos, hands, or other objects can also anally penetrate them.
3. In a study of male college students, 24% reported coercive sexual contact, with 24% from women, 4% from men, and 6% from both women and men, and 20% of the men experiencing strong negative reactions.
4. The majority of male rapes by women involve psychological or pressured contact such as verbal persuasion or emotional manipulation.

B. **Rape of Men by Men**

1. Research indicates that between 2.5% and 5% of men are sexually victimized before the age of 13, and 84% of sexually assaulted boys reported that they had a male assailant. However, the rape of men by men remains largely underreported.
2. As in the case of female victims, male rape is an expression of power, a show of strength and masculinity.
3. Many victims of male rape question their sexual orientation and feel that the rape makes them less of a "real man."
4. The risk of suicide in men who have been raped has been found to be higher than women.

C. **Prison Rape**

1. Research has found that both men and women are raped in prison today, although the majority of the research has been focused on male victims of rape.
2. Prison rape has been found to have a significant role in the development of **post-traumatic stress disorder**: a psychiatric disorder where a person relives the experience through nightmares and flashbacks and may have difficult sleeping.
3. Women who are raped in prison in the U.S. might also be victims of sexual harassment, molestation, coercive sexual behaviors, and forced sexual intercourse. Prison staff perpetrates the majority of this abuse.

4. Prison rape of men has been found to be an act of asserting one's own masculinity in an environment that rewards dominance and power.
5. Although prison rape occurs in the male population most frequently, it also occurs between female inmates using a variety of different objects.

VII. THE RAPIST

- Research has shown that rapists are primarily from younger age groups, between the ages of 15 and 30, single, and tend to reduce their rape behavior as they get older.
- Other research indicates that men who rape generally have sexist views about women, accept myths about rape, have low self-esteem, and are politically conservative.

A. Treating the Rapist

1. Many different therapies have been tried, including shock treatment, psychotherapy, behavioral treatment, support groups, and the use of Depo Provera.
2. For many men in treatment, the most important first step is to accept responsibility for their actions.
3. Although attitudes about rape myths appear to change after these education programs about rape, research has yet to show that these attitude changes result in changes in sexually coercive behavior.

VII. REPORTING A RAPE

A. Telling the Police

1. It is estimated that about 1 in 3 rapes are reported, and the likelihood of reporting is increased if the assailant was a stranger, if there was violence, or if a weapon was involved.
2. Women who report rape to the police have been found to have a better adjustment and fewer emotional symptoms than those who do not report.
3. Society's victim precipitated view of rape also affects the attitudes of police, although police officers have become more sensitive in the past few years.

B. Pressing Charges

1. The decision to press charges is difficult because a victim often feels as if he or she is on trial along with constant delays in the process and the public ordeal.
2. Reasons for refusing to press charges included being afraid of revenge, wanting to just forget, feeling sorry for the rapist, or feeling like it would not matter anyway because nothing will be done.
3. Victims of rape can also file a civil lawsuit and sue the assailant for monetary damages.

C. Going to Court

1. Sitting on a rape trial can prepare a victim for the process.
2. If a victim does decide to proceed with legal action, he or she must also be prepared for the possibility that the rapist may be found not guilty.

VIII. AVOIDING RAPE

A. Avoidance Strategies

1. Rape is the only violent crime in which society expects a person to fight back.
2. Research has studied victim's deciding to submit to an attack, which may increase post-assault guilt or self-blame.
3. Escape is the first strategy, followed by screaming, dissuasive techniques, or negotiation.
4. One study found that women who had take a self-defense class felt less scared and angrier during the rape than women who had never taken such a class.

B. **Rapist Typology and Avoiding a Rape:** the psychological traits of a rapist can affect the strategies for avoiding a rape.

IX. SEXUAL ABUSE OF CHILDREN

- Incest refers to sexual contact between a child, adolescent, or adult and a child they are closely related to, although specific definitions for incest vary from state to state.
- Child sexual abuse ranges from genital touching or oral and genital stimulation to penetration and the involvement of children in prostitution and pornography.
- The incest taboo (the prohibition of sex between family members) is universal across cultures.
- There is evidence that brother-sister sexual relationships are more common than father-daughter incest, but disagreement exists as to the amount of trauma involved, which suggestions that it is not traumatic unless there is force or exploitation.
- Men who have been sexually abused by their mothers often experience more trauma symptoms than do other sexually abused men.
- The Internet has given rise to another avenue for child sexual abuse.

A. **Incidence of Child Sexual Abuse**

1. Recent research suggests that the incidence of child sexual abuse has been increasing over the past 30 years.
2. It is estimated that 1 of every 4 girls and 1 of every 10 boys experiences sexual abuse as a child, although it remains underreported.
3. The increase in the incidence of child sexual abuse may be a reflection of changing social views rather than an actual increase in the number of sexual assaults on children.

B. **Victims of Sexual Abuse**

1. Boys are more likely to be abused by strangers, while girls are more likely to be abused by family members, with the median age of 8-9.
2. Victims of incest with a biological father delay reporting the longest, while those who are victims of stepfather or live-in partners told more readily.

X. HOW SEXUAL ABUSE AFFECTS CHILDREN

- Research suggests that the greatest trauma of sexual abuse occurs when it exists over a long period of time, the offender is a person who is trusted, penetration occurs, and there is aggression.
- If a family handles the sexual abuse in a caring and sensitive manner, the effects on the child are often reduced.

A. **Psychological and Emotional Reactions**

1. Children who hide their sexual abuse often experience shame and guilt and fear the loss of affection from family and friends.
2. Whether they tell someone about the abuse or not, many victims experience psychological symptoms such as depression, increased anxiety, nervousness, emotional problems, low self-esteem, and personality and intimacy disorders.
3. Many girls and women develop a tendency to blame themselves for sexual abuse, most likely due to more internal **attributional styles** (pattern of internal or external styles of attributing meaning to various events).
4. In its extreme form, **dissociative disorder** may result in **multiple personality disorder (MPD)** in which there is actually two or more distinct personalities in one body.
5. Antisocial behavior and promiscuous sexual behavior are also related to a history of childhood sexual behavior.

B. **Long-Term Effects**

1. It is not uncommon for children who are sexually abused to display what researchers call **traumatic sexualization**, where children begin to exhibit compulsive sex play or masturbation and show an inappropriate amount of sexual knowledge.
2. When they enter adolescence they may begin to show promiscuous and compulsive sexual behavior, which may lead to sexually abusing others in adulthood.
3. Recent research reveals a connection between eating disorders and sexual abuse.
4. Research hypothesized that children who were abused incorporate "badness" into their self-concept, explaining why they are more likely to develop drug and alcohol problems and engage in prostitution.

XI. WHO ABUSES CHILDREN AND WHY?

A. **Who Are the Sexual Abusers?**

1. Research comparing child molesters to nonmolesters has shown us that molesters tend to have poorer social skills, lower IQs, unhappy family histories, lower self-esteem, and less happiness in their lives.
2. Many abusers have strict religious codes yet still violate sexual norms.
3. Denying responsibility for the offense and claiming they were in a trancelike state is also common.
4. Those who abuse children often report disdain for all sex offenders.

B. **The Development of a Sexual Abuser**

1. Learning theorists believe that what children learn from their environment or those around them contributes to their behavior later in life, such as learning that sexual behavior is how adults show love and affection to children.
2. Gender theories recognize that sexual abusers are overwhelmingly male and that sex-role stereotyping can lead men to abuse.
3. Biological theories suggest that physiology, such as hormonal differences of neurological differences, contributes to the development of sexual abusers.

XII. TREATING SEXUAL ABUSE

A. **Helping the Victims Heal**

1. Currently, the most effective treatments for victims of sexual abuse include a combination of cognitive and behavioral psychotherapies, which teach victims how to understand and handle the trauma of their sexual assaults more effectively.
2. Being involved in a relationship that is high in emotional intimacy and low in expectations for sex is beneficial.

B. **Treating Abusers**

1. In addition to decreasing sexual interest in inappropriate sexual objects, other treatment goals for abusers include teaching them to interact and relate better with adults, assertiveness skills training, empathy and respect for others, and increasing sexuality education.
2. Since recidivism is high, it is also important to find ways to reduce the incidence of engaging in these behaviors.

C. **Preventing Child Sexual Abuse**

1. Teaching children to say "no" to inappropriate touching is one educational campaign.
2. Increasing the availability of sexuality education has also been cited as a way to decrease the incidence of child sexual abuse.
3. Telling children that abuse is not typical and where they can go to get assistance is important.
4. Other prevention areas include funding child welfare agencies and training physicians and educators to recognize signs.

XIII. DOMESTIC VIOLENCE

A. Defining Domestic Violence and Coercion

1. Domestic violence is coercive behavior that is done through the use of threats, harassment, or intimidation that may include physical, emotional, or sexual abuse.
2. Most women in this situation begin to believe that the problems are their fault so they stay in the abusive relationship.
3. Finances, low self-esteem, fear, and isolation are factors that make it difficult to leave an abusive relationship, and 75% are at greater risk of being killed by the partner if they leave.
4. Domestic violence in same-sex relationships looks similar to domestic violence in heterosexual relationships.

B. Preventing Domestic Violence

1. Battered women's shelters across the U.S. can provide women with information and a safe haven.
2. The Violence Against Women Act (VAWA) requires that cases of domestic violence be prosecuted by the Department of Justice.

XIV. SEXUAL HARASSMENT

- Sexual harassment is a very broad term that includes anything from looks, jokes, unwanted sexual advances, a "friendly" pat, an "accidental" brush on a person's body, or an arm around a person.
- One definition of sexual harassment sees it as a sexual pressure imposed on someone who is not in a position to refuse it.

A. Incidence and Reporting of Harassment

1. It is estimated that 50% to 85% of American women will experience some form of sexual harassment during their professional life.
2. Sexual harassment creates a hostile and intimidating environment.

B. Preventing Sexual Harassment

1. The first step in reducing the incidence of sexual harassment is to acknowledge the problem.
2. Workplaces need to design and implement strong policies against sexual harassment.

TEST YOURSELF

Below you will find fill-in-the-blank and short answer essay questions for topics covered in chapter 17. Check your answers at the end of this chapter.

Theories, Culture, Campus, and Effects

1. Physically or psychologically forcing sexual relations on another person is usually referred to as ________________________.

2. Although the word ________________________ emphasizes the person's lack of responsibility for a rape, it may also imply that the person was a passive recipient of the attack.

3. On average, a rape occurs every ________________________ minutes in the United States.

4. It is difficult to assess the actual incidence of rape because forcible rape is one of the most ________________________ crimes in the United States.

5. ______________________ theorists believe that sex-role stereotyping—which reinforces the idea that men are supposed to be strong, aggressive, and assertive while women are expected to be slim, weak, and passive—encourages rape in our culture.

6. ______________________ believe that rape is an expression of power differentials in society and explore the ways people guard their interests in society.

7. It has been found that people who believe in the victim-precipitation model tend to have more ______________________ attitudes in general.

8. ______________________ use is one of the strongest predictors of acquaintance rape.

9. One study found that ____ % of college men reported giving women alcohol or drugs in an attempt to obtain sex.

10. During the ______________________ Phase of Rape Trauma Syndrome, most victims feel a fear of being alone, fear of strangers, or even fear of their bedroom or their car if that is where the rape took place.

11. ______________________ ______________________ is stage two of RTS and involves restoring order in the victim's lifestyle and reestablishing control.

12. Some victims never discuss their rape with anyone and carry the burden of the assault alone within themselves. This is called the ______________________ ______________________ Reaction.

13. It has been estimated that their husbands rape ____ % to ____ % of all married women, although this number is much higher in battered women.

14. Students often laugh at the idea that a man could be ______________________ by a woman because they believe the myth that men are always willing to have sex, and so a woman would never need to rape a man.

15. Research estimates that between ____% and ____% of men are sexually victimized before the age to thirteen.

16. What is sexual assault?

17. Why are women so unlikely to report being raped?

18. What is the victim precipitation theory?

19. What is Rape Trauma Syndrome?

20. Describe how emotional reactions to rape vary depending on whether or not the victim knew her assailant.

The Rapist, Reporting Rape, and Other Types of Abuse

21. Research indicates that men who rape generally have ________________________ views about women, accept myths about rape, have low self-esteem, and are politically conservative.

22. Correlations have also been found between being the victim of past ________________________ ________________________ and raping behavior, and between the use of violent and degrading pornography and a negative view of women.

23. Many feminists argue that because ________________________, not sexual desire, causes rape, taking away sexual desire will not decrease the incidence of rape.

24. It is estimated that about one in ________________________ rapes are reported.

25. Women who ________________________ rape have been found to have a better adjustment and fewer emotional symptoms than those who do not report.

26. If you are confronted with a potential or attempted rape, the first and best strategy is to try to ________________________.

27. ________________________ refers to sexual contact between a child, adolescent, or adult and a child they are closely related to (parent, stepparent, uncle, cousin, caretaker).

28. The ________________________ ________________________ – the absolute prohibition of sex between family members–is universal.

29. Sibling incest is more likely to occur in families where there is a(n) ________________________ father, a(n) ________________________ mother and a dysfunctional home life.

30. It is estimated that 1 out of every ________________________ girls experiences sexual abuse as a child.

31. Children who ________________________ their sexual abuse often experience shame and guilt and fear the loss of affection from family and friends.

32. It is not uncommon for children who are sexually abused to display what is referred to as ________________________ ________________________, in which children may begin to exhibit compulsive sex play or masturbation and show an inappropriate amount of sexual knowledge.

33. ________________________ theorists believe that what children learn from their environment or those around them contributes to their behavior later in life. For example, many child abusers were themselves sexually abused as children.

34. Currently, the most effective treatments for victims of sexual abuse include a combination of ________________________ and ________________________ psychotherapies, which teach victims how to understand and handle the trauma of their assaults more effectively.

35. Increasing the availability of sex ________________________ has been cited as a way to decrease the incidence of child sexual abuse.

36. What are some therapies that have been used to try and treat rapists so that they lose their desire to rape?

37. What is Depo Provera?

38. What is the difference found between men and women in the reasons for not reporting rape?

39. If you cannot escape a potential rape situation, what may be some effective strategies to use?

40. What is incest?

POST TEST

Below you will find true/false, multiple-choice and matching quiz items covering the entire chapter. Check your answers at the end of this chapter.

True/False

1. In research on rape on college campuses, it is estimated that 90% of rapists were known to the women they raped.

2. Men have been found to attribute more responsibility to a rape victim than women.

3. Although the majority of rape victims report experiencing physical, psychological, behavioral problems after a rape, sexual problems are rarely an issue.

4. Women with disabilities are assaulted, raped, and abused at a rate two times greater than women without disabilities.

5. Men raped in prison have fewer negative effects compared to men raped outside of prison.

6. Children who hide their sexual abuse often fear the loss of affection from family and friends.

7. Many abusers have strict religious codes yet still violate sexual norms.

8. Battered women's shelters across the U.S. are required to notify a husband of his wife's whereabouts if he requests the information.
9. Sexual harassment may be nonverbal.

10. Most victims of sexual harassment never say anything about it.

Multiple-Choice

11. What element is generally included in most definitions of rape?
    a. force or threat of force
    b. lack of consent
    c. vaginal penetration
    d. All of the above
    e. None of the above

12. Research suggests that what percent of women experience a completed and/or attempted rape during a typical college year?
    a. 1%
    b. 3%
    c. 28%
    d. 37%
    e. 51%

13. What is the theory that suggests rapists rape because they are mentally ill or intoxicated?
    a. the feminist theory
    b. the sociological theory
    c. the victim theory
    d. the disease model
    e. the psychotic model

14. What statement reflects the victim precipitation model?
    a. She shouldn't have been walking in the neighborhood.
    b. Why was she wearing that tight skirt?
    c. She was drinking too much that night.
    d. Why did she go up to his apartment?
    e. All of the above

15. Which statement below would reflect a feminist theory of rape attitudes?
    a. Women who maintain separate circles from men, especially at night, can prevent themselves from getting raped.
    b. The social, economic, and political separation of the genders has encouraged rape, which is viewed as an act of domination of men over women.
    c. Women who are acting friendly in a bar are usually interested in having sex.
    d. Some women make themselves vulnerable to rape, such as by how they dress, act, or where they walk.
    e. It is either mental illness that forces men to rape and that if they did not have this problem, they would not rape

16. What would sociologists point to as something that encourages rape in society?
    a. provocative clothing
    b. power differentials
    c. the need to procreate
    d. rape paraphilias
    e. mental illness

17. What have researchers used to measure attitudes about rape and rape victims?
    a. videotaped scenarios
    b. mock trials
    c. newspaper reports
    d. written vignettes
    e. All of the above

18. According to the research of Peggy Sanday, what aspect of a society leads to higher incidences of rape across cultures?
    a. high status of women
    b. egalitarian relationships between men and women
    c. the promotion of male violence
    d. less access to food
    e. All of the above

19. How does the use of alcohol affect a rape case that goes to court?
    a. Women who have been drinking or were drunk are more likely to be given compassion.
    b. Women who have been drinking or were drunk are more likely to be discredited.
    c. Men who have been drinking or were drunk are more likely to be convicted.
    d. All of the above
    e. None of the above

20. What percent of college rapes are estimated to happen in fraternities?
    a. 10%
    b. 25%
    c. 50%
    d. 75%
    e. 95%

21. What does research suggest about college athletes who participate on teams that make money for their college?
    a. They are less likely to engage in sexually abusive behavior.
    b. They are more likely to invite guest speakers to speak against rape.
    c. They are more likely to turn in team members who rape.
    d. They are more likely to engage in sexually abusive behavior
    e. None of the above

22. What is the first stage of Rape Trauma Syndrome when most victims feel a fear of being alone, fear of strangers? Feelings may also include anger, anxiety, depression, confusion, shock, disbelief, incoherence, guilt, humiliation, shame, and self-blame.?
    a. Primary Stage
    b. Intense Phase
    c. Denial Period
    d. Acute Phase
    e. Shock Stage

23. What percent of women will eventually talk to someone about being raped?
    a. 4%
    b. 18%
    c. 60%
    d. 80%
    e. 97%

24. Sylvia remains fearful, angry, and depressed after the rape she experienced 2 years ago, but is unable to tell her family. What is the term for this type of response?
    a. Muted Rape Response
    b. Silent Rape Reaction
    c. Unspoken Suffering Experience
    d. Quiet Post-Rape Period
    e. Hushed Rape Condition

25. Which of the following statements regarding the research on marital rape is FALSE?
    a. The more often a woman experiences marital rape, the more emotional and physical symptoms she experiences.
    b. Both men and women report that marital rape is more psychologically damaging to the victim compared to other types of rape.
    c. Women who stay with their husbands often endure repeated attacks.
    d. There is little social support for women who are raped by their husbands.
    e. None of the above

26. What is a psychological reaction experienced by partners of rape survivors?
    a. feelings of revenge
    b. guilt
    c. jealousy
    d. sense of loss
    e. All of the above

27. What can make male rape more psychologically painful for many men who are raped by women?
    a. the myth that men cannot be raped by a woman
    b. having an erection
    c. society's disbelief or laughter at the idea
    d. All of the above
    e. None of the above

28. When it comes to men being raped by other men, which of the following statements is FALSE?
    a. The risk of suicide in men who have been raped has been found to be higher than women who were raped.
    b. Most men raped by men report the crime.
    c. Male rape is an expression of power, a show of strength and masculinity.
    d. Male rape victims may increase their subsequent sexual activity.
    e. Many heterosexual victims of male rape question their sexual orientation.

29. What attitudes about rape have been reflected by police responses to a victim of rape?
    a. feminist
    b. rapist psychopathology
    c. victim precipitation
    d. evolutionary
    e. sociological

30. Why do victims of rape report that they pressed charges against the rapist?
    a. to experience the trial process
    b. to protect others
    c. to be able to go public about the experience
    d. to discuss their sexual history in court
    e. All of the above

31. Men who have been sexually abused by their ______________ often experience more trauma symptoms than do other sexually abused men.
    a. mothers
    b. sisters
    c. brothers
    d. fathers
    e. cousins

32. What percent of girls are estimated to have been sexually abused?
    a. 2%
    b. 10%
    c. 25%
    d. 55%
    e. 75%

33. Victims of what type of incest delay reporting the abuse the longest?
    a. biological mother
    b. biological father
    c. stepmother
    d. stepfather
    e. sibling

34. What psychiatric disorder has been linked to child sexual abuse that occurs when victims try to cut themselves off from a painful or unbearable memory?
    a. schizophrenia
    b. obsessive compulsive disorder
    c. histrionic personality disorder
    d. multiple personality disorder
    e. body dysmorphic disorder

35. What characterizes traumatic sexualization?
    a. Children begin to exhibit compulsive masturbation.
    b. Children begin show an inappropriate amount of sexual knowledge.
    c. Children begin to exhibit compulsive sex play.
    d. All of the above
    e. None of the above

36. An abuser who says that he abused children because he has neurological differences in his brain is reflecting what theory of how sexual abusers develop?
    a. gender theory
    b. humanistic theory
    c. personality theory
    d. learning theory
    e. biological theory

37. What has NOT been recognized as a way to decrease the incidence of child sexual abuse?
    a. teaching children to say "no" to inappropriate touching
    b. training physicians to recognize signs of child abuse
    c. discouraging children from learning about their bodies
    d. increasing the availability of sexuality education
    e. funding child welfare agencies

38. What is the term for coercive behavior that is done through the use of threats, harassing, or intimidation that may include physical, emotional, or sexual abuse?
    a. sexual harassment
    b. sexual assault
    c. rape
    d. stalking
    e. domestic violence

39. What is an additional issue that can make domestic violence in same-sex relationships different than in heterosexual relationships?
    a. the threat of outing
    b. financial concerns
    c. feelings of responsibility
    d. fear
    e. None of the above

40. What separates sexual harassment from other sexual behaviors?
    a. the humorous nature of an action
    b. It always occurs between a man and a woman.
    c. the use of power or status
    d. It never involves physical contact.
    e. It always occurs in a workplace.

Matching

Column 1		Column 2
A. silent rape reaction	____	41. suggests that it is either disease or intoxication that forces men to rape and that if they did not have these problems, they would not rape.
B. victim precipitation theory	____	42. suggests that rape is an expression of power differentials in society.
C. domestic violence	____	43. the first stage of the response to rape when most victims feel a fear of being alone, fear of strangers, and may also include anger, anxiety, depression, confusion, shock, disbelief, incoherence, guilt, humiliation, shame, and self-blame.
D. sociological theory	____	44. second stage of the response to rape that involves restoring order and control in the victim's lifestyle.
E. rape trauma syndrome	____	45. focuses on the ways that victims make themselves vulnerable to rape, such as how they dress, act, or where they walk
F. rapist psychopathology theory	____	46. a two-stage stress response pattern characterized by physical, psychological, behavioral, and/or sexual problems, and it occurs after forced, non-consent sexual activity.
G. feminist theory	____	47. victims not discussing a rape with anyone and carrying the burden of the assault alone within themselves
H. sexual harassment	____	48. coercive behavior that is done through the use of threats, harassing, or intimidation that may include physical, emotional, or sexual abuse.
I. long-term reorganization	____	49. contends that rape and the threat of rape are tools used in our society to keep women in their place, keeping women in traditional sex roles that are subordinate to men's.
J. acute phase	____	50. unwanted attention of a sexual nature from someone in school or workplace that may include the use of status and/or power to coerce

Test Yourself Answer Key

Theories, Culture, Campus, and Effects

1. rape (p. 528)
2. victim (p. 529)
3. two (p. 529)
4. underreported (p. 530)
5. Feminist (p. 531)
6. Sociologists (p. 531)
7. conservative (p. 532)
8. Alcohol (p. 535)
9. 75 (p. 535)
10. Acute (p. 537)
11. Long-Term Reorganization (p. 538)
12. Silent Rape (p. 539)
13. 10, 14 (p. 539)
14. raped (p. 541)
15. 2.5, 5 (p. 541)
16. It is defined as the unwanted touching of an intimate part of another person, including the genitals, buttocks, and/or breasts, for sexual arousal. (p. 529)
17. Some women do not report it because they do not think that they were really raped, because they think that it was their fault because they did something to put themselves at risk, or because they feel shame and humiliation over the rape. Others fear no one will believe them or that nothing will be done legally. (p. 530)
18. It explores the ways victims make themselves vulnerable to rape, such as how they dress, act, or where they walk. By focusing on the victim and ignoring the motivations of the attacker, many have labeled this a "blame the victim" theory. (p. 531)
19. RTS is a two-stage stress response pattern characterized by physical, psychological, behavioral, and/or sexual problems, and it occurs after forced, non-consenting sexual activity. (p. 537)
20. Women who report being raped by strangers experience more anxiety, fear, and startle responses, while those raped by acquaintances usually report more depression, guilt, and a decrease in self-confidence. (p. 537)

The Rapist, Reporting Rape, and Other Types of Abuse

21. sexist (p. 543)
22. sexual abuse (p. 543)
23. violence (p. 543)
24. three (p. 543)
25. report (p. 543)
26. escape (p. 545)
27. Incest (p. 546)
28. incest taboo (p. 546)
29. dominating, passive (p. 546)
30. 4 (p. 547)
31. hide (p. 548)
32. traumatic sexualization (p. 548)
33. Learning (p. 550)
34. cognitive, behavioral (p. 550)
35. education (p. 551)
36. shock treatment, psychotherapy, behavioral treatment, support groups, the use of Depo Provera (p. 543)

37. It is a drug (progestin) that can diminish a man's sex drive. The idea behind it is that if the sex drive is reduced, so too will the likelihood of rape. (p. 543)
38. Women are less likely to report a rape if it does not fit the stereotypical rape scenario, while men are less likely to report if it jeopardizes their masculine self-identity. (p. 543)
39. verbal strategies such as screaming, dissuasive techniques, empathy, negotiation, and stalling for time (p. 545)
40. Incest refers to sexual contact between a child, adolescent, or adult and a child they are closely related to (parent, stepparent, uncle, cousin, caretaker). (p. 546)

Post Test Answer Key

	True/False		Multiple Choice		Matching
1.	T (p. 529)	11.	d (p. 529)	41.	F (p. 530)
2.	T (p. 532)	12.	b (p. 529)	42.	D (p. 531)
3.	F (p. 539)	13.	d (p. 530)	43.	J (p. 537)
4.	T (p. 540)	14.	e (p. 531)	44.	I (p. 538)
5.	F (p. 542)	15.	b (p. 531)	45.	B (p. 531)
6.	T (p. 548)	16.	b (p. 531)	46.	E (p. 537)
7.	T (p. 550)	17.	e (p. 532)	47.	A (p. 539)
8.	F (p. 553)	18.	c (p. 534)	48.	C (p. 551)
9.	T (p. 553)	19.	b (p. 535)	49.	G (p. 531)
10.	T (p. 554)	20.	a (p. 536)	50.	H (p. 553)
		21.	d (p. 536)		
		22.	d (p. 537)		
		23.	c (p. 537)		
		24.	b (p. 539)		
		25.	b (p. 539)		
		26.	e (p. 540)		
		27.	d (p. 541)		
		28.	b (pp. 541-542)		
		29.	c (p. 543)		
		30.	b (p. 544)		
		31.	a (p. 546)		
		32.	c (p. 547)		
		33.	b (p. 547)		
		34.	d (p. 548)		
		35.	d (p. 548)		
		36.	e (p. 550)		
		37.	c (p. 551)		
		38.	e (p. 551)		
		39.	a (p. 552)		
		40.	c (pp. 553-554)		

Chapter 18 – Sexual Images and Selling Sex

CHAPTER SUMMARY

By the dawn of the great ancient civilizations such as Egypt, people were drawing erotic images on walls or pieces of papyrus just for the sake of eroticism. The story of pornography is not just about publishing erotic material but also about the struggle between those who try to create it and those who try to stop them. Over the last 25 years, representations in the mass media have become more explicitly erotic. Pornography in the modern sense began to appear when printing became sophisticated enough to allow fairly large runs of popular books, beginning in the 16th century. Due to fears that people would turn away from religious teachings, by the 17th century, the Church was pressuring civic governments to allow them to inspect bookstores, and soon forbidden books, including erotica, were being removed.

The debate over pornography is particularly active today because sexual images have become so common on television, the Internet, and in advertising. In 1967 President Lyndon Johnson set up a commission to study "a matter of national concern": the impact of pornography on American society. The commission concluded that no reliable evidence was found to support the idea that exposure to elicit sexual materials is related to the development of delinquent or criminal sexual behavior among youth or adults, so adults should be able to decide for themselves what they will or will not read. The 1986 Attorney General's commission on pornography, the Meese Commission, came to the opposite conclusion. The issue of pornography has been divisive among feminist scholars, splitting them into two general schools.

Definitions of prostitution vary, with some state penal codes defining prostitution as the act of hiring out one's body for sexual intercourse and other states defining it as sexual intercourse in exchange for money or as any sexual behavior that is sold for profit. The most common predisposing factor for prostitutes is an economically deprived upbringing. Streetwalkers are the most common type of prostitute, and streetwalking is considered the most dangerous because they are often victims of violence, rape, and robbery. Like women, men become prostitutes primarily for the money. More than half of male prostitutes report that they are afraid of violence while they are hustling.

Men visit prostitutes for a variety of reasons including guaranteed sex, to eliminate rejection, greater control, companionship, undivided attention, adventure, and curiosity. More recent research supports Kinsey's 1948 findings that the majority of men who visit prostitutes are middle-aged and most often married and tend to be regular or repeat clients. Even though prostituting and engaging in sex with prostitutes are both illegal activities, arrests of the prostitute are 100 times greater than arrests of clients.

Prostitution could remain a criminal offense, or it could be legalized and regulated, which would subject it to government regulation including licensing, location, health standards, and advertising. Those who feel that prostitution should be legalized believe that it would result in lower levels of sexually transmitted infections and less disorderly conduct and the government could collect taxes. Eighty percent of prostitutes report using condoms with their clients during intercourse and 33% during oral sex.

LEARNING OBJECTIVES

After studying Chapter 18, you should be able to:

1. Provide examples of sexual images in history.
2. Explain how society began controlling sexually explicit media.
3. Discuss the role of sexual images on television and in advertising.
4. Summarize the historical court decisions on obscenity.
5. Compare and contrast the presidential commissions on pornography from 1970 and 1986.
6. Explain the views of antipornography groups.
7. Compare and contrast the views of the antipornography and anticensorship groups.
8. Summarize some of the research on pornography and harm.
9. Discuss the difficulties of defining prostitution.
10. Define some of the major terms related to prostitution.
11. List some predisposing factors related to becoming a female and male prostitute.
12. List some types of female and male prostitutes.
13. Identify some characteristics of adolescent prostitutes.
14. Describe the characteristics of the majority of men who visit prostitutes.
15. Discuss the positions for and against legalizing prostitution in the United States.
16. Identify how prostitutes try to minimize their risks of sexually transmitted infections.
17. Discuss issues relevant to the lives of people who leave prostitution.
18. Describe examples of prostitution in other cultures.

CHAPTER OUTLINE

I. SEXUAL IMAGES

A. Images have become more sexual in recent years.
B. There is a tendency to overemphasize explicit sexual images and to neglect the sexualized images that appear almost everywhere in modern society.

II. EROTIC REPRESENTATIONS IN HISTORY

A. By the dawn of the great ancient civilizations such as Egypt, people were drawing erotic images on walls or pieces of papyrus just for the sake of eroticism.
B. Greece is famous for the erotic art that adorned objects like bowls and urns.
C. **The Invention of Pornography**
1. Pornography, which tends to portray sexuality for its own sake, did not emerge as a distinct category until the middle of the 18th century.
2. **Obscenity** was illegal among the Puritans because it was an offense against God.
3. Historians of pornography trace the modern pornographic novel back to Renaissance Italian writer Pietro Aretino (1492-1556).
4. The story of pornography is not just about publishing erotic material but also about the struggle between those who try to create it and those who try to stop them.
5. The term **erotica** is often used to refer to sexual representations that are not pornographic, but it means only that a viewer or society considers it as within the acceptable bounds of decency.
6. Michel Foucault, a French philosopher and historian of sexuality, has referred to the constant sexuality in the media as a modern compulsion to speak incessantly about sex.

III. SEXUALITY IN THE MEDIA AND THE ARTS

A. Over the last 25 years, representations in the mass media have become more explicitly erotic.
B. **Erotic Literature: The Power of the Press**
1. Although the portrayal of sexuality is as old as art itself, pornography and censorship are more modern concepts.
2. Pornography in the modern sense began to appear when printing became sophisticated enough to allow fairly large runs of popular books, beginning in the 16th century.
3. Due to fears that people would turn away from religious teachings, by the 17th century, the Church pressured civic governments to allow them to inspect bookstores, and soon forbidden books, including erotica, were being removed.
4. It was the struggle between the illicit market in sexual art and literature and the forces of censorship that started what might be called a pornographic subculture that still thrives today.
5. Although modern debates about pornography tend to focus more on explicit pictures and movies, it was the erotic novel that first established pornographic production as a business in the Western world, provoking a response from religious and governmental **authorities.**

C. **Television and Film: Stereotypes, Sex, and the Decency Issue**
1. In 2003, Sex on TV: the Henry J. Kaiser Family Foundation published Content and Context, the largest study ever of sexual content on television.
a. Sixty-four percent of all shows included some sexual content.
b. Thirty-three percent were found to include sexual behaviors.
c. Fourteen percent included sexual intercourse.
d. It is estimated that 15% of shows today with sexual contact have references to safer sex.
2. Television, Film and Minority Sexuality

a. Popular sitcoms address may issues related to same-sex relationships.
b. African-American sexuality is still a taboo subject.

3. Television, Film, and Gender
a. While the portrayal of women is changing and improving on television today, men still outnumber women in major roles, and the traditional role of woman as sex object still predominates.
b. In the last 30 years, sexual stereotyping has been one of the most researched areas of media studies.

4. Television and Children
a. Research shows that when children are exposed to books or films that portray nonstereotyped gender behaviors, their gender stereotypes are reduced.
b. Today, more children's shows and cartoons show female stars.

5. The Movement Against the Equalization of the Visual Media
a. Portrayal of sexuality in movies has long been a source of controversy.
b. A backlash does seem to be developing, and Hollywood has been reducing the sexual explicitness of its general release movies.

D. **Advertising: Sex Sells and Sells**

1. Advertising and Gender Role Portrayals
a. In his groundbreaking book *Gender Advertisements*, Erving Goffman (1976) used hundreds of pictures from print advertising to show how men and women are positioned or displayed to evoke sexual tension, power relations, or seduction.
b. Research from the 80s and 90s suggested that gender biases existed in advertising.

2. Advertising and Portrayals of Sexuality
a. Sexually explicit advertising has become more common in the last 30 years.
b. Some researchers claim that advertisements have tried to use subliminal sexuality in advertisements.

E. **Other Media: Music Videos, Virtual Reality, and More**

1. Sexuality pervades other types of media including sex-advice columns, 900 number telephone lines, and the Internet.
2. Virtual reality producers have been making sexually explicit movies that are coordinated with vibrators.

IV. GRAPHIC IMAGES: PORNOGRAPHY AND THE PUBLIC'S RESPONSE

A. The debate over pornography is particularly active today because pornography has become so widely available.

B. **Defining Obscenity: "Banned in Boston"**

1. Court Decisions
a. The legal definition of obscenity dates back to the 1868 case of *Regina v. Hicklin* in England where the court defined obscenity as material that tended "to deprave and corrupt those whose minds are open to such immoral influences."
b. The Hicklin decision permitted the confiscation of obscene materials due only to their sexual content, which remained the American standard until the 1930s.
c. Court cases in the U.S. have established the 3-part definition of obscenity.

- It must appeal to the prurient interest
- It must offend contemporary community standards
- It must lack serious literary, artistic, political, or scientific value

d. Obscenity laws have been used in the 20th century to control many fiction and nonfiction books.

2. Presidential Commissions

a. In 1967, President Lyndon Johnson set up a commission to study "a matter of national concern": the impact of pornography on American society.
b. The commission studied four areas: pornography's effects, trafficking and distribution of pornography, legal issues, and positive approaches to cope with pornography.
c. The commission concluded that no reliable evidence was found to support the idea that exposure to elicit sexual materials is related to the development of delinquent or criminal sexual behavior among youth or adults, so adults should be able to decide for themselves what they will or will not read.
d. The 1986 Attorney General's commission on pornography (the Meese Commission)

- In 1985 President Ronald Reagan set out to overturn the 1970 ruling on pornography.
- Those who did not support the Commission's positions were treated with hostility.
- The Meese Commission came to the opposite conclusion of the 1970 ruling, dividing pornography into 4 categories: violent, degrading, nonviolent/nondegrading, and nudity.
- The reaction was strong and mixed.

C. **The Pornography Debates: Free Speech and Censorship**

- The religious-conservative opposition to pornography is based on a belief that people have an inherent human desire to sin and that pornography reinforces that tendency and so undermines the family, traditional authority, and the moral fabric of society.
- The issue of pornography has been divisive among feminist scholars, splitting them into two general schools.
  - The antipornography feminists, led by Catharine MacKinnon and Andrea Dworkin, see pornography as an assault on women that silences them, reinforces male dominance, and indirectly encourages sexual and physical abuse against women.
  - The other perspective includes group such as the Feminist Anticensorship Taskforce and argues that censorship of sexual materials will eventually be used to censor such things as feminist writing and gay erotica and would endanger women's rights and freedoms of expression.

1. Antipornography Arguments

a. MacKinnon argues that pornography is about power and cannot be separated from the long history of male domination of women and that it reinforces women's second-class status.
b. Andrea Dworkin sees pornography as a central aspect of male power, which she sees as a long-term strategy to elevate men to a superior position in society by forcing even strong women to feign weakness and dependency.

2. Anticensorship Arguments

a. Many critics argue that a restriction against pornography cannot be separated from a restriction against writing or pictures that show other oppressed minorities in subordinate positions.
b. Once sexually explicit portrayals are suppressed, anticensorship advocates argue, so are the portrayals that try to challenge sexual stereotypes.

D. **Studies on Pornography and Harm**

1. Societywide Studies

a. Correlations that sex offenders have large amounts of pornography have been used since the early 19th century to justify attitudes toward pornography.

b. Researchers have found higher rates of rape in states with the highest circulation of sex magazines, yet in countries where pornography is common and/or unregulated there are low rates of rape relative to the United States.
c. One longitudinal research study could find no increase in rape relative to other crimes in four countries as the availability of pornography increased dramatically.

2. Individual Studies
a. While little evidence indicates that nonviolent, sexually explicit films provoke antifemale reactions in men, many studies have shown that violent or degrading pornography does influence attitudes.
b. Other studies show that men's aggression tends to increase after seeing any violent movie, even if it's not sexual.

3. What is Harm?
a. A variety of studies show that women tend to have negative reactions to viewing pornography, but women's reactions are not generally considered relevant to the discussion.
b. Questions of harm focus on whether pornography induces sexual violence in men.

E. **What the Public Thinks About Pornography**
1. The majority of the general public wants to ban violent pornography and feels that pornography can lead to a loss of respect for women, acts of violence, and rape.
2. Research suggests that even people who felt that pornography had negative effects on others were opposed to regulating it.

V. SELLING SEX

A. **Defining Prostitution**
1. Definitions of prostitution vary, with some state penal codes defining prostitution as the act of hiring out one's body for sexual intercourse and other states defining it as sexual intercourse in exchange for money or as any sexual behavior that is sold for profit.
2. Prostitution is defined in the textbook as the act of a male or female engaging in sexual activity in exchange for money or other material goods.
3. A **pimp** is the term for a person who may act as a protector or business manger for prostitutes.
4. A **madam** is in charge of managing a home, or **brothel**, or group of prostitutes.
5. A **john** is a person who hires a prostitute, and a **trick** is the service that the prostitute performs.
6. Historically, most prostitutes worked in brothels, although with the exception of certain areas of Nevada, few brothels remain in the United States.

B. **Sociological Aspects of Prostitution**
1. Prostitution has existed as long as marriage has existed, which has led some people to argue that it is necessary.
2. Some sociologists suggest that prostitution developed out of the **patriarchal** nature of most societies.
3. The degree to which men govern a society has an influence over the type and degree of prostitution that exists in that society; however, prostitution is also linked to other economic, sociological, psychological, and religious factors.

VI. THE PROSTITUTE

A. The majority of prostitutes do not enjoy their work, with 24% of prostitutes reporting that they like it for the financial and personal freedom it offers.

B. **Female Prostitutes**

1. One study found that 75% of prostitutes were less than 25 years old and began during adolescence.
2. Female prostitutes may live in a home with several other prostitutes and a pimp, known as a **pseudofamily**.
3. One study found that 95% of prostitutes used drugs since many women were drug addicts first and had to find a way to pay for their addiction.
4. Predisposing Factors
    a. The most common predisposing factor for becoming a prostitute is an economically deprived upbringing, although the research may focus more on poor women since they're more likely to get involved with the law.
    b. Prostitutes are more often victims of sexual abuse, initiate sexual activity at a younger age, experience a higher frequency of rape, experience intrafamilial violence, and have past physical abuse histories.
    c. Sexually abused children who ran away from home were more likely to become prostitutes than those who did not run away.
5. Types of Female Prostitution
    a. Streetwalkers
        - They are the most common type of prostitute.
        - It is considered the most dangerous type of prostitution because they are often victims of violence, rape, and robbery.

    b. Bar Prostitutes: They work in bars for the owner.
    c. Hotel Prostitutes: They work in hotels and give some money to the hotel.
    d. Brothel Prostitutes: They work out of a home or apartment that is shared by a group of prostitutes and run by a madam who shares in the earnings.
    e. Massage Parlor Prostitutes
    f. Call Girls and Courtesans: This often refers to higher-class prostitution.
    g. Other Types of Prostitutes
        - **Bondage and discipline prostitutes** engage in sadomasochistic services.

C. **Male Prostitutes**

1. Male prostitutes who service women are referred to as **gigolos**, whereas male prostitutes who service other men are referred to as hustlers or boys.
2. Ninety-nine percent of male prostitutes say they perform fellatio.
3. Predisposing Factors
    a. Like women, men become prostitutes primarily for the money.
    b. Early childhood experiences, such as coerced sexual behavior, increase the chance of becoming a prostitute.
    c. More than 50% of male prostitutes report using alcohol and drugs with their clients and commonly accept drugs as a trade for sex.
    d. More than half of male prostitutes report that they are afraid of violence while they are hustling.
4. Types of Male Prostitution
    a. Street and Bar Hustlers
    b. Escort Prostitution
    c. Call Boys
    d. Transsexual and Transvestite Prostitutes

D. **Adolescent Prostitutes**

1. Adolescent prostitution can have long-term psychological and sociological effects on the adolescents and their families.

2. It is estimated that between 750,000 and 1,000,000 minors run away from home each year in the U.S., and that more than 85% of these minors become involved in prostitution.

VII. OTHER PLAYERS IN THE BUSINESS

A. The Pimp

1. Pimps take all of the prostitute's earnings and manage the money, providing her with clothes, jewelry, food, and sometimes, a place to live.
2. Factors that attract men to pimping include money, feelings of power and prestige within their peer group, and the lack of stress in the job.

B. The Client

1. Men visit prostitutes for a variety of reasons including guaranteed sex, elimination of rejection, greater control, companionship, undivided attention, no other sexual outlets, physical or mental handicaps, adventure, curiosity, etc.
2. Fellatio is the most commonly requested sexual behavior.
3. Clients believed that oral sex had a lower risk of STI or HIV transmission than other sexual behaviors.
4. Sadomasochistic behavior is the most common form of atypical sexual behavior.
5. More recent research supports Kinsey's 1948 findings that the majority of men who visit prostitutes are middle-aged and most often married and tend to be regular or repeat clients.
6. The majority of clients are not concerned with the police because enforcement of the law is usually directed at prostitutes rather than clients.
7. Even though prostituting and engaging in sex with prostitutes are both illegal activities, arrests of the prostitute are 100 times greater than arrests of clients.

C. The Government: Prostitution and the Law

1. Prostitution could remain a criminal offense, or it could be legalized and regulated, which would subject it to government regulation including licensing, location, health standards, and advertising.
2. The biggest roadblock to legalized prostitution in the U.S. is that the majority of the people views prostitution as an immoral behavior.
3. Those who feel that prostitution should be legalized believe that it would result in lower levels of sexually transmitted infections and less disorderly conduct and the government could collect taxes.
4. COYOTE (Call Off Your Old Tired Ethics) is regarded as the best-known prostitutes' rights groups in the U.S., and their mission is to repeal all laws against prostitution, to reshape prostitution into a credible occupation, and to protect the rights of prostitutes.

VII. PROSTITUTION AND SEXUALLY TRANSMITTED INFECTIONS

A. Prostitutes try to minimize their risks of sexually transmitted infections by using condoms, rejecting clients with STIs, and routinely taking antibiotics.
B. Condoms are used less frequently with their own sexual partners than with their clients.
C. Eighty percent of prostitutes report using condoms with their clients during intercourse and 33% during oral sex.
D. Rates of STIs in Europe were found to decrease when prostitution was legalized and to increase when it was illegal.

VIII. LIFE AFTER PROSTITUTION

A. Research suggests that female prostitutes practice prostitution for a relatively short time, usually 4 or 5 years.
B. There is a lot of disagreement about whether or not mandatory treatment programs should exist for prostitutes.

IX. PROSTITUTION IN OTHER CULTURES

A. During World War II, it is estimated that between 70,000 and 200,000 women (**comfort girls**) from Japan, Korea, China, the Philippines, Indonesia, Taiwan, and the Netherlands were forcibly taken by the Imperial Japanese Army from their homes and put in brothels for Japanese soldiers.
B. Recently a group called GABRIELA (General Assembly Binding Women for Reforms, Integrity, Equality, Leadership, and Action) has formed in the Philippines in an attempt to fight prostitution, sexual harassment, rape, and battering of women.
C. Thailand has a thriving prostitution industry.
D. In Amsterdam, Holland, there is a strip known as the Red Light District, which has legal window prostitutes.
E. In Cuba, male prostitutes who solicit tourists are known as jineteros who exchange sex for clothing or other luxuries brought over from other countries.

TEST YOURSELF

Below you will find fill-in-the-blank and short answer essay questions for topics covered in chapter 18. Check your answers at the end of this chapter.

History, Media, and Pornography

1. ______________________, which tends to portray sexuality for its own sake, did not emerge as a distinct, separate category until the middle of the eighteenth century.

2. The term ______________________ is often used to refer to sexual representations that are not pornographic.

3. Although the portrayal of sexuality is as old as art itself, ______________________ and ______________________ are more modern concepts, products of the mass production of erotic art in society.

4. The ______________________ of the nudes on Michelangelo's Sistine Chapel were painted over with loincloths and wisps of fabric by clerics.

5. Pornography in the modern sense began to appear when ______________________ became sophisticated.

6. ______________________ has been used to sell products since the turn of the century.

7. ______________________ laws have been used in the 20th century to control almost anything of a sexual nature.

8. The 1986 Attorney General's Commission on Pornography is also known as the ______________________ Commission.

9. Nowhere has the issue of pornography been as divisive as it has been among ______________________ scholars, splitting them into two general schools.

10. ______________________ ______________________ argues that in the case of pornography it is a mistake to distinguish speech from action; pornography is itself an act, linked to the general disempowerment of women.

11. While little evidence indicates that nonviolent, sexually explicit films provoke antifemale reactions in men, many studies have shown that _______________________ or _______________________ pornography does influence attitudes.

12. Some studies show that men's _______________________ tends to increase after seeing any violent movie, even if it is not sexual, and so the explicit sexuality of the movies may not be the important factor.

13. Pornography, Lahey argues, is the form that female _______________________ takes in American culture.

14. Many who defend sexually explicit materials that show consensual sex abhor the _______________________ and _______________________ pornography.

15. Unlike many other businesses today, the American _______________________ industry continues to do well.

16. Hundreds of acts of sexual intercourse are portrayed or suggested on television shows and in the movies every day. Name some things related to sexuality that we rarely see a couple do or discuss.

17. Court cases in the United States have established the three-part definition of obscenity. Name theses three requirements.

18. What are the four areas that were studied by the 1970 Commission on Obscenity and Pornography?

19. What was the official charter of the Meese Commission?

20. How does Catharine MacKinnon define pornography?

Selling Sex and the Prostitute

21. Some state penal codes define ______________________ as the act of hiring out one's body for sexual intercourse.

22. A(n) ______________________ is a person who hires a prostitute.

23. A(n) ______________________ is the service that the prostitute performs.

24. Some sociologists suggest that prostitution developed out of the ______________________ nature of most societies.

25. The majority of prostitutes do not ______________________ their work.

26. Sexually abused children who were ______________________ were more likely to become prostitutes than those who were not.

27. In the United States, ______________________ is the only state with counties in which brothels are legal.

28. Bondage and discipline prostitutes engage in ______________________ services.

29. ______________________ are young men who are hired by older women and they have an ongoing sexual relationship with her.

30. Male prostitutes who service other men are referred to as ______________________ or ______________________.

31. Male prostitutes who have sex with men may be otherwise ______________________.

32. Like the pimp for female prostitutes, many male prostitutes also have mentors, or ______________________.

33. Like females, males become prostitutes mainly for the ______________________.

34. More than ______% of male prostitutes report using alcohol and a variety of drugs with their clients and commonly accept drugs or alcohol as a trade for sex.

35. It is estimated that between 750,000 and 1,000,000 minors run away from home each year in the United States and that more than ______% of these minors eventually become involved in prostitution.

36. What is a pseudofamily?

37. What are some common threads that run through the lives of many prostitutes?

38. What is a courtesan?

39. When male prostitutes are asked what types of sexual behavior they engage in with their clients, what do they report?

40. From the male prostitutes' perspective, what are the differences between street hustling, bar hustling, and escort prostitution?

Other Players in the Business

41. The majority of clients of prostitutes are ________________________.

42. The majority of clients are not concerned with the ________________________ because enforcement of the law is usually directed at prostitutes rather than clients.

43. Even though prostituting and engaging in sex with prostitutes are both illegal activities, arrests of the prostitute are ________________________ times greater than arrests of the client.

44. The biggest roadblock to legalized prostitution in the United States is that prostitution is viewed as an ________________________ behavior by the majority of people.

45. ________________________ is the acronym for the organization whose mission is to repeal all laws against prostitution, to reshape prostitution into a credible occupation, and to protect the rights of prostitutes.

46. Among gay male prostitutes, receptive ________________________ intercourse without a condom is the most common mode of HIV transmission.

47. Among female prostitutes, ________________________ ________________________ ________________________ is the most common mode of HIV transmission.

48. Rates of STIs in Europe were found to ________________________ when prostitution was legalized and to ________________________ when it was illegal.

49. Life after prostitution is often grim, because most prostitutes have little ________________________ and few ________________________.

50. There is a lot of disagreement about whether or not mandatory ________________________ programs should exist for prostitutes.

51. In 1993, Japan finally admitted to having forced women to prostitute themselves as ________________________ ________________________, and now these women are demanding to be compensated for the suffering they were forced to endure.

52. A group with the acronym ________________________ has formed in the Philippines in an attempt to fight prostitution, sexual harassment, rape, and battering of women.

53. ________________________ is often referred to as the prostitution capital of the world.

54. The Red Light District in ________________________ is crowded with sex shops, adult movie and live theater shows, and street and window prostitutes.

55. In Cuba, male prostitutes who solicit tourists are known as ________________________.

56. What are some reasons that males visit prostitutes?

57. What are two ways that the government could address the issue of prostitution?

58. What is COYOTE?

59. How do some prostitutes try to minimize their risks of getting STIs?

60. What country has instituted a "100% condom use" program targeted at the prostitution industry?

POST TEST

Below you will find true/false, multiple-choice and matching quiz items covering the entire chapter. Check your answers at the end of this chapter.

True/False

1. Erotic representations have appeared in most societies throughout history.

2. Sexual stereotyping is one of the most researched areas of media studies.

3. The 1970 Presidential Commission on Pornography under President Lyndon Johnson found that exposure to elicit sexual materials is related to the development of delinquent sexual behaviors.

4. Many feminists believe that pornography should not be prohibited since some pornography challenges sexual stereotypes.

5. Most of the research on the effects of viewing pornography focuses on viewers' attitudes about safer sex.

6. The majority of prostitutes were sexually abused as children.

7. Unlike female prostitutes, male prostitutes have few psychological or emotional problems.

8. Men who visit prostitutes are more likely to be arrested than the prostitutes.

9. People who support the legalization of prostitution suggest that legalization would decrease prostitution by taking away the taboo nature.

10. Rates of STIs in Europe were found to decrease when prostitution was legalized and to increase when it was illegal.

Multiple-Choice

11. Who in society has struggled to determine which types of sexual images should be acceptable and which might be harmful to society?
    a. presidential commissions
    b. scholars
    c. the Supreme Court
    d. All of the above
    e. None of the above

12. What was the state of hard-core sexual representations before the 19th century?
    a. They were openly displayed and distributed.
    b. They were prevalent but hidden.
    c. They were extremely rare.
    d. They were popular among people in upper classes.
    e. They were popular among people in lower classes.

13. What is the term for sexual representations that are considered by a viewer or society as within the acceptable bounds of decency?
    a. pornography
    b. blasphemy
    c. obscenity
    d. erotica
    e. archetypes

14. What is the name of the French philosopher and historian of sexuality that referred to the constant sexuality in the media as a modern compulsion to speak incessantly about sex?
    a. Marquis de Sade
    b. Michel Foucault
    c. Jean Cleland
    d. Francois Perrier
    e. Pierre Herbaud

15. During the 17th century, who pressured civic governments to allow them to inspect bookstores to remove forbidden books?
    a. women's rights activists
    b. artists and authors
    c. parent groups
    d. physicians
    e. the Church

16. In 2003, "Sex on TV: Content and Context," the largest study ever of sexual content on television, was published by what organization?
    a. The Alan Guttmacher Institute
    b. Henry J. Kaiser Family Foundation
    c. Public Broadcasting Service
    d. American Family Association
    e. The Gallop Association

17. According to research, how are children affected when they are exposed to books or films that portray nonstereotyped gender behaviors?
    a. Their gender stereotypes increase.
    b. Their gender stereotypes are reduced.
    c. Boys' gender stereotypes increase, while girls' are reduced.
    d. Girls' gender stereotypes increase, while boys' are reduced.
    e. There is no effect.

When was the movie rating system introduced by movie industry?
    a. 1890s
    b. 1930s
    c. 1950s
    d. 1970s
    e. 1980s

18. How has advertising changed in the last 30 years?
    a. Women are shown in more positions of authority.
    b. Nudity is more common.
    c. Men are shown in roles considered traditionally female.
    d. Sexual illustrations have become more overt and visually explicit.
    e. All of the above

19. What new computer technology has contributed to increased sexual images in the media?
    a. ROC servers
    b. line-up systems
    c. virtual reality
    d. internal erotica
    e. intra-vivo keyboards

21. What have been the ramifications of obscenity laws in the 20th century?
    a. Children have been allowed to buy pornography.
    b. Fiction and nonfiction books have been banned.
    c. Poor people have been prohibited from viewing pornography.
    d. The development of sexually explicit photographs has decreased.
    e. All of the above

22. What is another name given to the 1986 Attorney General's commission on Pornography set up by President Ronald Reagan?
    a. The Meese Commission
    b. The Censorship Commission
    c. The Morality Commission
    d. The Reagan Commission
    e. The Freedom Commission

23. What group takes the position that pornography reinforces an inherent human desire to sin?
    a. antipornography feminists
    b. religious civil libertarians
    c. anticensorship feminists
    d. religious conservative individuals
    e. None of the above

24. What is a key argument of antipornography feminists?
    a. Pornography undermines the family, traditional authority, and the moral fabric of society.
    b. Pornography is immoral and degrading to women, but the government shouldn't control it.
    c. Pornography is about power and cannot be separated from the long history of male domination of women.
    d. Once sexually explicit portrayals are suppressed, so are the portrayals that try to challenge sexual stereotypes.
    e. All of the above

25. Correlational data between pornography and rates of sexual violence have been used to research the effects of pornography. In Japan, pornography is sold freely and tends to be dominated by rape and bondage scenes. What is the incidence of reported rapes?
    a. Rates of rape have increased as rape scenes have increased.
    b. Rates of rape have always been high in Japan, even before rape scenes were common.
    c. There are low rates of reported rapes in Japan relative to the United States.
    d. Rates of reported rapes have recently increased as pornography has gone more mainstream.
    e. None of the above

26. What is the definition of prostitution?
    a. sexual intercourse in exchange for money
    b. sexual behavior that is sold for profit
    c. the act of a male or female engaging in sexual activity in exchange for money or other material goods
    d. the act of hiring out one's body for sexual intercourse
    e. There is no one definition since the definition changes from state to state.

27. Donald regularly hires prostitutes. What is a term for him?
    a. john
    b. token
    c. pimp
    d. frank
    e. hustler

28. Historically, where did most prostitutes work?
    a. on the street
    b. in hotels
    c. out of their homes
    d. in bars
    e. in brothels

29. According to research, what percent of prostitutes were less than 25 years old?
    a. 15%
    b. 38%
    c. 52%
    d. 75%
    e. 95%

30. What type of female prostitute is considered to be the most common?
    a. bar prostitutes
    b. call girls
    c. streetwalkers
    d. brothel prostitutes
    e. massage parlor prostitutes

31. What is a gigolo?
    a. male prostitutes who pass as female
    b. male prostitutes who work with both men and women
    c. male prostitutes whose clients are women
    d. male prostitutes who accept drugs for sex
    e. male prostitutes who engage in bondage and discipline

32. What percent of male prostitutes report using alcohol and drugs with their clients?
    a. 10%
    b. 25%
    c. 50%
    d. 75%
    e. 90%

33. What leads many adolescents into prostitution?
    a. sexual abuse
    b. need for money
    c. running away from home
    d. psychological problems
    e. All of the above

34. What is the most commonly requested sexual behavior from female prostitutes?
    a. vaginal intercourse
    b. anal intercourse
    c. fellatio
    d. sadomasochistic behavior
    e. fetish behavior

35. What is a characteristic of most men who visit prostitutes?
    a. married
    b. lower-income
    c. in their twenties
    d. first-time clients
    e. All of the above

36. How do crackdowns on prostitution affect the practice?
    a. reduces the number of men seeking prostitutes
    b. drives it further underground
    c. reduces the likelihood that women will engage in prostitution
    d. decreases the transmission of sexually transmitted infections
    e. None of the above

37. What is the name of a prostitutes' rights groups whose mission is to repeal all laws against prostitution?
    a. PONY
    b. NAPRW
    c. PRAL
    d. COYOTE
    e. TRICKS

38. Prostitutes try to minimize their risks of sexually transmitted infections by ______________________.
    a. rejecting clients with sexually transmitted infections
    b. using condoms
    c. routinely taking antibiotics
    d. All of the above
    e. None of the above

39. Which of the following statements about life after prostitution is TRUE?
    a. Prostitutes usually leave because they made enough money to do other things.
    b. Most women leave prostitution after an average of 6 months.
    c. There are many treatment programs available to prostitutes who seek training in an effort to stop engaging in prostitution.
    d. All of the above
    e. None of the above

40. What country forced between 70,000 and 200,000 women from many different countries to work as "comfort girls," or prostitutes, for soldiers during World War II?
    a. France
    b. Japan
    c. Great Britain
    d. Italy
    e. Germany

Matching

Column 1		Column 2
A. obscenity	____	41. country that has "window prostitutes" that are legal and regulated by the government through taxes, regular physical checkups, and government insurance plans
B. comfort girls	____	42. antipornography crusader who was in charge of the 1985 Presidential Commission that determined that violent and degrading pornography should be banned in the U.S.
C. Catharine MacKinnon	____	43. a legal term for materials that are considered offensive to standards of sexual decency in a society
D. Holland	____	44. the term for a person who may act as a protector or business manger for prostitutes
E. erotica	____	45. a higher-class female prostitute who is often contacted by telephone and may either work by the hour or the evening
F. gigolo	____	46. a prostitute in Japan or the Philippines who was forced into prostitution by the government to provide sex for soldiers
G. call girls	____	47. a man who is hired to have a sexual relationship with a woman and receives financial support from her
H. Edwin Meese	____	48. sexual representations that are considered by a viewer or society as within the acceptable bounds of decency
I. pimp	____	49. antipornography crusader who sees pornography as an assault on women that silences them, reinforces male dominance, and indirectly encourages sexual and physical abuse against women
J. Thailand	____	50. country that has instituted a "100% condom use" program targeted at the prostitution industry

Test Yourself Answer Key

History, Media, and Pornography

1. Pornography (p. 561)
2. erotica (p. 562)
3. pornography, censorship (p. 562)
4. genitals (p. 562)
5. printing (p. 563)
6. Sexuality (p. 569)
7. Obscenity (p. 571)
8. Meese (p. 571)
9. feminist (p. 574)
10. Catherine MacKinnon (p. 575)
11. violent, degrading (p. 577)
12. aggression (p. 577)
13. victimization (p. 577)
14. violent, degrading (p. 577)
15. pornography (p. 577)
16. contraception, the morality of their action, contract an STI, worry about AIDS, experience an unwanted pregnancy, experience erectile dysfunction, or regret the act afterward (p. 563)
17. it must appeal to the prurient interest, offend contemporary community standards, and lack serious literary, artistic, political, or scientific value (p. 570)
18. pornography's effects, traffic and distribution of pornography, legal issues, and positive approaches to cope with pornography (p. 571)
19. to "find more effective ways in which the spread of pornography could be contained" (p. 571)
20. "the sexually explicit subordination of women, graphically depicted, whether in pictures or words," which is "central in creating and maintaining the civil inequalities of the sexes" (p. 575)

Selling Sex and the Prostitute

21. prostitution (p. 578)
22. john (p. 579)
23. trick (p. 579)
24. patriarchal (p. 579)
25. enjoy (p. 579)
26. runaways (p. 581)
27. Nevada (p. 582)
28. sadomasochistic (p. 582)
29. Gigolos (p. 583)
30. hustlers, boys (p. 583)
31. heterosexual (p. 583)
32. sugardaddies (p. 583)
33. money (p. 584)
34. 50 (p. 584)
35. 85 (p. 585)
36. a type of family that develops when prostitutes and pimps live together; rules, household responsibilities, and work activities are agreed on by all members of the family (p. 580)
37. an economically deprived upbringing, early sexual contact with many partners in superficial relationships, past history of being sexually abused, the initiation of sexual activity at a younger age, a higher frequency of rape (p. 580)
38. a prostitute who often interacts with men of rank or wealth (p. 582)

39. 99% say that they perform fellatio, 80% say that they engage in anal sex, and 63% participate in rimming (p. 583)
40. The main differences between these types of prostitution are in income potential and personal safety. (p. 584)

Other Players in the Business

41. male (p. 586)
42. 75 (p. 587)
43. police (p. 587)
44. immoral (p. 587)
45. COYOTE (p. 589)
46. anal (p. 589)
47. intravenous drug use (p. 589)
48. decrease, increase (p. 589)
49. money, skills (p. 590)
50. treatment (p. 590)
51. comfort girls (p. 590)
52. GABRIELA (p. 590)
53. Thailand (p. 590)
54. Amsterdam (p. 591)
55. jineteros (p. 592)
56. for guaranteed sex, to eliminate the risk of rejection, for greater control in sexual encounters, for companionship, to have the undivided attention of the prostitute, because they have no other sexual outlets, because of physical or mental handicaps, for adventure, curiosity, or to relieve loneliness (pp. 586-587)
57. Prostitution could remain a criminal offense, or it could be legalized and regulated. (p. 587)
58. COYOTE is regarded as the best-known prostitutes' rights groups in the United States. (p. 589)
59. by using condoms, rejecting clients with STIs, and routinely taking antibiotics (p. 589)
60. Thailand (p. 591)

Post Test Answer Key

True/False		Multiple Choice		Matching	
1.	T (p. 561)	11.	d (p. 561)	41.	D (p. 591)
2.	T (p. 566)	12.	c (p. 561)	42.	H (p. 571)
3.	F (p. 571)	13.	d (p. 562)	43.	A (p. 561)
4.	T (p. 576)	14.	b (p. 562)	44.	I (p. 579)
5.	F (p. 577)	15.	e (p. 563)	45.	G (pp. 581-582)
6.	T (p. 580)	16.	b (p. 564)	46.	B (p. 590)
7.	F (p. 584)	17.	b (p. 567)	47.	F (p. 583)
8.	F (p. 587)	18.	b (p. 567)	48.	E (p. 562)
9.	F (pp. 587-588)	19.	e (p. 569)	49.	C (pp. 574-575)
10.	T (p. 589)	20.	c (p. 570)	50.	J (p. 590)
		21.	b (p. 571)		
		22.	a (p. 571)		
		23.	d (p. 574)		
		24.	c (p. 575)		
		25.	c (p. 577)		
		26.	e (p. 578)		
		27.	a (p. 579)		
		28.	e (p. 579)		
		29.	d (p. 580)		
		30.	c (p. 581)		
		31.	c (p. 583)		
		32.	c (p. 584)		
		33.	e (p. 585)		
		34.	c (p. 587)		
		35.	a (p. 587)		
		36.	b (p. 589)		
		37.	d (p. 589)		
		38.	d (p. 589)		
		39.	e (p. 590)		
		40.	b (p. 590)